ION CHROMATOGRAPHY

CHROMATOGRAPHIC SCIENCE

A Series of Monographs

Editor: JACK CAZES
Silver Spring, Maryland

Volume 1: Dynamics of Chromatography
J. Calvin Giddings

Volume 2: Gas Chromatographic Analysis of Drugs and Pesticides
Benjamin J. Gudzinowicz

Volume 3: Principles of Adsorption Chromatography: The Separation of Nonionic Organic Compounds (out of print)
Lloyd R. Snyder

Volume 4: Multicomponent Chromatography: Theory of Interference (out of print)
Friedrich Helfferich and Gerhard Klein

Volume 5: Quantitative Analysis by Gas Chromatography
Joseph Novák

Volume 6: High-Speed Liquid Chromatography
Peter M. Rajcsanyi and Elisabeth Rajcsanyi

Volume 7: Fundamentals of Integrated GC-MS (in three parts)
Benjamin J. Gudzinowicz, Michael J. Gudzinowicz, and Horace F. Martin

Volume 8: Liquid Chromatography of Polymers and Related Materials
Jack Cazes

Volume 9: GLC and HPLC Determination of Therapeutic Agents (in three parts)
Part 1 edited by Kiyoshi Tsuji and Walter Morozowich
Part 2 and 3 edited by Kiyoshi Tsuji

Volume 10: Biological/Biomedical Applications of Liquid Chromatography
Edited by Gerald L. Hawk

Volume 11: Chromatography in Petroleum Analysis
Edited by Klaus H. Altgelt and T. H. Gouw

Volume 12: Biological/Biomedical Applications of Liquid Chromatography II
Edited by Gerald L. Hawk

Volume 13: Liquid Chromatography of Polymers and Related Materials II
Edited by Jack Cazes and Xavier Delamare

Volume 14: Introduction to Analytical Gas Chromatography: History, Principles, and Practice
John A. Perry

Volume 16: Steroid Analysis by HPLC: Recent Applications
Edited by Marie P. Kautsky

Volume 17: Thin-Layer Chromatography: Techniques and Applications
Bernard Fried and Joseph Sherma

Volume 18: Biological/Biomedical Applications of Liquid Chromatography III
Edited by Gerald L. Hawk

Volume 19: Liquid Chromatography of Polymers and Related Materials III
Edited by Jack Cazes

Volume 20: Biological/Biomedical Applications of Liquid Chromatography IV
Edited by Gerald L. Hawk

Volume 21: Chromatographic Separation and Extraction with Foamed Plastics and Rubbers
G. J. Moody and J. D. R. Thomas

Volume 22: Analytical Pyrolysis: A Comprehensive Guide
William J. Irwin

Volume 23: Liquid Chromatography Detectors
Edited by Thomas M. Vickrey

Volume 24: High-Performance Liquid Chromatography in Forensic Chemistry
Edited by Ira S. Lurie and John D. Wittwer, Jr.

Volume 25: Steric Exclusion Liquid Chromatography of Polymers
Edited by Josef Janča

Volume 26: HPLC Analysis of Biological Compounds: A Laboratory Guide
William S. Hancock and James T. Sparrow

Volume 27: Affinity Chromatography: Template Chromatography of Nucleic Acids and Proteins
Herbert Schott

Volume 28: HPLC in Nucleic Acid Research: Methods and Applications
Edited by Phyllis R. Brown

Volume 29: Pyrolysis and GC in Polymer Analysis
Edited by S. A. Liebman and E. J. Levy

Volume 30: Modern Chromatographic Analysis of the Vitamins
Edited by Andre P. De Leenheer, Willy E. Lambert, and Marcel G. M. De Ruyter

Volume 31: Ion-Pair Chromatography
Edited by Milton T. W. Hearn

Volume 32: Therapeutic Drug Monitoring and Toxicology by Liquid Chromatography
Edited by Steven H. Y. Wong

Volume 33: Affinity Chromatography: Practical and Theoretical Aspects
Peter Mohr and Klaus Pommerening

Volume 34: Reaction Detection in Liquid Chromatography
Edited by Ira S. Krull

Volume 35: Thin-Layer Chromatography: Techniques and Applications, Second Edition, Revised and Expanded
Bernard Fried and Joseph Sherma

Volume 36: Quantitative Thin-Layer Chromatography and Its Industrial Applications
Edited by Laszlo R. Treiber

Volume 37: Ion Chromatography
Edited by James G. Tarter

Other Volumes in Preparation

ION CHROMATOGRAPHY

Edited by

James G. Tarter

NORTH TEXAS STATE UNIVERSITY
DENTON, TEXAS

MARCEL DEKKER, INC.　　　　　　New York and Basel

Ion Chromatography

(Chromatographic science ; v. 37)
Includes bibliographies and index.
1. Ion exchange chromatography. I. Tarter, James G.,
. II. Series.
QD79.C453164 1987   543'.0893   86-23934
ISBN 0-8247-7634-8

COPYRIGHT © 1987 by MARCEL DEKKER, INC.   ALL RIGHTS RESERVED

Neither this book nor any part may be reproduced or transmitted in any form or by any means, electronic or mechanical, including photocopying, microfilming, and recording, or by any information storage and retrieval system, without permission in writing from the publisher.

MARCEL DEKKER, INC.
270 Madison Avenue, New York, New York 10016

Current printing (last digit):
10 9 8 7 6 5 4 3 2 1

PRINTED IN THE UNITED STATES OF AMERICA

Preface

Ion chromatography is the kind of invaluable technique that is all too rare in the field of chemical analysis. It is similar to other liquid chromatographic techniques in that movement between the various techniques does not require acquiring an extensive new body of knowledge. However, its difference appears in its ability to provide many novel and challenging problems just waiting for solutions. One of the key advantages and uses of IC is the analysis of inorganic anions, an area of chemical analysis in which a relative dearth of information exists.

The aim of this book is to explain and discuss ion chromatography in such a manner as to engender an appreciation for the merits of ion chromatography. It is hoped that this book will serve as a handy review and reference book for the experienced ion chromatographer and as a valuable teaching aid for the person just beginning to work in the field, and that it will provide the information necessary to determine the potential usefulness of IC for a given situation.

Chapers 1 and 2 discuss the fundamental operating premises of the two different approaches to ion chromatographic analysis. The third chapter condenses the appropriate information from the first two chapters, and presents additional information, to compare the two techniques. The purpose of the third chapter is to present as much relevant information as possible to help others decide which type of ion chromatography best suits their particular situation. Chapter 4 discusses from both a theoretical and a practical viewpoint the various types of detectors that can be used in ion chromatography. The fifth and sixth chapters discuss various modifications of the basic ion chromatographic process that have extended the applicability of IC even further.

The seventh chapter is a bibliographic review of the journal literature concerning ion chromatography. This chapter is divided into three nonexclusive sections. The first section covers the general types of analytical analyses that have been performed using IC. The second part covers the various kinds of ions that have been analyzed by IC. The third part covers the fundamental aspects of ion chromatography that have led to a better understanding of this technique. Only published works are included in Chapter 7, since it is assumed that the works of major importance presented at symposia and meetings have been published in either journals or conference proceedings. In addition, a detailed appendix is included for assisting in the use of this book as a reference.

The material presented in this volume will not become incorrect or useless over time, merely incomplete. The foundation upon which ion chromatography is built will not change; only the direction and extent of its growth are unknown.

James G. Tarter

Introduction

Ion chromatography is a technique that has its origins in a paper by Small et al. in 1975 [1]. From this relatively recent beginning, ion chromatography has grown and become an accepted method of analysis for many diverse applications. The acceptance and diversity of ion chromatography are illustrated by the escalating number of published papers. Table 1 delineates the increase in both the total number of journal articles and the international component of that growth.

Ion chromatography is a form of high-performance liquid chromatography (HPLC) that operates at pressures ranging from several hundred to several thousand pounds per square inch. The analyte species are nonvolatile and essentially ionic in nature. In contrast, typical HPLC and reverse phase HPLC are used with nonpolar to moderately polar compounds. The primary differences between ion chromatography and HPLC can be attributed to the ionic nature of the analyte in ion chromatography. The ion chromatographic eluants are frequently aqueous buffer solutions, as opposed to the nonaqueous media common in HPLC, and the separation of the analyte ions is effected using ion exchange resins. In many cases, the same chromatographic components (pumps, injection valves, etc.) can be employed in both HPLC and ion chromatography. Many of the basic chromatographic principles developed and used in HPLC can be applied to ion chromatographic analysis with only minor modifications.

No major scientific advance is made without many years of careful, competent research to provide a solid working foundation for the advances. The background of ion exchange prior to 1975 involved many

Table 1 Increase in Ion Chromatography Journal Articles

Year	Journal articles	Not written in English	Languages used
1975	1	0	1
1976	1	0	1
1977	7	0	1
1978	13	5	4
1979	29	6	4
1980	46	12	6
1981	57	12	5
1982	97	29	6
1983	124	58	8
1984[a]	105	19	6

[a]As listed in *Chem. Abstracts* through December 1984.

scientists from many nations working in many different fields of research. The German scientist Gans [2], after observing the ion exchange characteristics of natural zeolites, developed a process for preparing artificial zeolites for use as ion exchangers for cation exchange. This interest in the inorganic ions eventually led to one of the more significant events in the early chromatographic literature: the separation of radioisotopes and rare earths by Tompkins et al. in 1947 [3]. Tompkins and co-workers used a synthetic resin, dilute citric acid buffers, and radiochemical detection in the separation of the species using column chromatography. The work of Tompkins and co-workers was part of the Manhattan Project, which ultimately produced the atomic weapons that helped bring an end to the second World War.

While some scientists were interested in the separation of inorganic ions using ion-exchange column chromatography, other scientists were interested in separating species of biological importance. In 1948, Stein and Moore [4] used column chromatography with starch packing material to separate a series of amino acids. In 1951, these same two scientists used a synthetic sulfonated polystyrene resin to increase the versatility of the amino acid separation and to minimize unwanted

Introduction

reactions [5]. In both cases, Stein and Moore used photometric ninhydrin detection of the amino acids in the fractionally collected eluant from the column. This work on amino acids eventually culminated in the 1972 Nobel Prize.

The early work in ion exchange chromatography, as it was originally called, required that the analyte species be detected by some property other than that of their ionic nature. Such detection methods included colorimetry and radiochemical methods. One of the early attempts at the use of conductance as a means of detection for ion exchange work was that of Duhne and Sanchez de Ita in 1962 [6]. These scientists used a four-cell conductivity bridge to detect the separated ions from an anion exchange column. The eluants used were water and dilute hydrochloric acid. The types of species investigated included the sodium salts of citric, oxalic, nitric, boric, acetic, and hydrochloric acids. Problems inherent in this work included the relatively slow process of column chromatography and the relatively large cell volume, over 0.14 ml, for each of the four cells. Kambara and Tachikawa [7] used derivative electrical conductivity to quantitate separated zinc and lanthanum ions using ethylenediammonium chloride as the eluant in column chromatography. The derivative technique measured the difference in electrical conductivity between the pure eluant and the eluant plus analyte. Neither the procedure of Duhne and Sanchez de Ita nor that of Kambara and Tachikawa was readily amenable to use with the new HPLC techniques and instruments then being developed. The main problem in using conductometric detection for the HPLC detection of ions was the necessity of using high-conductivity eluants in the separation process. These high-conductivity eluants produced a high background conductance, rendering the measurement of trace levels of ions virtually impossible and drastically limiting the usefulness of ion exchange chromatography for trace analysis.

The beginnings of modern ion chromatography can be traced to the work of Small et al. in 1975 [1]. Small and co-workers used a second column, positioned in line after the first or separator column, to chemically suppress the background conductance. This suppression was carried out by converting the buffer solution into the weakly ionized acid prior to measurement of the separated anions. As a result of this pioneering work, Small and co-workers were awarded the Pittsburgh Conference Award for Applied Analytical Chemistry.

One of the concerns with the method of Small and co-workers was the potential for band broadening of the separated ions as these ions passed through the second column. In addition, the extra column increased the complexity of the instrumentation necessary to perform the analysis. Gjerde and co-workers [8] developed an alternative method of ion chromatographic analysis based on the use of eluants with low initial conductance. Low-conductivity eluants have, in many cases, negated the need for the use of an eluant suppression device. The resins and eluants used in the two different ion chromatography techniques are different, leading to specific advantages and disadvantages for each of the techniques.

Ion chromatography now encompasses an ever-widening range of samples and applications. The main thrust of ion chromatography in the early days of its development was anion analysis due to the near nonexistence of rapid multielement anion analysis methodology. The analysis of inorganic cations by ion chromatography has been a little slower to develop because of the acceptability and accessibility of atomic spectroscopy. In recent years, however, the ion chromatographic analysis of inorganic cations has established a firm place in analytical chemistry methodology. An extension of ion chromatographic analysis of inorganic ions involves the analysis of transition metal ions, including those with multiple oxidation states. The analysis of these transition metals has rapidly become routine. The ion chromatographic analysis of organic species and species of biological and biochemical interest, along with inorganic ions, has become an area of active research and application. All of these topics are expounded on by the contributing authors to this volume.

The growth of ion chromatography, both the eluant suppressed system of Small et al. [1] and the single-column system of Gjerde et al. [8], has been dramatic in the first decade of its existence, as is evident in this book. In fact, one of the main attractions of ion chromatography to many researchers seems to be that ion chromatography emphasizes the *chemistry* in analytical chemistry. The nature of future applications is limited only by the ingenuity of those performing the research.

REFERENCES

1. H. Small, T. C. Stevens, and W. C. Bauman, Novel ion exchange chromatographic method using conductometric detection. *Anal. Chem.* 47, 1801-1809 (1975).
2. R. Gans, German Patent #174,097 (1906).
3. E. R. Tompkins, J. X. Khym, and W. E. Cohn, Ion-exchange as a separations method. I. The separation of fission-produced radioisotopes, including individual rare earths, by complexing elution from amberlite resin. *J. Am. Chem. Soc.* 69, 2769-2777 (1947).
4. W. H. Stein and S. Moore, Chromatography of amino acids on starch columns. The separation of phenylalanine, leucine, isoleucine, methionine, tyrosine, and valine. *J. Biol. Chem.* 176, 337-365 (1948).

5. S. Moore and W. H. Stein, Chromatography of amino acids on sulfonated polystyrene resins. *J. Biol. Chem. 192*, 663-681 (1951).
6. C. Duhne and O. Sanchez de Ita, Use of a four-cell conductivity bridge as a continuous detector for ion exchange chromatography. *Anal. Chem. 34*, 1074-1076 (1962).
7. T. Kambara and T. Tachikawa, A derivative method in ion exchange chromatography, *J. Chromatogr. 32*, 728-731 (1968).
8. D. T. Gjerde, J. S. Fritz, and G. Schmuckler, Anion chromatography with low-conductivity eluents. *J. Chromatogr. 186*, 509-519 (1979).

Contents

Preface	iii
Introduction	v
Contributors	xv

1 Eluant-Suppressed Ion Chromatography   1
   Edward L. Johnson

	I. Introduction	1
	II. Principles of Eluant Suppression	1
	III. Anion Analysis	6
	IV. Cation Analysis	13
	V. Conclusion	20
	References	20

2 Single-Column Ion Chromatography   23
   Thomas Jupille

	I. Overview	23
	II. Electrical Conductivity Detection	26
	III. Other Detection Techniques	54
	IV. Ancillary Techniques and Applications	59
	References	78

3 Comparison of Eluant-Suppressed and Single-Column Ion Chromatography   83
   James G. Tarter, Edward L. Johnson, and Thomas Jupille

4	Detection Methods in Ion Chromatography Paul R. Haddad and Petr Jandik	87
	I. Introduction	87
	II. Electrochemical Methods of Detection	88
	III. Spectroscopic Methods of Detection	126
	IV. Detection via Post-Column Reactions	146
	References	151
5	Ion Chromatography Exclusion Phyllis E. Buell and James E. Girard	157
	I. Introduction	157
	II. The Theory of Ion Chromatography Exclusion	158
	III. Background	159
	VI. ICE/IC	166
	V. Experimental Conditions for ICE	169
	VI. Detection Limits	174
	VII. Applications of ICE	175
	VIII. Conclusion	186
	References	186
6	Approaches to Ionic Chromatography Purnendu K. Dasgupta	191
	I. Introduction	191
	II. Membrane-Based Eluant Counterion Suppression Systems	194
	III. Indirect Detection Methods	236
	IV. Ion Chromatography with Ion Interaction Reagents	253
	V. Chromatographic Analysis of Metal Species	272
	VI. Improving Detectability	310
	VII. Conclusions	337
	References and Notes	338
7	A Review of Ion Chromatography: A Bibliography James G. Tarter	369
	I. Introduction	369
	II. Books and Dissertations	370
	III. Review Articles	371
	IV. Applications of Ion Chromatography: Sample Type	372
	V. Applications of Ion Chromatography: Ions Analyzed, Anions	375

VI	Applications of Ion Chromatography: Ions Analyzed, Cations	385
VII.	Instrumental, Procedural, and Theoretical Considerations	387
	References	390

Index *417*

Contributors

Phyllis E. Buell Department of Chemistry, The American University, Washington, D.C.

Purnendu K. Dasgupta Department of Chemistry and Biochemistry, Texas Tech University, Lubbock, Texas

James E. Girard Department of Chemistry, The American University, Washington, D.C.

Paul R. Haddad Department of Analytical Chemistry, University of New South Wales, Sydney, Australia

Petr Jandik Waters Chromatography Division, Millipore, Ltd., Milford, Massachusetts

Edward L. Johnson Department of Technology, Dionex Corporation, Sunnyvale, California

Thomas Jupille* Wescan Instruments, Inc., Santa Clara, California

James G. Tarter Department of Chemistry, North Texas State University, Denton, Texas

---

**Current affiliation*: Pi Technologies, Orinda, California.

ION CHROMATOGRAPHY

1
Eluant-Suppressed Ion Chromatography

Edward L. Johnson
Dionex Corporation
Sunnyvale, California

I. INTRODUCTION

The development of eluant-suppressed ion chromatography in the mid-1970s marked a renaissance in the chromatography of inorganic anions and cations. The pioneering work first begun by Small, Stevens, and Bauman in the Dow Laboratories resulted in the development of both (a) a unique detection mode that could be applied to nearly all ionic species; and (b) new ion exchange resins that featured high efficiencies and sample loading capacities [1]. This work has been significantly refined and enhanced within the research laboratories of the Dionex Corporation with the development of mobile-phase ion chromatography (MPIC) [2] and ion exclusion chromatography (ICE) [3]. The aim of this chapter is to describe via discussion and illustrations both the unique resins and the detection system. The principles behind the detection will be illustrated with specific examples. The columns used for the separation of the ions, as well as the more common eluants, will be described.

II. PRINCIPLES OF ELUANT SUPPRESSION

Chromatographic separations utilizing ion exchange techniques require the use of eluants that consist of some mixture of an ionic species. The eluant may or may not contain other constituents such as organic solvents. The purpose of the ionic species added to the eluant is to act as the "pusher" of the analyte species. For simplicity, we shall call this "pusher" the *eluant*, it being understood that other constituents are often present in the ultimate eluant employed during the chromatography. The eluant competes with the analyte species for the ion exchange sites in the analytical column. The analyte species migrate differentially down the column, i.e., are separated into discrete bands. Once the separation has been achieved, the eluant is then directed to a suitable detector.

If the effluent from the typical analytical column is directed to a conductivity detector, one observes that:

1. The conductivity of the eluant is somewhat larger than zero.
2. The conductivity of the analyte represents only a small percentage of the total observed conductivity.
3. Drift and noise are a problem.

In this mode, the conductivity detector is serving as a bulk property detector; i.e., detecting all of the ions present in the effluent. Note that in this situation the analyte ions are, in fact, being detected indirectly by the presence or absence of additional conductivity (above or below that of the eluant). The result is that peaks can appear as either positive or negative, depending upon the eluant conductivity and the concentration of the analyte ion. The ideal situation would be to have the conductivity detector functioning as a solute-specific detector, i.e., see only the ions of interest. This would result in:

1. Easier, more reliable operation
2. Increased stability
3. Superior detection limits

How can the conductivity detector be made to operate in such a manner?

We have already pointed out that all chromatographic separations employing ion exchange require an ionic species in the eluant. The only way that a conductivity detector can be made insensitive to this analyte constituent is to somehow convert the eluant to a nonconductive species or remove it from the solution without losing or destroying the analyte ions. If this can be accomplished, then indeed the conductivity detector becomes solute specific, i.e., the detector detects the analyte ions directly, because they are the only ions in solution. The result is a detector that is specific to ions, has a low background (thus being less sensitive to environmental effects), and has a wide linear dynamic range (an analogous detector would be the fixed UV detector of high-performance liquid chromatography, HPLC). Thus, the Dow researchers developed a unique postcolumn detection scheme which not only succeeded in removing or converting the eluant to a less conductive form but also converted the species of interest to a more

Eluant-Suppressed Ion Chromatography

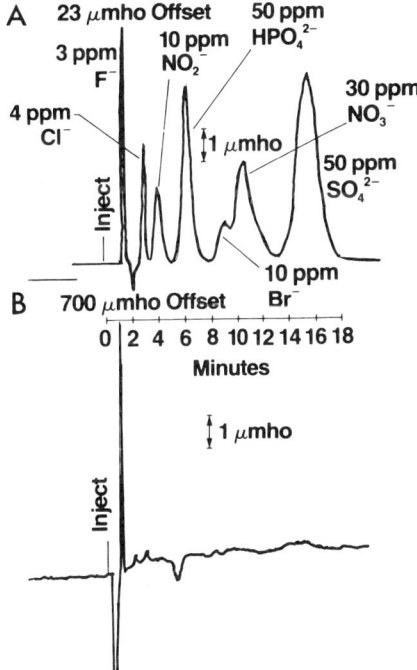

Figure 1. Effect of high-conductance background on sensitivity.

conductive form. This postcolumn scheme succeeded in lowering the background noise while simultaneously increasing the analyte response!

Let us turn to a simple example to illustrate this unique postcolumn reaction.

Consider the separation of the common anions shown in Figure 1. This separation is easily accomplished with a carbonate/bicarbonate eluant. If a conductivity detector were placed directly at the end of the analytical column, one would observe a result similar to that shown in Figure 1B. If, however, the effluent were directed to a device that contained a cation exchange material in the H^+ form, the following reactions would occur:

1. The cation exchanger would exchange its H^+ for Na^+ (reaction 1).
2. The carbonate/bicarbonate of the eluant would be neutralized to poorly conducting carbonic acid (reaction 2).
3. The chloride would now exist in the effluent as hydrochloric acid (its most conductive form) (reaction 3).

(R-1): Na^+ (eluant and sample) + H^+ ~~ Resin $\longrightarrow$ H^+ (eluant) + Na^+ ~~ Resin

(R-2): CO_3^{2-} and HCO_3^- (eluant) + H^+ (eluant) $\longrightarrow$ H_2CO_3 (removal of anions of eluant)

(R-3): Cl^- (sample) + H^+ (eluant) $\longrightarrow$ Hydrochloric acid

(Note: Ion sources are in parentheses.)

If we now direct the effluent from this device to a conductivity detector, we would obtain a chromatogram as shown in Figure 1A. Note the significantly improved signal-to-noise ratio as well as the reduction in the baseline noise. The background conductivity in this simple case has been lowered from 700 µS to 23 µS and the analyte signal enhanced by a factor of 16.

In an analogous manner, it is possible to design a postcolumn reaction device that can be used for cations as well. A simple example would be the separation of sodium and potassium with hydrochloric acid as the eluant. In this case, the suppressor device would contain an anion exchange material in the hydroxide form. As the effluent from the column enters the device, the anion exchange material would be converted to the chloride form, with the concurrent conversion of the hydronium ions to water. In effect, the eluant ions have been removed from the flowing stream. Meantime, the sodium and potassium are now present as their corresponding bases which are highly conductive.

As is the case for most postcolumn reactions, there are restrictions that are placed upon both the column capacity and the eluant. In the case of anions, these restrictions are only a minor limitation. Unfortunately, in the case of cations, the technique is limited to the alkali metals, alkaline earths, and organic amines because of the poor solubility of most metal hydroxides.

General restrictions include the following:

1. The eluant must be convertible to a less conductive form.
2. The converted form of the analyte must remain soluble and conductive.
3. The eluant concentration must not exceed the "suppressing" capacity of the postcolumn device.

Examples of species that can be used for the elution of anions in this manner are carbonate, bicarbonate, borate, phenate(s), cyanide, amino acids, quaternary amine hydroxides, and hydroxide. Examples of species that can be used for the elution of cations in this manner are hydrochloric acid, nitric acid, barium chloride, alkyl and aryl sulfonic acids, amino acids, and silver nitrate.

As was mentioned earlier, the converted form of the analyte species must remain soluble and be conductive. In the case of anions, solubility is not a significant problem because the anions are converted

Eluant-Suppressed Ion Chromatography

to their corresponding acids, which generally have sufficient solubility in the aqueous eluants employed. Solubility is a more significant problem in the case of cations. When the usual acid eluants are employed, the postcolumn device generally converts the analytes to their corresponding bases. Only the bases of the alkali metals and the alkaline earths have reasonable solubility in water, and thus the technique is limited to these inorganic species. Organic amine salts and ammonium ion are converted to the free amine or ammonia and remain soluble, making the technique applicable to them as well. (Other postcolumn techniques have been developed for transition and heavy metals. See later chapters of this book.)

Because the postcolumn device essentially converts the eluant to its weakly conductive form, a converted analyte species must be more conductive than the weakly conductive form of the eluant to be detected. In general terms, this means that the pK (a or b) must be greater than 7. Examples of species detectable and their detection limits are listed later in this discussion. An interesting modification of the general postcolumn reaction permits the detection of such weak acid species as cyanide, borate, and carbonate via the formation of conductive ion-pairs in the ion exclusion technique [4].

$H^+ + A^-$ (eluant) $+ R^+ + OH^-$ (regenerant) $\rightarrow H_2O + (R^+)(A^-)$
(low conducting ion pair)

$H_3BO_3 + R^+ + OH^-$ (regenerant) $\rightarrow H_2O + (R^+)(H_2BO_3^-)$
(higher conducting ion pair)

(Note: A^- is usually a large organic anion such as octane sulfonate.)

In the first implementation of the postcolumn reaction, Small et al. employed a second column packed with ion exchange resin. If anions were to be determined, the resin was a cation resin in the hydrogen ion form. For cations, an anion resin in the hydroxide form was employed. During a series of analyses, the second column would slowly be converted to its expended form, and at some point it would be necessary to regenerate it. Although instrumentation was designed that permitted a fresh column to be easily placed in line without significantly delaying analyses, there were other factors that caused the development of even better devices for performing the postcolumn reaction. Chief among these factors were the problems caused by the Donnan exclusion properties of the resins employed in the postcolumn reactor. These properties are the chief cause of the movement of the so-called water dip and the apparent changing response of weak acids such as nitrite and acetate.

Consider now the usual anion case, which employs a carbonate eluant and a cation exchange resin in the hydrogen form. If a mixture containing fluoride and chloride dissolved in water is injected, one obtains a chromatogram that has a dip between the fluoride and chloride peaks. The "dip" is caused by the water of the sample, which is less conductive than the carbonic acid formed from the eluant. Such upsets caused by the sample solvent are common in chromatographic techniques. What makes it unusual is that the dip's retention time is not constant, but related to the degree of regeneration of the postcolumn device. Thus, as the device became more expended, the "dip" would elute at a shorter time since the Donnan exclusion effects became less significant. This moving of the dip caused problems in the routine quantitation of early-eluting components such as fluoride and chloride. The Donnan exclusion effects also caused similar quantitation problems with weak acid or base species, such as nitrous acid, acetic acid, and ammonia.

It is a tribute to the power of this unique postcolumn reaction system that in spite of the limitations listed above, the analytical technique was the fastest-growing in the past decade.

As is the case in most new technologies, significant improvements have been made in the manner in which the postcolumn reaction is performed. A significant improvement was achieved by the development of fiber-based devices (Figure 2). These devices eliminated the troublesome Donnan exclusion effects and greatly simplified the operation of the instrument because they did not require periodic regeneration. In an extension of this technology, membrane-based devices were recently introduced [5,6] which not only have all the advantages of the fiber-based devices, but also make the technique applicable to wider concentration ranges of the eluant. Indeed, these new devices make gradient chromatography feasible with the conductivity detector.

However, all of these improvements in the postcolumn devices would be of little value if the analytical columns were not compatible with them.

Although often overlooked, the anion columns first described by Small and co-workers were unique from the usual ion exchange resins commonly employed in chromatography. These resins led directly to the development of improved resins with wide ranges in selectivity. New columns for improved ion exclusion and ion pair separations were also developed. We turn our discussion to the columns used for the separation of the various analyte species. We limit our discussion primarily to those resins used in the more traditional ion exchange mode. Other chapters will discuss the additional separation techniques in greater detail. For convenience, the anion and cation techniques are discussed separately.

III. ANION ANALYSIS

Although pellicular packings were widely used in the early development of HPLC, they quickly gave way to the totally porous types currently

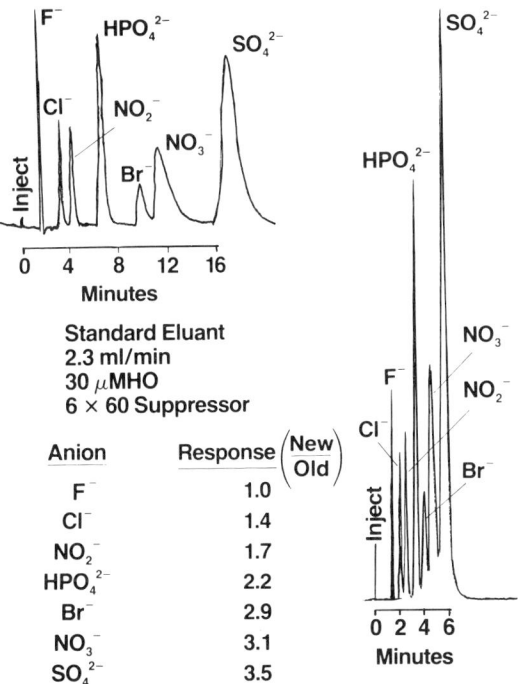

Figure 2. Fiber suppressor (on the right) versus packed-bed suppressor (on the left).

used. The shift to the totally porous materials was primarily due to their much higher surface areas and consequently their higher sample loading capacity. The need for high surface area was due to the fact that the common bonded-phase techniques rely upon relatively weak interactions with the solute species. Thus, to obtain reasonable sample capacity, a high surface area is required. In contrast, ion exchange interactions are quite strong. These strong interactions allow one to utilize pellicular packings with their inherently high efficiencies (and low back pressures) for common ion exchange chromatography. Although silica-based pellicular packings were available, they proved to be incompatible with the highly alkaline eluants needed in most separations. Thus, the Dow researchers searched for techniques to develop pellicular packings from resin-based materials, which were widely known to have excellent stability in a wide range of pHs. They discovered that excellent columns could be obtained from resins with the unique structure shown in Figure 3. The large substrate bead acts as support for the smaller ion exchange beads. To "anchor" the

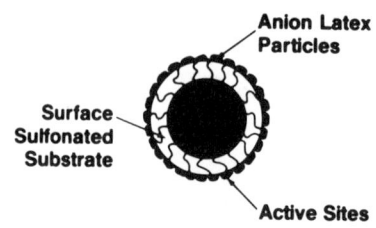

Figure 3. IC separator resins.

small beads in place, the substrate was partially functionalized to contain surface sulfonate groups. The smaller beads, being anion exchange material (formed from grinding larger beads or from a latex), are held in place by multiple ion interactions with the surface. These ion interactions are undoubtedly reinforced by Van der Waals forces as well. Indeed, the beads are so strongly attached to the surface that eluants as strong as 5 M sulfuric acid cannot displace them. The smaller beads are similar in structure to the more common anion exchange resins, such as Dowex 1. Several important advantages are derived from this unique structure:

1. Column packings have high permeability, i.e., they operate at high flow rates with low back pressures.
2. Swelling or shrinking problems associated with the usual ion exchange resins are significantly decreased.
3. Because high flow rates can be employed, equilibration times can be significantly shortened.
4. Because the active portion of the packing is similar in structure to the common ion exchange resins, a large portion of existing knowledge dealing with separations can be directly transferred to this new approach.

5. The unique structure permits unusual variations in crosslinkage, functional group, capacity, and size to be quickly and easily assessed for their problem-solving ability.

Research within both the Dionex and Dow laboratories has established that the observed chromatographic selectivity can be varied in rather dramatic manners by subtle changes in various properties of the smaller resin beads agglomerated to the substrate. The ability to alter the observed selectivity by changing the column packing rather than the eluant makes the analytical technique operationally easier for the user.

Table 1 lists the types of columns available for the ion exchange separation of the common inorganic and organic anions via eluant-suppressed ion chromatography. The table is arranged in descending order of hydrophobicity of the packing. For comparative purposes, the relative resolution for bromide and nitrate is also indicated. For convenience, this selectivity is normalized such that fluoride is arbitrarily defined as unity. Note that as the hydrophobicity of the packing decreases, the selectivity of bromide and nitrate also decreases. In fact, as the resin becomes more hydrophilic, it becomes very difficult to separate these two species. This illustrates that most common ion-exchange separations involve a complex mechanism of which ion exchange is only one part. With the ability to vary the composition of the active portion of the resin, it is possible to exploit the many components of the complete ion-exchange mechanism in such a manner as to optimize a column to a specific application problem.

Table 1. Columns Used in Anion Eluant-Suppressed Ion Chromatography

Column	Latex[a] XL	Latex[a] Size	Substrate Size (μm)	Rel. Separation of Br/NO_3
AS2	1	1	20	+++
AS3	1	0.5	25	++
AS4a	0.2	1.7	15	++
AS4	0.7	0.3	15	++
AS1	1	1	25	+
AS5	0.2	0.3	15	-

[a]XL, crosslinks. All values for crosslinks are relative to the AS1 material, which is arbitrarily defined as unity.

Table 2. Other Anion Separation Techniques Used in Eluant-Suppressed Ion Chromatography

---

Ion exclusion:
  Utilizes Donnan exclusion effects
  Typical eluants: dil. HCl, octanesulfonic acid
  Typical regenerants: dil. quaternary amine hydroxides
  Best for weak organic acids such as those of the Krebs cycle

Ion pair (MPIC):
  Utilizes reverse-phase ion-pair effects
  Typical eluants: dil. quaternary amine bases with organic solvents
  Typical regenerants: dil. sulfuric acid
  Ideal for hydrophobic ions such as many aryl organics, perchlorate, and iodide

---

The columns used for ion exclusion are typically totally sulfonated resins with different crosslinking and thus pore volumes. The columns used for ion pair separation are macroporous resins with large surface areas. Reverse-phase silica columns have been used with marginal success because of their poor stability in alkaline eluants. Table 2 lists the characteristics of these techniques.

A. The Eluants

The eluants most commonly employed in the case of anion exchange are inexpensive, simple mixtures of carbonate, bicarbonate, and hydroxide. These salts are readily available, and easily suppressed to a non- or low-conducting form. Table 3 lists the most commonly employed combinations of these salts. Other eluants that have been reported are phenate, glutamate, and borate. In some cases, additives such as p-cyanophenol have been used to lessen absorption properties of the resins. However, the need for such additives is less important since the development of resins with significantly greater hydrophilic properties.

B. The Regenerants

The regenerant must supply a source of hydrogen ions to convert the eluant anions to a less conductive form. The most common regenerant

Table 3. Typical Eluants for Anion Separation[a]

Eluant	Typical column
3.0 mM $NaHCO_3$; 2.4 mM Na_2CO_3	AS1
3.0 mM Na_2CO_3; 2.0 mM NaOH	AS2
4.5 mM Na_2CO_3; 2.0 mM NaOH; 0.7 mM 4-cyanophenol; 5% (v/v) acetonitrile	AS2
2.8 mM $NaHCO_3$; 2.2 mM Na_2CO_3	AS3, AS4
3.4 mM Na_2CO_3; 4.3 mM $NaHCO_3$; 0.8 mM 4-cyanophenol; 2% (v/v) acetonitrile	AS5
0.7 mM $NaHCO_3$	AS3, AS4

[a]Other combinations may be used. Those listed represent typical starting points. Eluants derived from phenate and borate salts have been reported in the literature.

is dilute sulfuric acid. Again, it is readily available and easily prepared in the dilute solutions required. Acids such as hydrochloric and nitric are not usually employed because they tend to cause high backgrounds due to inability of the membrane material to completely exclude the anion of the acid. Large organic acids typified by dodecylbenzenesulfonic acid have been used with good success. Care must be taken in the preparation of solutions of these organic acids since they are excellent surfactants and thus produce rather significant foaming problems. Table 4 lists typical regenerants most commonly employed.

Table 4. Typical Regenerants Used in Anion Eluant-Suppressed Ion Chromatography

0.025 N Sulfuric acid

0.025 N Dodecylbenzene sulfonic acid

C. Detectable Species

It is difficult to tabulate all of the species detectable by this analytical technique. Table 5 lists many of the more common species with an estimation of detection limits. The detection limits assume ideal conditions and a 50-μl direct injection. Values are those that can be easily determined; i.e., three times the noise. In some cases, lower values can be obtained by appropriate concentration techniques and/or large injection volumes. The table is limited to those species detectable via eluant-suppressed techniques. Species such as $HS-$, $CN-$, $RS-$, etc., are usually separated by the columns discussed earlier; however, other detection means are employed (usually amperometric). The published literature contains several reviews that will provide the reader with more information [7,8]. Dionex Corporation maintains an extensive bibliographic library that lists most published papers dealing with all forms of ion chromatography. Examples of various separations are shown in Figures 4 through 7.

Table 5. Anion Species Detectable via Eluant-Suppressed Ion Chromatography

Species	Detection limit (50-μl injection)
Fluoride	5 ppb
Chloride	5-10 ppb
Nitrite	5-10 ppb
Phosphate	5-10 ppb
Nitrate	5-10 ppb
Bromide	5-10 ppb
Sulfate	5-10 ppb
Thiosulfate	20-40 ppb
Tetrathionate	1 ppm
Sulfite	10-20 ppb
Tungstate	1 ppm
Molybdate	1 ppm
Chromate	1 ppm
Acetate	1 ppm
Formate	0.5 ppm
Carbonate	1 ppm
Perchlorate	0.5 ppm
Chlorate	0.5 ppm
Phosphite	0.1 ppm
Hypophosphite	0.1 ppm
Arsenate	0.1 ppm

Table 5. (Continued)

Species	Detection limit (50-µl injection)
Selenate	0.1 ppm
Iodide	0.1 ppm
Ferrocyanide	1 ppm
Ferricyanide	1 ppm
Gold (I) cyanide	1 ppm
Gold (III) cyanide	1 ppm
Cobalt (III) cyanide	1 ppm
Benzoate	1 ppm
Alkyl sulfonates	0.5-1 ppm
Aryl sulfonates	0.5-1 ppm
Alkyl sulfates	0.5-1 ppm
Lactate	0.5 ppm
Tartrate	0.5 ppm
Citrate	1 ppm
Succinate	0.5 ppm
Glycolate	0.5 ppm
Pyruvate	0.5 ppm
Proprionate	1-2 ppm
Tetrafluoroborate	0.1 ppm
Monofluorophosphate	10 ppb
Oxalate	5-10 ppm
Hexafluorosilicate	5-10 ppm
Thiocyanate	0.5 ppm
Iodate	0.1 ppm

IV. CATION ANALYSIS

A. The Columns

The first columns developed for cation analysis were synthesized by lightly sulfonating a raw polystyrene-divinylbenzene polymer. These columns are ideally suited for the separation of the common alkali metals and the alkaline earths. Since these first columns, improvements have been made in both the efficiency and capacity. New cation columns that have been developed have been designed to separate transition metals and amino acids. Because these applications do not utilize eluant suppression, they will not be discussed in detail in this chapter. Analogous to the use of macroporous resins for the MPIC (ion pair) separation of anions, the same type of columns can be used to separate cations. Table 6 lists the columns used for the separation of cation by means of eluant suppression.

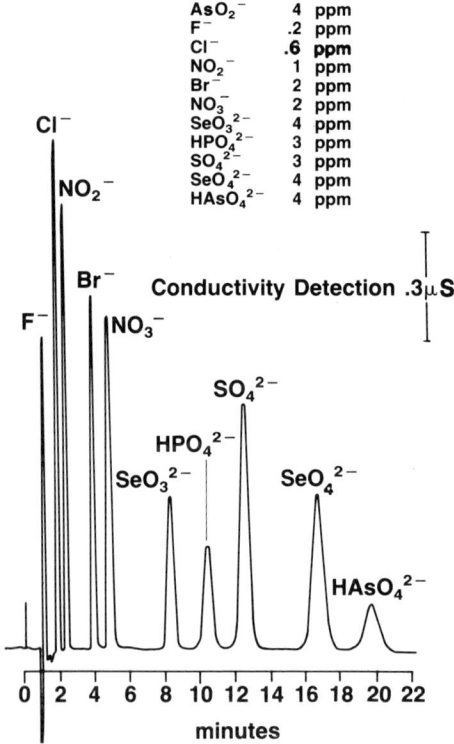

Figure 4. Separation of common anions. Column, HPIC-AS4A; eluant, carbonate/bicarbonate.

Table 6. Typical Columns and Eluants for Cation Separations

Ion exchange separations:

    CS-1   0.005 M HCl for monovalents

                0.002 M m-PDA-2HCl, 0.002 M HCl for divalents

MPIC separations:

    NS-1   Mixtures of hexane sulfonic acid or octanesulfonic acid with acetonitrile and water

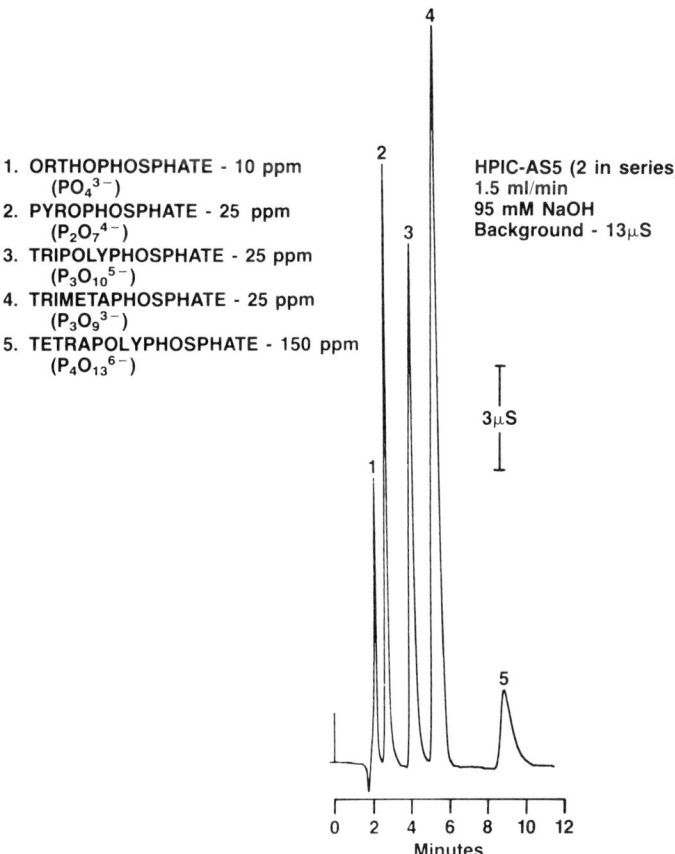

Figure 5. Separation of polyphosphates with a conductivity detector, using the anion membrane suppressor.

B. The Eluants

The separation of the common alkali metals and simple amines is usually accomplished with the use of dilute mineral acid solutions. The most common choice is HCl, but nitric acid may also be used with good success. The only limitation to the eluant choice is that it can be suppressed by the postcolumn device. The divalent alkaline earths, such as magnesium and calcium, require the use of eluants with much higher affinity for the ion exchange resin. Silver nitrate was described in the first publication by Small and co-workers [1]. However, this eluant

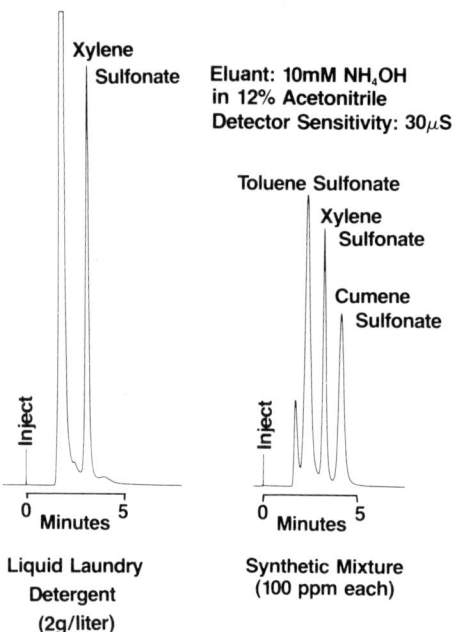

Figure 6. Analysis of hydrotropes by MPIC.

did not gain much popularity because the suppressors could not be easily regenerated. The best success was achieved when eluants derived from aromatic diamines were used. *Para*-phenylenediamine adjusted to a pH of 2-3 with HCl was found to provide adequate results. It was soon discovered that this eluant was not particularly stable with time and that air oxidation of the amine could lead to a "poisoning" of the analytical column. Experimentation has shown that the meta isomer yields equivalent chromatographic results (Figure 8) and is significantly more resistant to air oxidation. The separation of cations via the MPIC or ion-pair mode is usually accomplished by use of an eluant derived from a hydrophobic acid such as hexane or octane sulfonic acid in combination with organic modifiers such as acetonitrile or methanol. Typical eluants, along with the column types, are given in Table 6.

C. The Regenerants

Regardless of the type of suppressor used, the most common regenerant used is hydroxide. The early packed-bed devices employed sodium

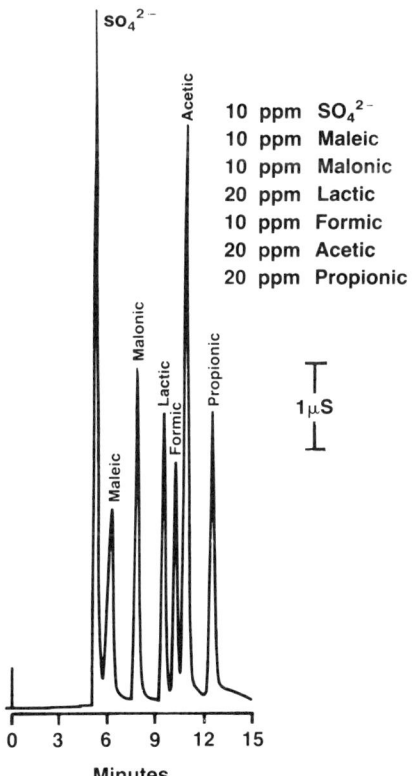

Figure 7. Ion exclusion separation.

hydroxide. The new fiber devices employ potassium or tetramethylammonium hydroxide. Leakage of the smaller cations, such as sodium and lithium, across the membrane causes a higher background, and thus these cations are usually not recommended. An important point must be remembered when one is determining amines. If the regenerant is a hydroxide, the amine cation is converted to the corresponding free amine. The free amine is, of course, nonconducting. The response observed is due to the hydrolysis of the amine to the amine hydroxide. If one desires higher sensitivity for amines, this can be achieved by the use of a carbonate-based regenerant. This regenerant will convert all of the amine to the corresponding amine carbonate salt, and a much larger peak will be observed. In general, the regenerants used for MPIC techniques are also derived from strong alkali

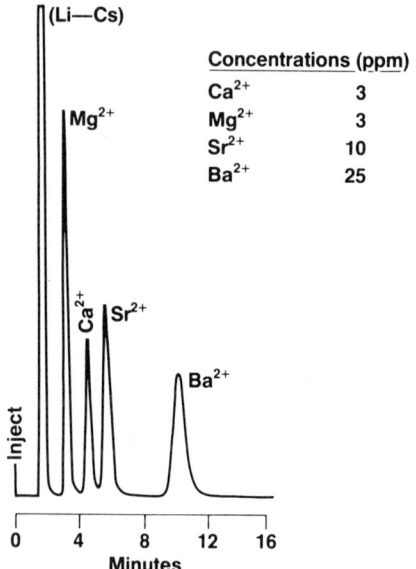

Figure 8. Separation of divalent cations. Eluant, 2.5 mM *meta*-phenylendiamine dihydrochloride, 2.5 mM HCl.

bases or quaternary amines. The common regenerants are listed in Table 7.

D. Detectable Species

The species detectable by this technique are listed in Table 8. The injection volume is 50 μl, and the figures given are three times the detection limit. Just as for the anion case, the limits can often be lowered by the use of appropriate concentration techniques. Examples of typical separations are shown in Figures 9, 10, and 11.

Table 7. Typical Regenerants for Cations

Packed-bed devices	0.5 M NaOH
Fiber devices	0.04 M Tetramethylammonium hydroxide
Suggested regenerant for best amine sensitivity	0.05 M Potassium carbonate

Table 8. Cations Detectable via Eluant Suppression

Sodium	5 ppb
Lithium	2 ppb
Ammonia	5 ppb
Potassium	5 ppb
Cesium	10 ppb
Magnesium	20 ppb
Calcium	20 ppb
Barium	30 ppb
Strontium	40 ppb
Alkyl amines (C1-C4)	100 ppb
Alkyl amines (>C4)	500 ppb
Aromatic amines	500 ppb

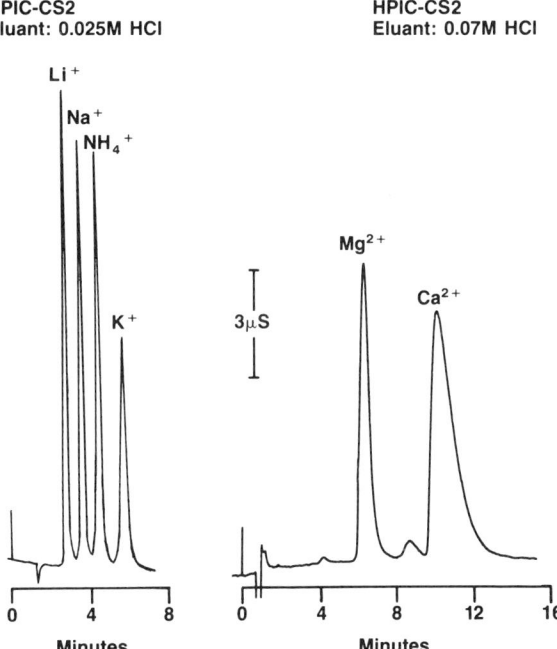

Figure 9. Rapid separation of alkali-alkaline earth metals, using the cation membrane suppressor.

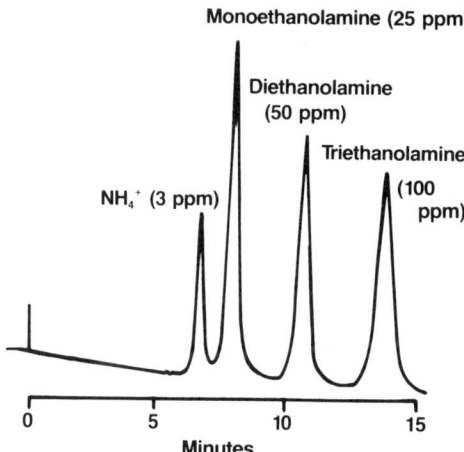

Figure 10. MPIC separation of ethanol amines (HSA = hexanesulfonic acid).

V. CONCLUSION

Eluant-suppressed ion chromatography is a powerful analytical technique, providing sensitivities comparable to the best spectroscopy methods. The combination of rugged high-capacity columns and post-column suppression provides the analyst with a technique that puts less demand on sample preparation and thus makes his/her job less complicated. Advances in the suppressor devices have developed an analytical technique utilizing a bulk property detector which is compatible with both step and continuous gradients. The application of the method continues to expand into new areas, particularly large organic molecules with little or no ultraviolet (UV) absorption.

REFERENCES

1. H. Small, T. Stevens, and W. Bauman, Novel ion exchange chromatographic method using conductimetric detection, *Anal. Chem.* 47, 1801 (1975).
2. C. Pohl, Chromatographic Separation and Quantitative Analysis of Ionic Species, U.S. Patent 4,265,634.

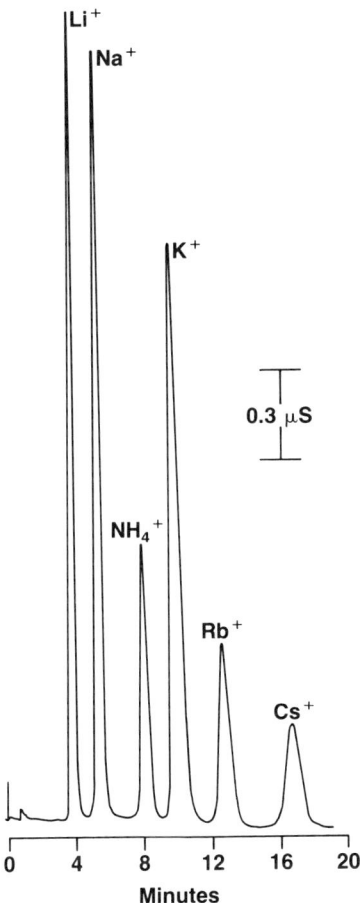

Figure 11. Rapid separation of all alkali metals, eluant, 0.005 M HCl.

3. W. Rich, E. Johnson, and T. Sidebottom, Method and Apparatus for Quantitative Analysis of Weakly Ionized Anions, U.S. Patent 4,242,097.
4. C. Pohl, R. Slingsby, E. Johnson, and L. Angers, Method and Apparatus for Ion Analysis and Detection Using Reverse Mode Suppression, U.S. Patent 4,455,233.
5. J. Stillian, R. Slingsby, and C. Pohl, A Revolutionary New Suppressor for Ion Chromatography, presented at 1985 Pittsburgh Conference, New Orleans, LA.

6. J. Stillian, An Improved Suppressor for Ion Chromatography, *LC, Liq. Chromatogr. HPLC Mag. 3*, 802-812 (1985).
7. P. R. Haddad and A. Heckenberg, *Chem. Aust. 50*, 275-278 (1983).
8. F. C. Smith, Jr., and R. C. Chang, *CRC Crit. Rev. Anal. Chem. 9*, 197-217 (1980).

2
Single-Column Ion Chromatography

Thomas Jupille*
Wescan Instruments, Inc.
Santa Clara, California

I. OVERVIEW

A. Introduction

In the late 1970s, a number of ion chromatographic techniques were developed that did not require the use of a suppressor column in contrast to the original technique developed in 1975 [1]. The various terms used to describe these approaches to ion chromatography (IC)-- "suppressorless" IC, "nonsuppressed" IC, "electronic suppression" IC, etc.--all leave something to be desired. Because these techniques have as a common characteristic the direct coupling of an appropriate detector to the outlet of a column, the term *single-column ion chromatography* (SCIC) has come to be generally accepted [2]. The alternatives to SCIC technology can be classified together as those involving postcolumn eluant modification before detection.

At one time or another, virtually every type of HPLC detector has been applied to SCIC. Specifically, SCIC applications have been reported with the use of (a) refractive index detection [3]; (b) UV absorbance detection, by both direct absorbance measurement [4-7] and indirect absorbance measurement [8-10]; and (c) electrochemical detection [11,12], in addition to the "workhorse" (d) electrical conductivity detection [13,14]. More specific information concerning the various detectors can be found in a later chapter.

Current affiliation: Pi Technologies, Orinda, California.

B. Separation Mechanisms

Along with all other aspects of ion chromatography, the variety of separation mechanisms has proliferated in recent years. To the original ion exchange technology have been added ion exclusion and ion pair chromatography, and also reverse phase and even normal phase chromatography. Each mechanism has its particular advantages and disadvantages and can be optimized for different sample types.

Ion Exchange Chromatography

Ion exchange chromatography was the first approach for both eluant-suppressed IC and SCIC [1,15-17]. Because the technique depends on competition between sample ions and buffer ions for active sites on the ion exchanger surface, ion concentration in the column effluent must remain constant; only the identity of the ions changes as sample ions replace buffer ions and vice versa. This confers a great predictability on ion exchange systems; their behavior is well defined.

Ion Pair Chromatography

Ion pair separation can be visualized as an ion exchange process wherein the ion exchanger (stationary phase) is "dynamically" maintained by the adsorption of relatively hydrophobic charged species from the mobile phase onto the hydrophobic surface of the reverse phase column. Viewed in this manner, ion pair chromatography confers a much greater degree of flexibility than does conventional "fixed-site" ion exchange [18-20]. Both the identity of the ion exchange sites and the capacity of the system can readily be modified as necessary to meet new analytical conditions. Because of this great degree of flexibility, however, optimization is often far from straightforward.

Ion Exclusion Chromatography

Ion exclusion (occasionally referred to as *ion-moderated partition* chromatography) actually involves a complex set of mechanisms: ion exclusion, partition, ligand exchange, and size exclusion [21]. They are exploited to separate relatively weakly charged species on a strongly ionized stationary phase of the same charge (thus, negatively charged anions are separated on a negatively charged cation exchange resin). As carried out in practice, ion exclusion allows only limited flexibility in selectivity. It does, however, separate all the ions in a given pK range; it is thus useful in providing a profile of ion concentrations in a sample.

Reverse Phase Chromatography

In many applications, the distinctions among reverse phase chromatography, ion pair chromatography, ion exclusion chromatography, etc., are only tenuous; separations actually arise from a mixture of mechanisms. What are traditionally thought of as "reverse phase" columns in HPLC--typically C18 or C8 alkyl functionality bonded to a silica gel substrate--are used in ion chromatography both for ion pair separations described above and for separations based clearly on the hydrophobicity of sample molecules for separations such as alkyl sulfate or sulfonate surfactants [22].

Normal Phase Chromatography

The degree to which ion chromatography and HPLC have converged is illustrated by the application of normal phase columns with nonaqueous eluants and conductivity detection for the analysis of long-chain alkyl quaternary ammonium surfactants [22]. The sample species have no chromophores and are thus difficult to detect by UV absorbance. The eluant used (chloroform/methanol) is not one that is often associated with conductivity detection. The low background conductance, however, allows conductivity measurements to be made with a high-sensitivity cell and adequate (microgram range) detection limits to be obtained.

C. Column Technology

Column technology for SCIC has generally followed the principles and practices associated with HPLC. Column dimensions are typically 5, 10, or 25 cm long by 2-5 mm internal diameter. Systems are typically operated at flow rates in the 0.5- to 8.0-ml/min range. Normal back pressures vary from a few hundred to a few thousand psi. With very few exceptions, columns for SCIC are compatible with HPLC instrumentation and hardware.

Injection volumes in ion chromatography (including SCIC) are generally somewhat larger than those normally associated with HPLC practice. "Typical" SCIC injection volumes are in the 100-µl range, with volumes as large as 2 ml not uncommon. This contrasts with HPLC, where 5-10-µl injections are the norm.

In most SCIC applications, the peak width is almost independent of injection volume over a wide range. The result can be the seeming paradox of peak widths being actually smaller than the volume injected. The phenomenon is explained by the "preconcentration" of sample ions at the head of the column. Because most samples have only a very low concentration of ions, there is little if any competition for ion exchange sites. Virtually all of the sample ions displace buffer ions and are concentrated at the column inlet. These concentrated species are displaced in their turn by the much higher concentration of buffer ions in the eluant.

II. ELECTRICAL CONDUCTIVITY DETECTION

A. Ion Exchange Separations

Effect of Equivalent Conductance on Detection Sensitivity

The key to the development of SCIC was the observation that differences in equivalent conductance determine sensitivity [15-17]. Although SCIC first became practical as a result of the development of sensitive conductivity detectors with a wide range zero offset or "electronic suppression" capability, successful implementation of the techniques depends primarily on the choice of an eluant buffer whose equivalent conductance differs greatly from that of the sample ions.

We should keep this in mind in formulating our first rule for eluant selection in SCIC with conductivity detection: *Wherever possible, choose an eluant buffer that maximizes the difference in equivalent conductance between the sample and the buffer.*

Because sample ions are apt to be intermediate in equivalent conductance, the most generally useful buffer ions have either very high or very low equivalent conductance. For anion analysis, this usually means big, bulky organic acids or their salts (low equivalent conductance) [15,17] or hydroxide ion (high equivalent conductance) [14]. For cation analysis, big, bulky quaternary ammonium compounds are, in principle, suitable as low-equivalent-conductance driving ions, but practical problems with adsorption onto ion exchange resins limit their utility. Most cation separations are carried out using high-equivalent-conductance driving ions such as hydronium (monovalent) or ethylenediammonium (divalent) [17].

Anion Exchange

Knowing that we want low equivalent conductance is not enough. In order to effectively optimize anion exchange separations in SCIC, we must also consider the factors that determine eluant driving strength.

In general, the driving strength of a buffer ion is closely related to its affinity for the ion exchanger used. Other things being equal, the order of affinity/driving strength is monovalent < divalent < trivalent [24]. Of course, nowhere is it written that only a single type of ion may be used in a single eluant buffer. In practice, a great deal of flexibility in system optimization is available if one chooses as a buffer ion a weak acid that is a difunctional (or trifunctional) with different pK_a values for each ionization [25]. This allows the relative concentration of monovalent (weak) and divalent (strong) driving ion to be controlled by changing pH. Figure 1, for example, shows retention for a number of inorganic ions as a function of sample pH with 4 mM phthalate buffer on the Wescan "standard" low-capacity, silica-based anion-exchange column.

Single-Column Ion Chromatography

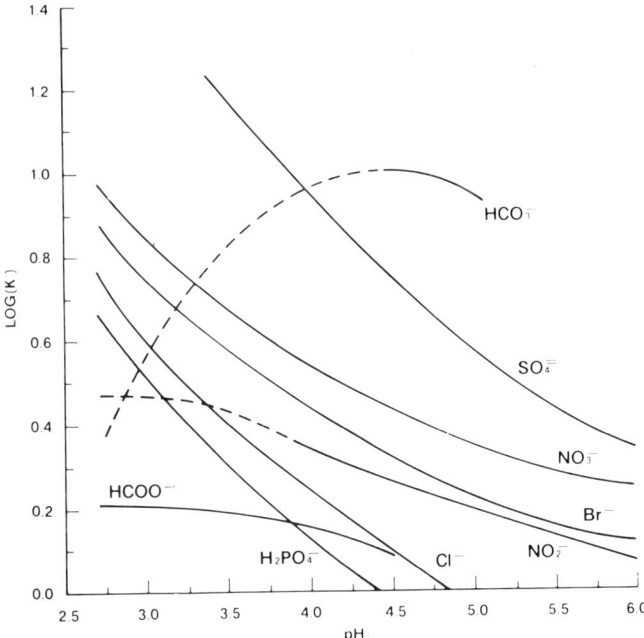

Figure 1. Retention behavior as a function of pH for common inorganic anions on a commercially available silica-based anion exchange column, using 4 mM phthalate buffers. Note that the peak mislabeled "HCO_3^-" is actually the *system peak*, representing adsorption of neutral phthalic acid.

Figure 1 illustrates a number of general patterns in retention behavior that should be considered when one is optimizing a buffer system:

1. *Other things being equal, the effect of buffer driving strength on sample ion retention depends on the charge of the sample ions.* Again, the order of response is monovalent < divalent < trivalent. Thus, the retention time of a divalent ion such as sulfate changes much more rapidly in response to changes in buffer composition than does the retention time of a monovalent ion such as chloride.
2. *The order of elution for ions of the same charge is more or less independent of the buffer composition.* In fact, if we assume that the buffer and sample ions do not interact, but only compete for ion exchange sites on the stationary phase, then selectivity should be uniquely a function of the stationary phase

chemistry. In practice, this means that we have little hope of making nitrate elute before, instead of after chloride by merely changing the buffer; we would have to find a different column packing that provides the appropriate selectivity.

3. *Buffer optimization can be complicated by the existence of a system peak in the chromatogram.* This peak may be either positive or negative, but elutes at a characteristic retention time for a given eluant buffer. The most convincing explanation of the genesis of system peaks is that given by Fritz, Gjerde, and Becker: A system peak results from the elution of an excess (or deficit) of the neutral form of the buffer ion [17].

Because of the importance of system peaks, a more detailed explanation is in order. Any chromatographic system involves a number of simultaneous equilibria. In the present case (an anion exchange column eluted with an organic acid buffer), we must consider three related equilibria (Figure 2):

1. Between buffer ions in the mobile phase and buffer ions bound to ion exchange sites on the stationary phase
2. Between buffer ions and un-ionized acid in the mobile phase
3. Between un-ionized acid in the mobile phase and un-ionized acid adsorbed onto the nonionic portions of the stationary phase surface

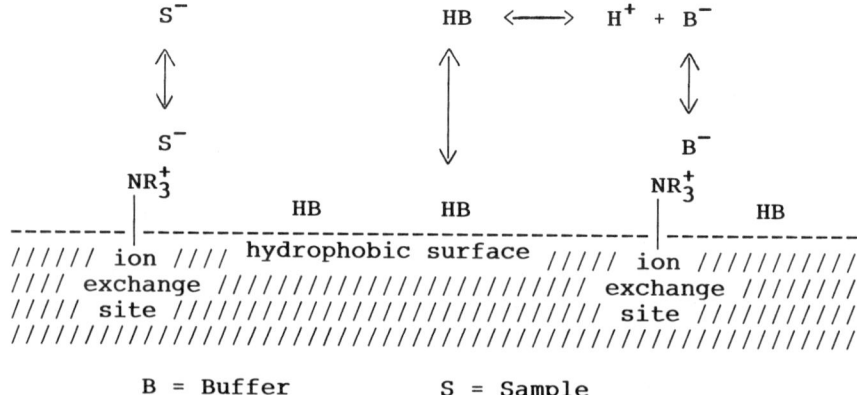

Figure 2. A number of equilibria affect single-column ion chromatography. Buffer ions (usually weak acid anions) equilibrate with the free acid in solution. Both of these species, in turn, equilibrate with their bound forms at the surface of the stationary phase.

Single-Column Ion Chromatography

Assuming initial equilibrium conditions, the injection of a sample whose pH is lower than that of the eluant buffer will tend to protonate buffer ions. This increases the concentration of un-ionized acid in the mobile phase and, in turn, increases the amount of acid adsorbed on the stationary phase surface. The result is the same as if a sample containing the acid has been injected. The "excess" acid is carried through the column at a characteristic fraction of the mobile phase velocity and elutes as a peak at a characteristic retention time: the *system peak*.

Injection of a sample whose pH is higher than that of the eluant will similarly shift the equilibrium in favor of ionized acid in the mobile phase. The effect is to "strip" some of the adsorbed acid from the stationary phase surface. The resulting deficit of un-ionized acid is made up from the mobile phase and propagates down the column, finally emerging as a negative peak at its characteristic elution time.

System peaks are not unique to silica gel-based columns or to phthalate eluants, nor is their detection limited to electrical conductivity [9]. The fact that system peaks can be either positive or negative suggests that the *absence* of a component normally present in the mobile phase of an equilibrium system chromatographs with the same retention time as the *presence* of the same component. This suggestion has been invoked to explain the presence of a negative peak at the retention time of chloride when deionized water was injected into a system in which the mobile phase had been contaminated with chloride from a pH electrode [26].

It should be stressed that system peaks are reproducible: For a given eluant buffer, the retention time of the system peak is predictable. When multiple driving ions are used in a mobile phase, a system peak can be expected for each ion (Figure 3). In practice, one need only make sure that the system peak (or peaks) do not coelute with sample peaks of interest.

The most common eluants for anion analysis in SCIC are based on organic acids and their salts. Organic acids tend to have a low equivalent conductance (the bigger and bulkier the acid, the better, from this point of view). Furthermore, the wide range of pK_a values available allows a great degree of flexibility in tailoring retention as a function of pH.

The difunctional phthalic acid (o-phthalic acid) has pK_a values of 2.8 and 5.5. This allows the ratio of monovalent (biphthalate) to divalent (phthalate) ions to be controlled by controlling eluant buffer pH over the range from about 2.5 to about 5. This range corresponds nicely to the range of stability of the silica-based columns, which were the first commercially available columns suitable for SCIC analysis of anions. Of course, not all pH values within this range are equally useful. The presence of a system peak from phthalic acid means that useful phthalate buffers for use with Wescan standard anion columns can be prepared in three pH ranges (Figure 4) [27]:

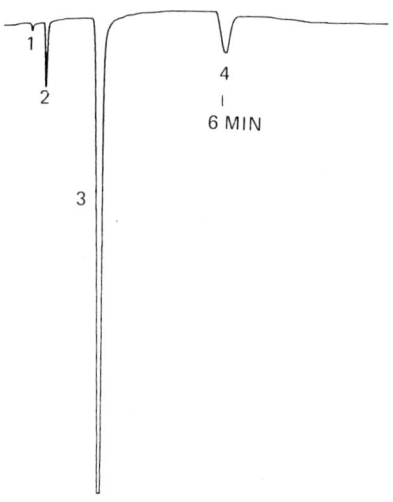

Figure 3. System peaks may occur even when no ions are injected with the sample (1% sugar solution). Furthermore, when multiple driving ions are used in the buffer, multiple system peaks will result. Peaks: (1) cations, (2) sugar, (3) phthalic system peak, (4) benzoic system peak.

1. pH 2.7. Essentially the pH of phthalic acid itself, this buffer places the system peak at the start of the chromatogram. Because only monovalent driving ions are present, phthalic acid is a relatively weak eluant. It provides very good resolution of such species as phosphate, chloride, or nitrate, but leaves sulfate and other di- or trivalent species essentially irreversibly retained.
2. pH 3.8. This intermediate-pH phthalate buffer provides good resolution of common inorganic ions, with the system peak typically located between nitrate and sulfate [13].
3. pH 4.5. Separations using standard silica gel-based anion exchange columns can also be carried out using 4 mM phthalate at pH 4.5. This eluant places the system peak after sulfate.

The analysis of inorganic acids in many real samples (foods, beverages, plant extracts, etc.) can be complicated by the presence of organic acids. The ability of SCIC to operate over a wide pH range provides the flexibility to minimize organic acid interference by carrying out the separation at low pH. Under these conditions, the weakly acidic organic acids are protonated and only weakly retained by the column. As a result, they elute early in the chromatogram, well away from the inorganic species of interest (Figure 5).

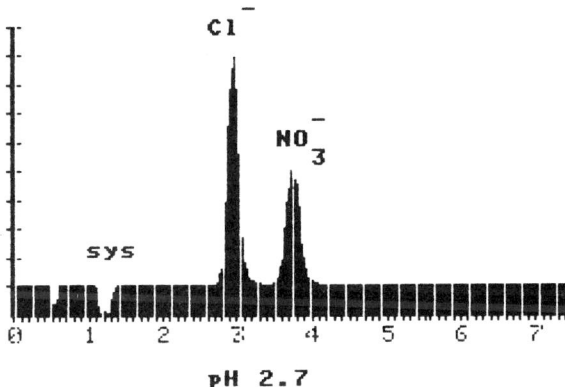

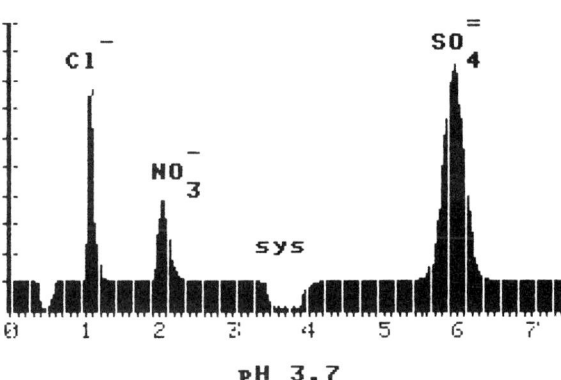

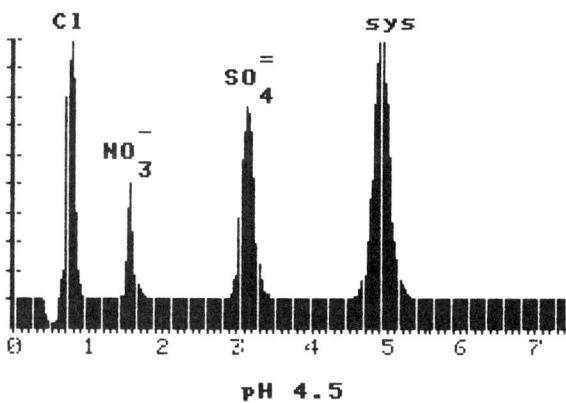

Figure 4. Four millimolar phthalate buffers can be used at three pH values on commercially available silica-based anion-exchange columns. At these pH values, the system peak is resolved from inorganic anions of interest.

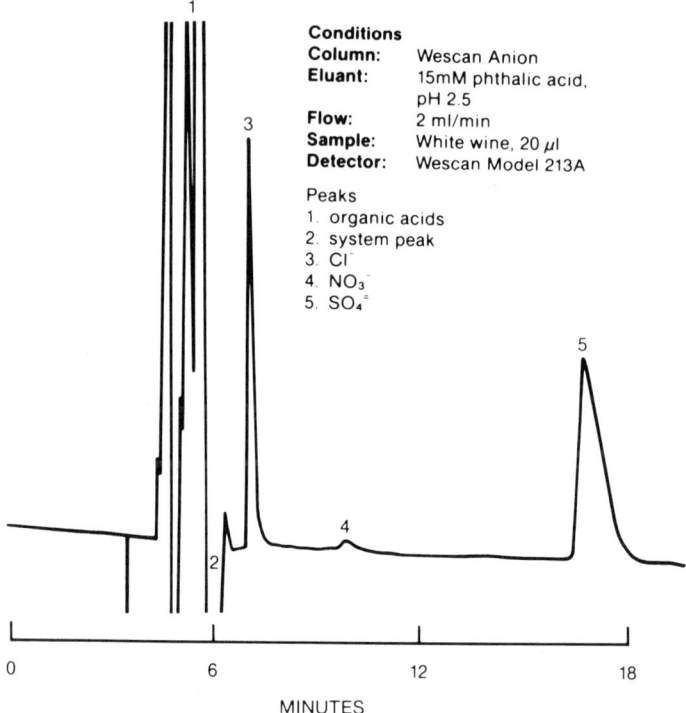

Figure 5. Low-pH eluants will tend to suppress ionization of weak acids in the sample. This allows easy analysis for inorganic ions in the presence of organic acids in samples such as wine.

A variant on this buffer can be used with extremely short columns (3-cm bed length) and high flow rates (8 ml/min) to provide rapid analysis of nitrate and/or sulfate (Figure 6) [28].

The same approach allows the analysis of low levels of fluoride in concentrated boric acid solutions. The eluant pH is low enough to suppress the ionization of the weakly acidic borate. This ensures that borate is essentially unretained by the column and does not constitute an interference.

Because the pK_a for the second ionization of phthalate is 5.5, raising the pH above 6 or so provides little further increase in driving strength (the phthalate is already essentially completely divalent). With the recent commercial introduction of resin-based anion exchange columns, the need has arisen for buffer ions that will provide pH and retention control in the pH range from 6 on up. A look at the pK_a values of common organic acids (Table 1) suggests *p*-hydroxybenzoate as a possibility.

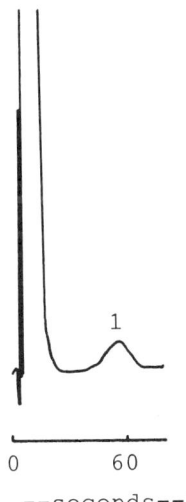

Figure 6. The use of a low-pH eluant coupled with a short column operated at a high flow rate allows for a rapid analysis of sulfate in serum without sample pretreatment. Conditions: column, Wescan Ion-Guard Anion; eluant, 20 mM phthalic acid; flow rate, 8 ml/min; sample, human serum, 100 µl; detector, Wescan ICM conductivity. Peaks: SO_4^{2-}, 2 ppm.

p-Hydroxybenzoic acid has pK_a values of 4.4 (carboxyl proton) and 9.3 (phenolic proton). These values not only make p-hydroxybenzoate a good buffer in the pH range from 4 to 10 or so, but also allow convenient control of retention as a function of pH (Figure 7). p-Hydroxybenzoic acid is typically used at pH 8.4-8.6 for the analysis of common inorganic ions. In contrast to the lower-pH phthalate eluants, p-hydroxybenzoate at pH 8.6 elutes phosphate between nitrate and sulfate, providing an elution sequence similar to that commonly encountered in eluant-suppressed IC with bicarbonate eluants.

Other organic acid systems have been used for anion analysis by SCIC: phenate [29], salicylate [30], and glycolate/borate [14]. Glycolate/borate eluants can be particularly useful because they are only weakly adsorbed on commercially available columns. As a consequence, the system peaks elutes early in the chromatogram (usually indistinguishable from the *water dip*) where it does not interfere with ions of interest (Figure 8).

The emphasis placed on difunctional acid buffers should not be overdone; monofunctional acids are still useful as driving ions.

Table 1. Organic Acid Dissociation Constants (Divalent)

Organic acid	pK_a 1	pK_a 2
Oxalic	1.23	4.19
Tartaric	2.98	4.34
Adipic	4.43	4.41
Fumaric	3.03	4.44
m-Phthalic	3.54	4.60
Mesaconic	3.09	4.75
meso-Tartaric	3.22	4.82
p-Phthalic	3.51	4.82
Malic	3.40	5.11
Glutaric	4.34	5.41
Itaconic	3.85	5.45
o-Phthalic	2.89	5.51
Succinic	4.16	5.61
Methylsuccinic	4.13	5.64
Malonic	2.83	5.69
Dimethylmalic	3.17	6.06
Maleic	1.83	6.07
Cyclohexane-1:1-dicarboxylic	3.45	6.11
Cyclopropane-1:1-dicarboxylic	1.82	7.43
p-Hydroxybenzoic	4.48	9.32
Aspartic	3.86	9.82
Cystine	7.85	9.85
m-Hydroxybenzoic	4.06	9.92
Cysteine	8.14	10.34
Ascorbic	4.10	11.79
o-Hydroxybenzoic	2.97	13.40

Weakly retained anions, for example, are well resolved by use of nicotinic acid (Figure 9) [31]. Nicotinic acid is used in preference to other monofunctional acids because of its low equivalent conductance and good solubility in water. In addition, nicotinic acid is only weakly adsorbed onto anion exchange resins, with the result that system peak formation is minimized.

A major advantage of the new resin-based anion exchange columns is their applicability to high-pH eluants for the analysis of weakly dissociated anions. Such species as cyanide, sulfide, silicate, or borate can be analyzed by SCIC via direct electrical conductivity detection [29,32]. For this type of analysis, the use of a high-pH buffer both ensures that sample species are ionized and provides good sensitivity

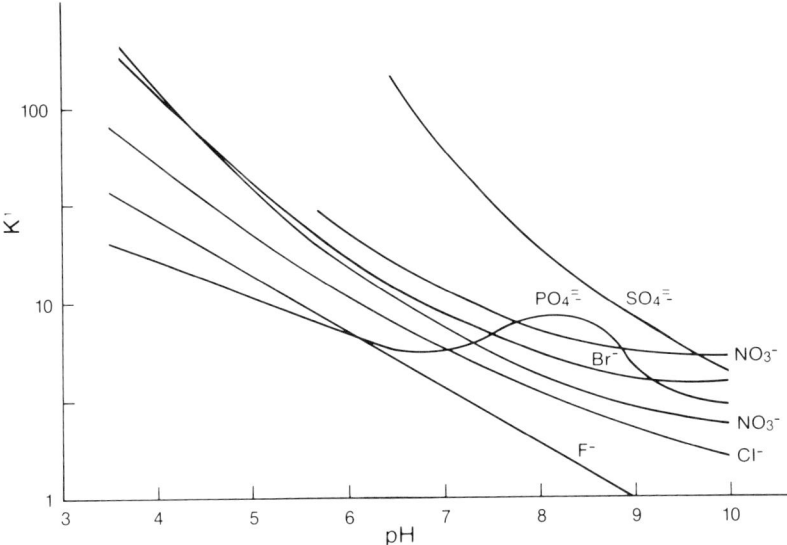

Figure 7. Retention behavior as a function of pH for common inorganic anions on a commercially available resin-based anion exchange column, using 5 mM p-hydroxybenzoate buffers.

by conductivity detection. Because the equivalent conductance of the sample ions is lower than that of the hydroxide used in the buffer, sample peaks in this type of system are *negative*.

Hydroxide itself is a relatively weak driving ion. In many cases, better sensitivity and retention control can be obtained by using eluant buffers containing multiple driving ions. The use of co-ions with hydroxide eluants, for example, allows for faster separations and (because peaks of interest elute earlier and are thus sharper) better sensitivity (Figure 10).

On a practical level, care should be taken in using basic eluants for SCIC to ensure that contact between the eluant and atmospheric carbon dioxide is minimized. This can be conveniently accomplished either by use of an adsorbent trap or an inert gas blanket over the eluant reservoir.

Detection limits for anions (expressed as *minimum detectable concentration*) vary with the sensitivity of the detector, with the volume of sample injected, and with the identity, concentration, and pH of the eluant buffer as well as with chromatographic factors such as column efficiency, elution volume, etc. As a result, absolute statements about detection limits are very difficult to come by. Detection

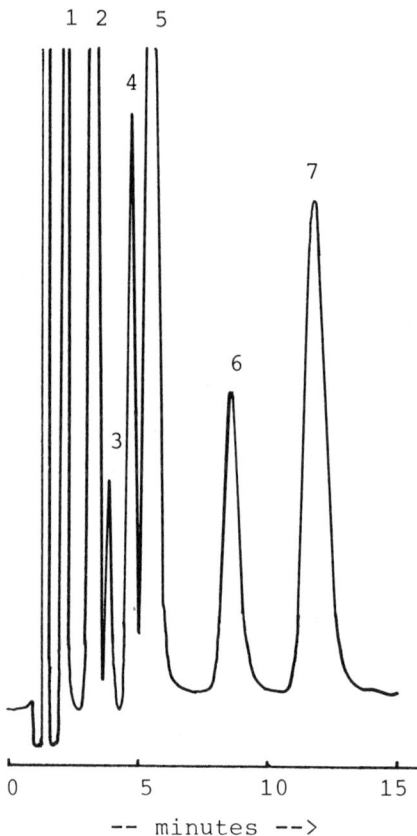

Figure 8. Glycolate-borate eluants are only weakly adsorbed. As a consequence, the system peak is not observed when these eluants are run on an acrylic resin-based anion exchange column. Conditions: column, Bio Rad IC Anion PW; eluant, 1.48 mM sodium gluconate, 5.82 mM boric acid, 1.30 mM sodium tetraborate, 12% acetonitrile, 0.25% glycerine; flow, 1 ml/min; sample, Anion Standard, 100 μl detector, Wescan ICM Conductivity. Peaks: (1) F^-, (2) Cl^-, (3) NO_2^-, (4) Br^-, (5) NO_3^-, (6) HPO_4^{2-}, (7) SO_4^{2-}.

limits for anion exchange SCIC with electrical conductivity detection have been quoted as:

1. One hundred to 200 ppb for 200-μl injection, using 4 mM phthalate at pH 4.5 [33].

Single-Column Ion Chromatography

Conditions
Column: Anion/R-HS
Eluant: 10 mM Nicotinic Acid
Flow: 2.7 ml/min
Sample: Anion Standard; 100 microliters
Detector: Wescan ICM Conductivity

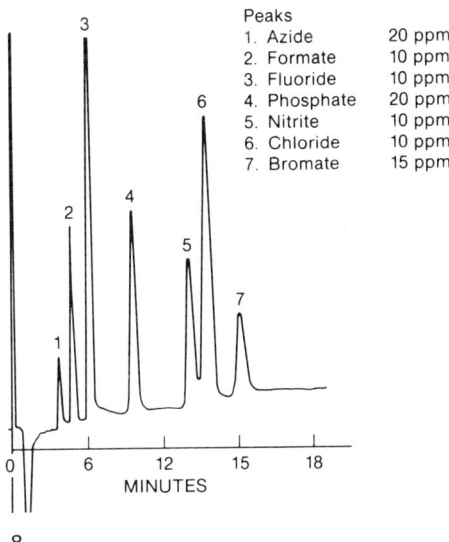

Peaks
1. Azide   20 ppm
2. Formate   10 ppm
3. Fluoride 10 ppm
4. Phosphate 20 ppm
5. Nitrite   10 ppm
6. Chloride 10 ppm
7. Bromate   15 ppm

Figure 9. Nicotinic acid is highly water soluble and only weakly adsorbed. As a strictly monovalent driving ion, it provides good resolution for weakly retained anions.

2. Forty to 300 ppb for 100-µl injection, using 5 mM p-hydroxybenzoic acid at pH 8.6 [34].
3. Five to 25 ppb for 100-µl injection, using 5 mM hydroxide eluant [14].

Much of the variation in detection limits can be accounted for on the basis of differences in equivalent conductance among the buffers used. A significant (and often overlooked) factor in SCIC sensitivity, however, is the influence of temperature control. The electrical conductivity of aqueous solutions typically varies by approximately 2% per degree Celsius temperature change. If we assume that a noise level of 0.001 µS or less must be maintained on a 100-µS background conductivity (these are typical values for today's detectors), then short-term temperature fluctuations at the detector cell must be limited to no more than 5×10^{-4} °C! In practice, temperature fluctuations are controlled by combination of heat exchangers (to equilibrate

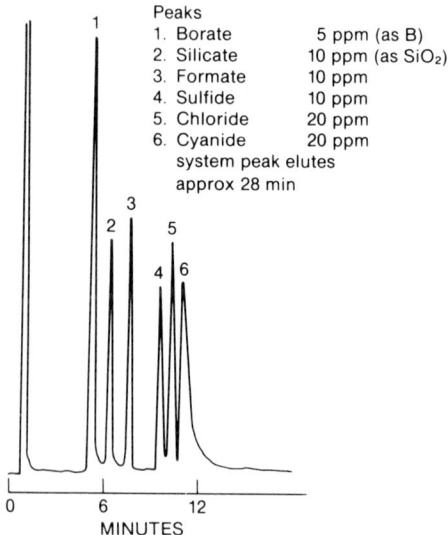

Figure 10. Co-ions may be used with hydroxide-based eluants to speed elution for sharper peaks and better sensitivity.

incoming and outgoing eluant), thermal mass (to ensure a stable thermal environment), and electronic temperature compensation (to adjust detector gain in response to residual temperature changes).

Cation Exchange

The basic rule for cation exchange SCIC is the same as that for anion exchange: Choose a buffer ion so as to maximize the difference in equivalent conductance with the sample ion.

The most commonly encountered approach to SCIC analysis of monovalent cations uses a dilute mineral acid buffer (typically 3 mM nitric acid) on a low-capacity, polystyrene resin-based cation exchange column [17] (Figure 11). Because hydronium ion is higher in equivalent conductance than, say, sodium or ammonium sample ions, the electrical conductivity signal *decreases* as sample ions elute from the column. In normal practice, recorder polarity is simply reversed to show these negative peaks in the "conventional" direction.

Single-Column Ion Chromatography

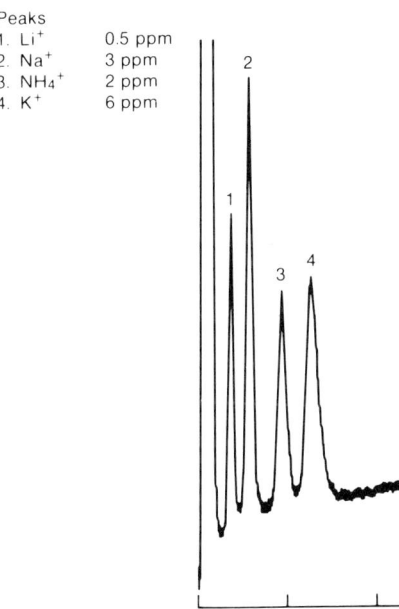

Conditions
Column: Wescan Cation/HS
Eluant: HNO_3, pH 2.0
Flow: 8 ml/min
Sample: Cation Std., 100 μl
Detector: Wescan Model 213A Conductivity Detector

Peaks
1. Li^+ — 0.5 ppm
2. Na^+ — 3 ppm
3. NH_4^+ — 2 ppm
4. K^+ — 6 ppm

Figure 11. Monovalent cations are separated by use of a nitric acid eluant on a resin-based cation exchange column.

The same column/eluant system can also be used for the analysis of simple amines. On commercially available columns, however, monomethylamine and potassium coelute when a 100% aqueous eluant is used. The addition of up to 40% methanol changes the selectivity of the columns to allow easy analysis of the simple amines (Figure 12) [35].

Additional differences in selectivity can accrue from the use of columns whose surface chemistry differs. Thus, whereas ethanolamines elute as a class from standard, low-capacity, polystyrene-based columns, they are readily separated on a silica gel-based cation exchange column of roughly the same capacity (Figure 13).

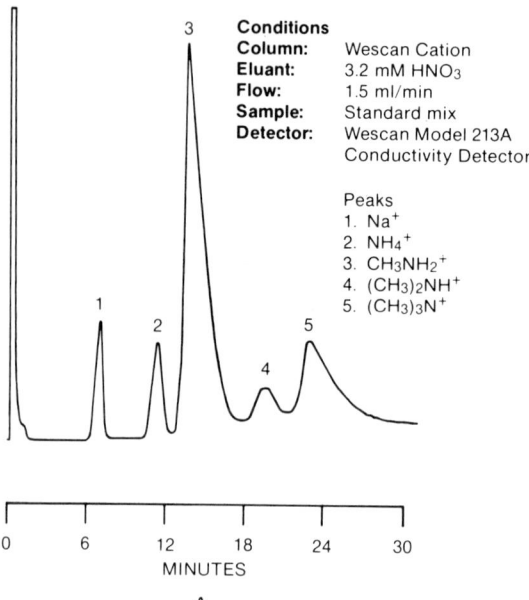

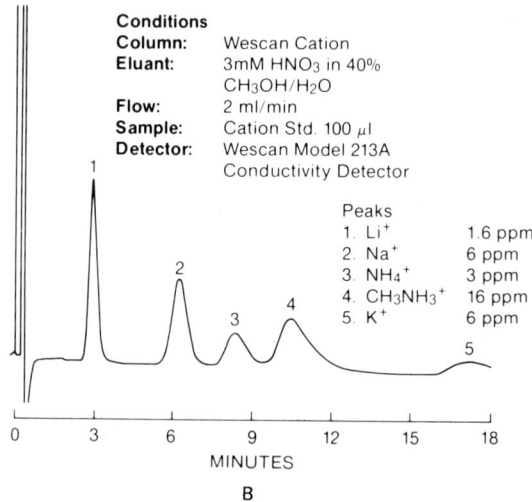

Figure 12. Methanol may be added as an organic modifier to improve the selectivity of SCIC cation exchange columns for simple amines. (A) Without methanol. (B) With methanol.

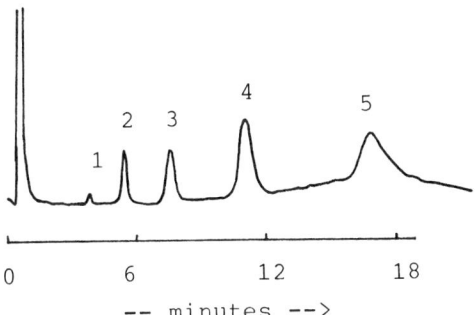

Figure 13. A silica-based cation-exchange column allows the separation of ethanolamines from one another (these species elute as a group on the resin-based column packing used in the previous figure). Conditions: column, Wescan Cation/S; eluant, 3.16 mM nitric acid; flow, 1.5 ml/min; sample, Ethanolamines Standard, 100 μl; detector, Wescan ICM Conductivity. Peaks: (1) sodium, trace; (2) ammonium, trace; (3) ethanolamine, 0.2 ppm; (4) diethanolamine, 7 ppm; (5) triethanolamine, 7 ppm.

Cation exchange columns used for alkali metal analysis are effective scavengers for traces of transition metals from both samples and the eluant buffer. As a consequence, column capacity (and retention time) gradually decreases as transition metal ions are tightly bound to exchange sites. As with many other problems, the best cure is prevention:

1. Use only "18 megohm" water and "Ultrex" grade or equivalent nitric acid in making up buffers.
2. Pretreat samples with chelating resins to remove high levels of transition metals, or with nonpolar adsorbents to remove nonpolar contaminants.

Contaminated columns can be quickly rejuvenated by injections of more concentrated nitric acid (5% or so works well) to remove calcium or magnesium or of complexing agents such as edetic acid (EDTA) to remove transition metal cations (Figure 14).

The comments made earlier about buffer-ion driving strength apply to cation analysis as well as to anion analysis. Divalent sample ions are generally too strongly retained to be eluted by a monovalent driving ion like hydronium. The analysis of divalent cations is typically carried out using a high-equivalent-conductance divalent driving ion such as ethylenediammonium (EDA) (Figure 15) [17]. Because the

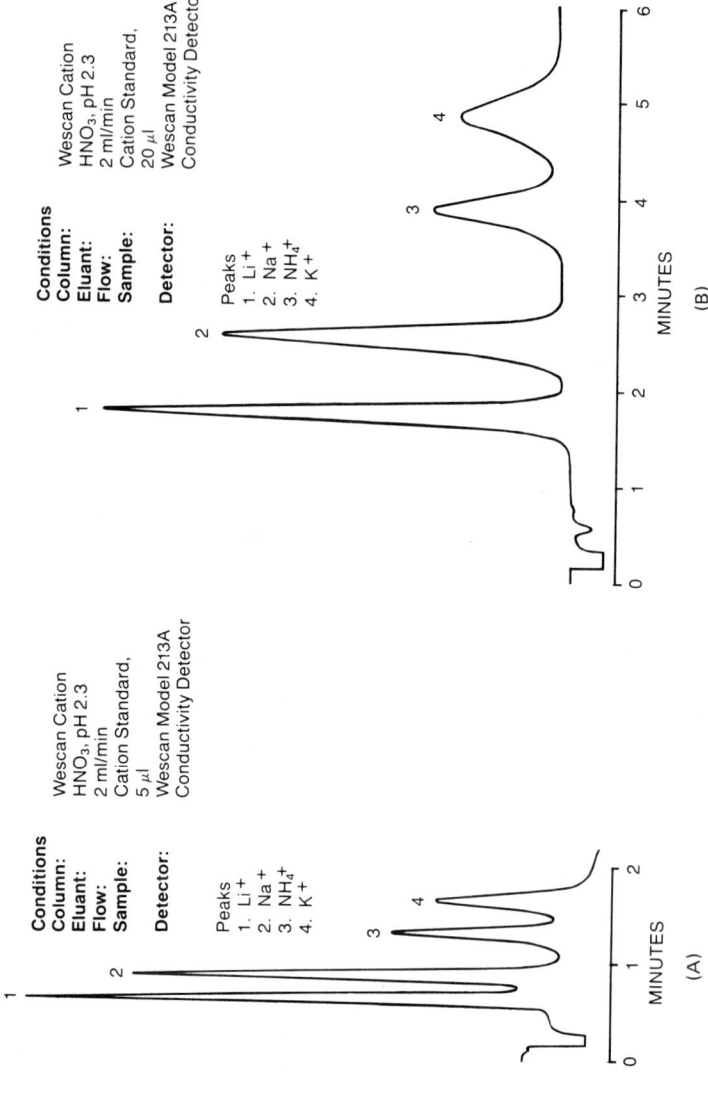

Figure 14. Cation column "poisoning" by trace heavy metal ions in the sample or the buffer can be cured by periodic cleaning with injected EDTA solutions. (A) Before cleaning. (B) After cleaning.

Single-Column Ion Chromatography

Conditions
Column: Cation/HS
Eluant: 1mM Ethylenediamine, pH 6.0 with HNO_3
Flow: 8 ml/min
Sample: Redwood City Tap Water, 100 µl
Detector: Wescan Model 213/HS Conductivity Detector

Peaks
1. Mg^{++} 2.4 ppm
2. Ca^{++} 11 ppm

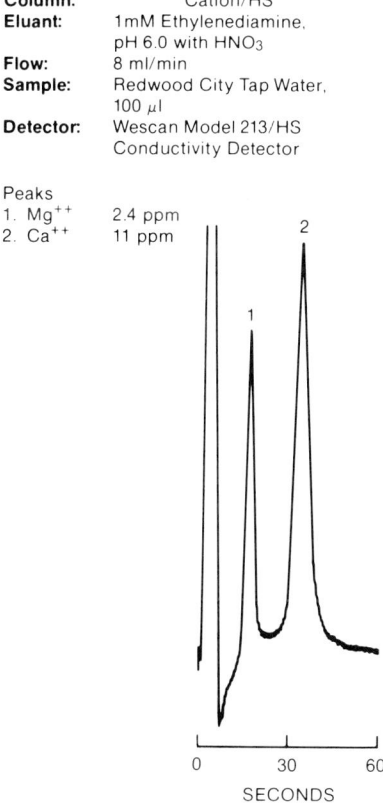

Figure 15. SCIC analysis for divalent cations is carried out with a divalent driving ion (ethylenediamine).

equivalent conductance of EDA is greater than that of the sample ions, the peaks are typically negative in divalent as in monovalent cation analysis.

Very strongly retained species such as transition metal cations can be separated by SCIC using eluants based on divalent EDA as the driving ion, but including complexing agents to reduce elution time and control selectivity (Figure 16) [36]. Tartaric acid and citric acid have both been used as complexing agents for this purpose. As is typical in cation analysis, peaks are negative because the sample species have a lower equivalent conductance than does the driving ion. Sensitivity for transition metals by SCIC employing conductivity detection is usually on the order of 10-100 ng.

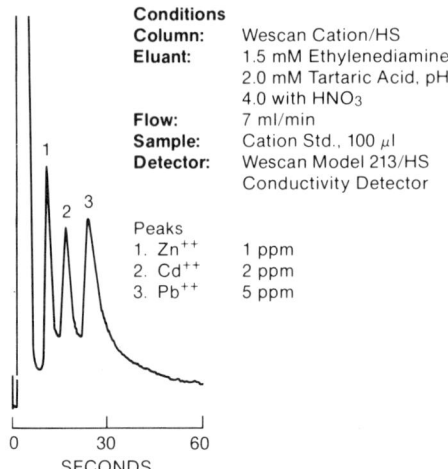

Figure 16. The addition of a complexing agent such as tartrate allows the separation of transition metal ions on a low-capacity cation exchange column.

Column Technology

The basic principles of column design are, of course, common to both SCIC and other forms of HPLC. In addition to the usual factors affecting efficiency, however, anion exchange SCIC columns must also meet constraints on capacity, selectivity, and adsorptivity.

The capacity of SCIC columns has generally been one to two orders of magnitude lower than that of "conventional" polystyrene-based anion exchangers of the "Dowex" type. The capacity of SCIC columns is, in fact, comparable to that of columns used for eluant-suppressed IC. In the latter case, low capacity is required in order that the capacity of the suppressor column or fiber not be exceeded. In the former case, low capacity is desirable in order to minimize the background signal. In both cases, the use of low-capacity columns maximizes the percentage change in conductivity resulting from replacement of a buffer ion by a sample ion and thus maximizes sensitivity.

The first commercially available column packings used for anion exchange SCIC were "pellicular" silica materials. Because the ion exchange sites are confined to a relatively thin porous layer on the surface of an impermeable glass core, the bulk capacity of the packing can be acceptably low for ion chromatography. The relatively large particle size (40 μm) of this type material, however, limits the attainable efficiency.

Single-Column Ion Chromatography

A similar selectivity is provided by totally porous packing material of 10-μm particle size based on a silica gel substrate. This material is made by covalently bonding an aromatic hydrocarbon stationary phase to the silica gel substrate by use of similar chemistry to that used in the preparation of reverse phase materials for HPLC. The aryl groups on the packing surface can then be further reacted to produce an anion exchanger. The selectivity of the packing material is determined by the ion exchange groups used, whereas the capacity is determined primarily by the pore size of the packing (which effectively controls the surface area available for bonding).

The totally porous silica packings provide good efficiency and selectivity, but are restricted to operation at relatively low pH. Although one would expect a silica gel substrate to limit pH stability to 7, in practice the weak link seems to be covalently bonded stationary phase. The author's experience with these packings has been that column lifetime decreases as the eluant pH is increased above 4.5.

Resin-based anion exchangers of sufficiently low capacity for ion chromatography have been described in the literature since 1978. Such packings have been produced by chloromethylating and then aminating finely ground samples of a commercially available macroporous polystyrene adsorbent. As with the silica-based packings described above, the selectivity of the packing is determined primarily by the chemistry of the ion exchange groups, whereas the packing capacity is determined by the pore size and surface area.

An alternative resin-based packing material is based on acrylate polymers. Because the surface of this type of material is more polar than that of polystyrene resins, it is somewhat less adsorptive toward organic acid molecules used in SCIC buffers. This tends to minimize the size of system peaks.

Polystyrene-based cation exchange resins for SCIC have been commercially available since 1980. The basic technology is similar to that described in 1980 [17]. A neutral polystyrene resin is partially sulfonated to produce a capacity in the 0.01-meq/g range (compared to 1-2 meq/g typical of fully sulfonated resins). Recently, a cation exchange resin based on a macroporous polystyrene has been described.

B. Ion Exclusion Separations

Principles

When the separation mechanism does not involve ion exchange, the requirement for electroneutrality associated with ion exchange no longer exists. Thus, the constraint on ion concentration discussed above no longer holds. In most cases, sample ions elute *in addition to* (*not instead of*) ions in the buffer. As a result, little or no attention need be paid to differences in equivalent conductance between the sample and the eluant. Buffers for non-ion-exchange separations using conductivity detection in SCIC are chosen to optimize selectivity and maximize the conductance of the sample ions. As a secondary consideration, the conductance of the eluant buffer should be minimized wherever possible, in the interests of maximizing signal-to-noise ratio.

Ion exclusion chromatography, sometimes referred to as ion-moderated partition (IMP) chromatography in fact involves a complex mixture of mechanisms [21]. A more detailed discussion can be found in Chapter 5. In the most common application, a high-capacity cation exchange resin is used to separate organic acids and other relatively weakly dissociated anions.

The early part of the chromatogram is dominated by *Donnan exclusion* or *charge exclusion*, in which strongly ionized sample species are excluded from the interior of the resin and hence elute quickly. The effect is exactly the same as though at the resin surface a membrane existed which was permeable to neutral or positive species but impermeable to negatively charged species. In point of fact, no physical membrane exists, and the phenomenon is better explained in terms of statistical thermodynamics: The free energy of the system will be minimized (entropy maximized) if the available negative charge is evenly dispersed throughout the system. Granted the existence of both fixed negative charge (the resin) and mobile negative charge (the sample species), this means that the mobile negative charge will not penetrate regions of fixed charge.

Practice

Although Donnan exclusion seems to predominate in the early part of the chromatogram (in which organic acids, for example, elute roughly in order of pK_a) (Figure 17), later-eluting peaks seem to be more strongly affected by hydrophobic interactions (short-chain fatty acids, such as formate, acetate, propionate, etc. elute in order of hydrocarbon chain length).

In practice, retention time for early-eluting species can be controlled to some extent by controlling the pH of the eluant buffer, thus controlling the degree of dissociation of the sample species. Because strong mineral acids are normally used as eluants, this typically means controlling the eluant acid concentration. In general, organic acids become more strongly retained as eluant acid concentration is increased. Although this behavior seems counterintuitive at first, it is consistent with the model for Donnan exclusion presented above: Decreasing eluant pH tends to protonate organic acids. As the neutral acids, they can penetrate the resin bed more readily. Changing the eluant pH can alter selectivity somewhat

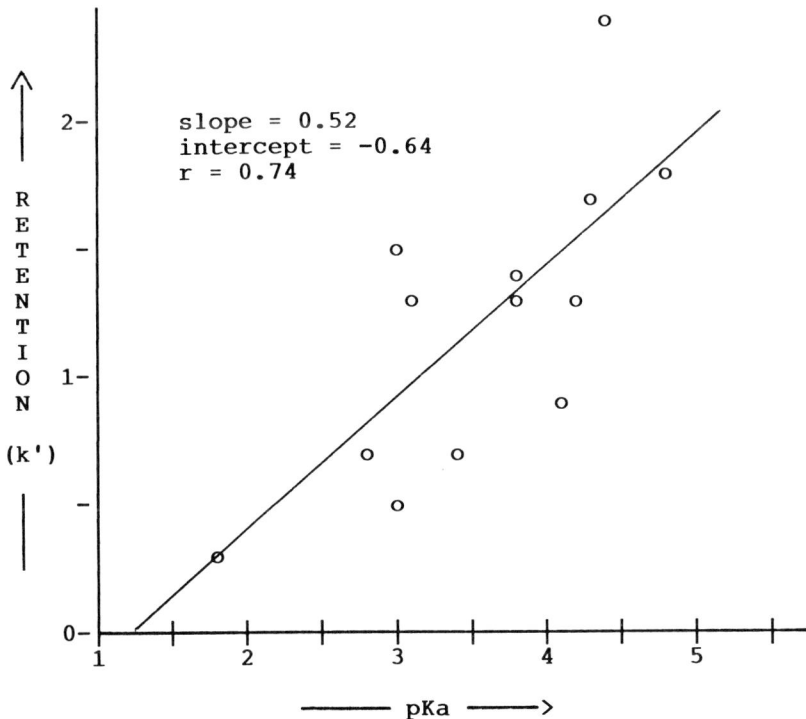

Figure 17. Organic acids elute from an anion exclusion column roughly in the order of pK_a. These data show the correlation between pK_a and retention for 13 mono- and dicarboxylic acids eluted from a standard anion exclusion column, with 1 mM sulfuric acid as the eluant.

(Figure 18), especially where polyfunctional acids are involved. In practice, however, the working pH range is from about 2 to about 4, because the resin must be kept in the hydrogen form.

Strongly retained species, such as aromatic acids, phenols, etc., can be eluted more quickly by the addition of an organic modifier to the mobile phase. This behavior is consistent with "reverse phase" adsorption as a mechanism for retention of those species that can penetrate the resin beads. Acetonitrile has proven to be the most useful organic modifier for use with commercially available anion exclusion columns.

Typical anion-exclusion eluant systems include water (for species such as fluoride, bicarbonate, or sulfide) (Figure 19), dilute mineral

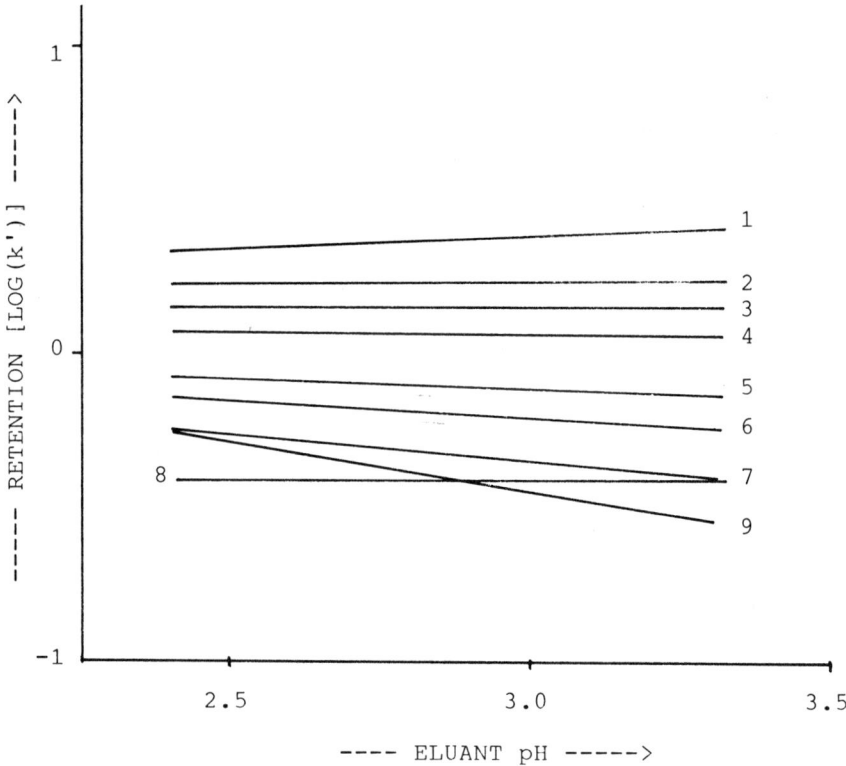

Figure 18. Changing the eluant pH has some effect on retention in anion exclusion chromatography. The effect is much less dramatic than that observed in ion exchange (see Figures 1 and 7 for comparison).

acids (for organic acids analysis) (Figure 20), or acid/acetonitrile mixtures (for C1-C8 fatty acids).

Compared to that of anion exchange columns, which can be operated over a very wide pH range, the selectivity of anion exclusion columns for organic acids is limited. Exclusion columns do, however, have the advantage of providing a complete profile of all anions in a sample with pK_a values in the range of, say, 2-5. This makes anion exclusion columns ideal for "fingerprinting" samples containing a variety of acids. Anion exchange columns, on the other hand, can usually be optimized for the analysis of any given organic acid, at the price of being "blind" to most of the rest.

Single-Column Ion Chromatography

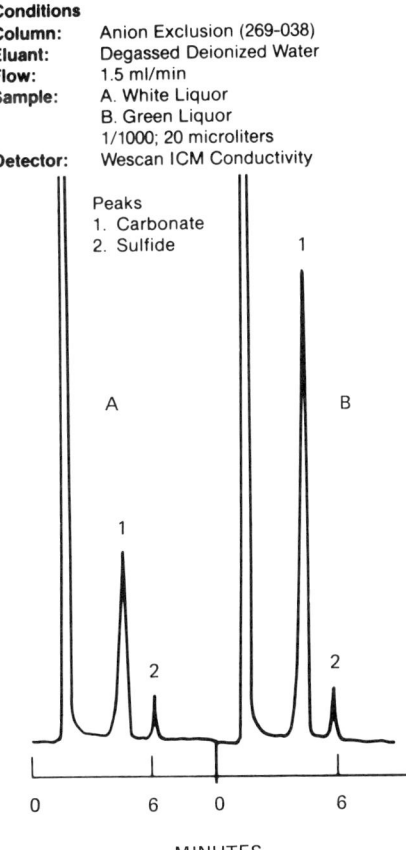

Figure 19. An anion exclusion column can be used with deionized water as the eluant to resolve weak acid anions such as sulfide or carbonate.

The sensitivity of electrical conductivity detection coupled with anion exclusion separation depends on the degree of ionization of the sample species (and thus, indirectly, on the pH of the eluant). In general, the detection limits for common organic acids (say, acetate) are on the order of 100 ng. This is comparable to the sensitivity attained with low-wavelength UV absorbance detection at 210 nm. Acids bearing conjugated double-bond systems will be detected with greater sensitivity by UV absorbance, whereas low-pK_a species will tend to be detected with greater sensitivity by electrical conductivity.

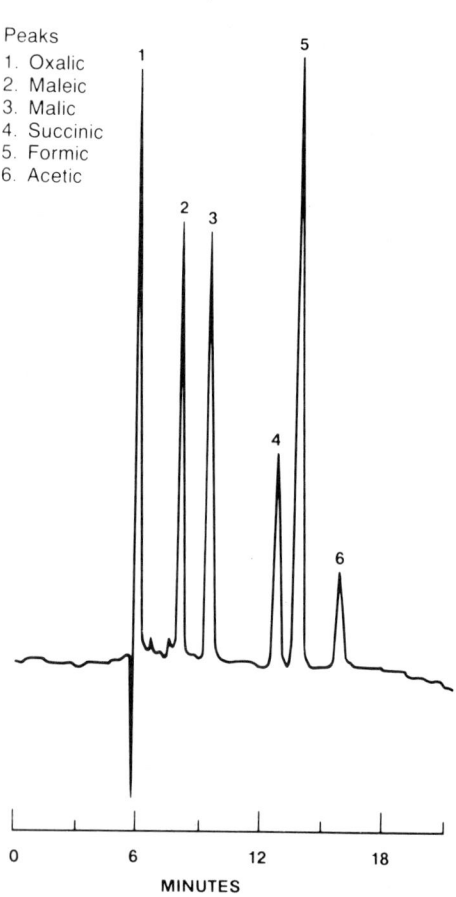

Figure 20. An anion exclusion column can be used with a mineral acid eluant to resolve organic acids.

In many cases, the use of both UV and conductivity detectors in series provides additional information about sample composition that is particularly useful if a "fingerprint" or profile is important for sample identifications. Although a number of the same peaks can be detected, the relative responses are different. In general, conductivity detection coupled with anion exclusion for the analysis of organic acids is less interference-prone than UV detection, which can respond to a variety of UV-absorbing but nonionized components such as sugars, aldehydes, ketones, etc. (Figure 21).

Column Technology

Columns used for ion exclusion chromatography are typically based on standard, high-capacity cation- or (less commonly) anion exchange resins. Anion exclusion separations using low-wavelength UV absorbance detection were first described by Turkelson and Richards in 1978 [37]. Their separations were carried out on fully sulfonated 4% crosslinked polystyrene-divinylbenzene resins. Today's typical column is packed with 8% or higher crosslinked resin with particles in the 9- to 10-µm average size range. These materials have a reasonable pressure resistance, although overpressurization (typically 1500 psi or so) can cause resin bed compression. Because the ion exchange resins used are somewhat elastic, catastrophic bed failure rarely occurs. Columns can typically be restored to their original condition by gently backwashing in order to "fluff" the resin bed.

Commercial columns are typically available in fairly large sizes. Because an exclusion mechanism is involved, many species of interest elute in less than one column volume as very narrow peaks. Because the columns are very efficient (60,000 theoretical plates/meter is not unusual), extracolumn dead volume plays a significant role in limiting overall system efficiency. Even with the large columns commercially available, efficiency losses become evident when sample volumes greater than about 20 µl are used (Figure 22).

C. Other Separation Mechanisms

Ion Pair Chromatography

Ion interaction or *ion pair chromatography* has been used in HPLC since the mid-1970s. (A further discussion is found in Chapter 6.) Although arguments abound as to the exact nature of the mechanism, in most applications to ion chromatography, it can be readily explained in terms of in situ formation of an ion exchange stationary phase by adsorption of an amphoteric moiety to the surface of a reverse phase column. Typical ion-pair reagents used for anion analysis include a variety of alkyl ammonium species (tetrabutyl ammonium being one of the most popular). Presumably, the "greasy" end of these species stick to the reverse-phase column surface, leaving the "charged" end protruding into the aqueous mobile phase to provide ion-exchange capacity.

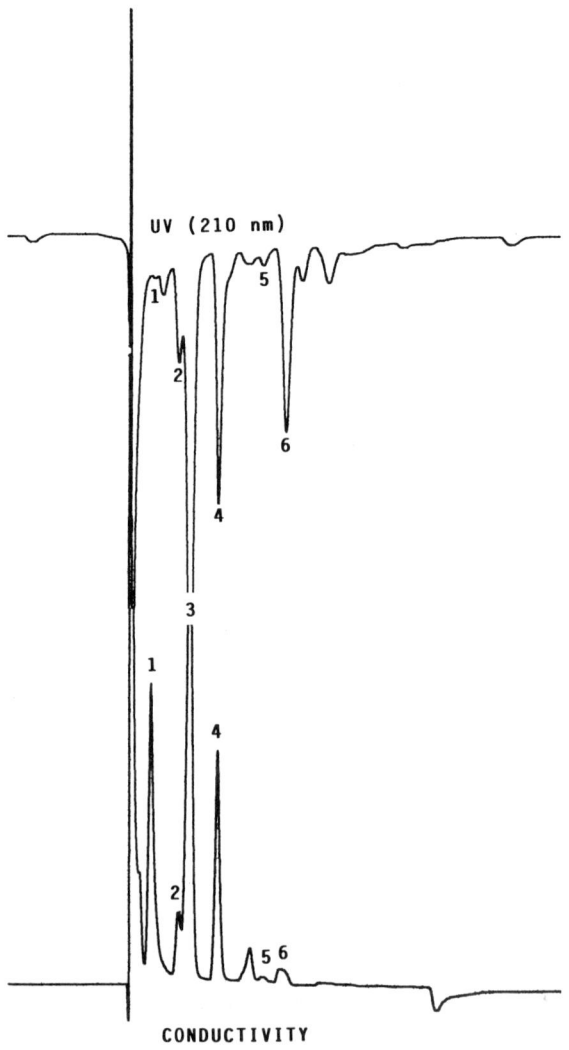

Figure 21. Conductivity detection in ion exclusion separations provides somewhat greater selectivity for organic acids than does UV detection because it does not respond to neutral carbonyl-bearing molecules. Conditions: column, Anion Exclusion; eluant, 0.002 N sulfuric acid; flow, 0.8 ml/min; sample, California "champagne," 20 μl; detector, Wescan 213A Conductivity (bottom); UV absorbance, 210 nm (top). Peaks: (1) phosphate, (2) citrate, (3) tartrate, (4) malate, (5) succinate, (6) fumarate.

Single-Column Ion Chromatography

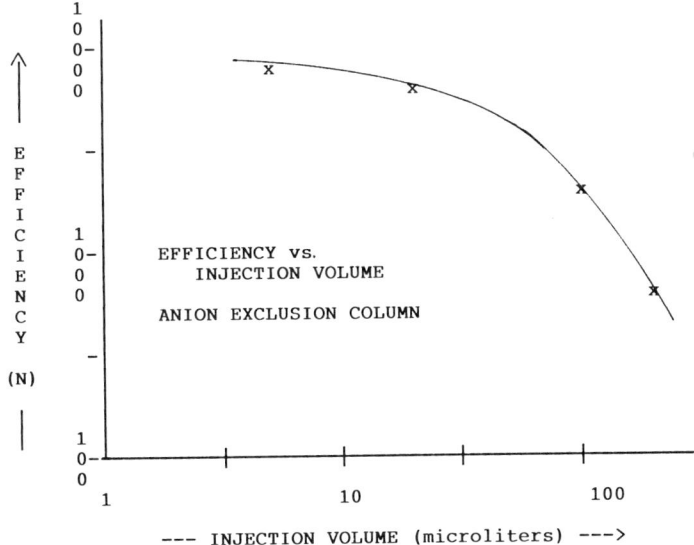

Figure 22. The efficiency of ion exclusion systems is quite sensitive to injection volume.

Because the ion-pair reagent is somewhat soluble in the mobile phase, it must be maintained in the mobile phase at an equilibrium concentration. This provides a very convenient way of controlling the ion exchange capacity of the column: Simply vary the concentration of ion-pair reagent in the mobile phase.

In a variant on this technique developed by Cassidy and Elchuk [38], a long-chain surfactant is coated onto a reverse phase column by use of an organic modifier. When the eluant is switched to an all-aqueous ion exchange system, the coated surfactant remains essentially irreversibly bound. The ion exchange capacity can be "tailored" by controlling the coating process. Perhaps more important, the coating can be stripped off by use of an organic phase such as acetonitrile, and the column recoated as required.

The flexibility of ion-pair/ion-interaction chromatography is at once its blessing and its curse. The ability to control ion exchange capacity adds another variable which must be optimized for each particular system. In addition, the multiple-equilibrium phenomena which give rise to system peaks in fixed-site ion-exchange chromatography are complicated further in ion pair chromatography, where not only sample ion, buffer ion, and buffer acid equilibria must be considered, but also ion-pair reagent equilibrium affects retention. The functional result is that ion-pair separation coupled with electrical conductivity detection is often plagued by system peaks and baseline upsets which can have long retention times.

The problem is less severe in the "permanently coated" type of columns, in which the ion-pair reagent is essentially insoluble in the mobile phase. Even here, however, some care must be taken to avoid "stripping" ion-pair reagent by the injection of samples containing organic solvents.

"Permanently coated" columns have been used to separate transition metal cations, with detection either by direct electrical conductivity or by postcolumn reaction [38]. More conventional ion-pair or ion-interaction chromatography based on dynamic equilibrium between the coating reagent in the mobile phase and on the column has been used for the analysis of both organic and inorganic anions [18-20].

Reverse Phase Chromatography

Probably the major commercial application of ion pair chromatography has been the analysis of anionic surfactants, such as alkyl sulfates or sulfonates. Because the ammonium acetate used in the mobile phase is a weak ion-pairing agent and, moreover, is used at low concentration, this application seems to cross over into the gray area between ion-pair and classical reverse-phase partition chromatography. The latter interpretation is reinforced by the elution order, which depends essentially on the total carbon number of the sample molecules (Figure 23).

Normal Phase Chromatography

Finally, the parallels between ion chromatography and HPLC are demonstrated by the coupling of normal phase separation, using a completely nonaqueous mobile phase, with electrical conductivity detection for the analysis of long-chain alkyl quaternary ammonium surfactants [23]. The conductivity detector can be considered essentially as a charge-specific detector in this application.

III. OTHER DETECTION TECHNIQUES

A. Refractive Index Detection

Refractive index detection is a "universal" detection technique; this property is simultaneously its great weakness and its great strength.

On the positive side, refractive index detection allows an extremely wide latitude in the selection of buffer driving ion and ionic strength. This confers great flexibility in column/eluant selection. In principle, refractive index detection can be substituted for electrical conductivity or UV detection in any of the separations described in the preceding sections.

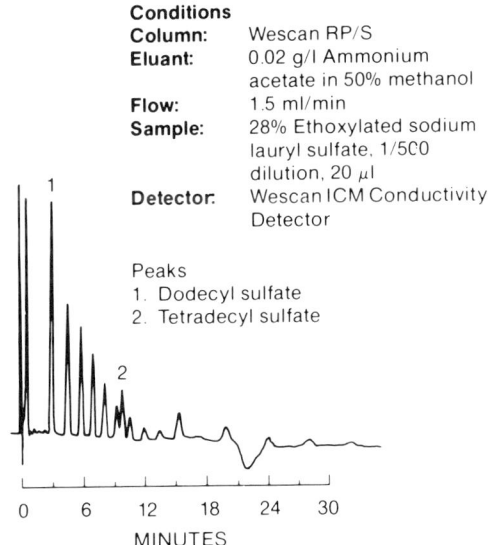

Figure 23. Retention of surfactants in reverse phase chromatography depends to a large extent on the hydrophobicity of the molecule. Reverse phase SCIC separates ethoxylated alkyl sulfonates first on the basis of hydrocarbon chain length (C_{10}, C_{12}, C_{14}, etc.). For a given chain length, a further separation occurs, based on the number of ethylene oxide groups (+EO_1, +EO_2, etc.), to produce a "picket fence" chromatogram.

On the negative side, refractive index detection is only moderately sensitive and is generally considered to be more than slightly interference prone [39].

The sensitivity of refractive index detection is somewhat poorer than that of other ion chromatographic detection techniques. Minimum detectable quantities for common anions such as chloride, nitrate, or sulfate are reported to be in the 20- to 50-ng range (compared with 1-5 ng for indirect UV or direct conductivity detection) [3]. Even these sensitivity levels can be reached only with careful control of temperature via a circulating water bath to maintain the temperature of both the column and the detector cell constant to within 0.1°C.

B. Electrochemical Detection

Although electrical conductivity detection can be classified as "electrochemical," in most applications that term is reserved for detectors in which a chemical reaction (oxidation or reduction) of the sample occurs at the electrode surface [2]. (This is a situation which both conductivity and potentiometric detectors are specifically designed to avoid.) A major distinction among electrochemical detector types is based on the fraction of the sample ions or molecules that are allowed to react. Amperometric detectors typically consume on the order of 3% to 10% of the sample [40]. Coulometric detectors, on the other hand, consume the entire sample. Coulometric detectors provide a larger signal than their amperometric cousins, as well as allowing a direct calibration of detector response based on Faraday's law. Paradoxically, however, amperometric detectors are generally considered to be more sensitive than coulometric types (the background noise increases faster than the signal as electrode surface area is increased). Applications of amperometric detectors in liquid chromatography far outnumber those of coulometric detectors. (A more detailed discussion can be found in Chapter 4.)

In principle, amperometric detection can be applied to any chemical species; one merely has to set the potential high (or low) enough to oxidize or reduce the sample species of interest. In practice, however, the working potential range of amperometric detection is limited at both ends of the potential range.

Because of the difficulties involved in removing dissolved oxygen from both sample and eluant solutions, most applications of amperometric detectors in ion chromatography have involved the detection of relatively easily oxidized species (cyanide, sulfide, sulfite, nitrite, or phenols, for example; see Figure 24).

In many respects, the characteristics of electrochemical detection are complementary to those of refractive index detection. Electrochemical detection is highly sensitive, but extremely selective in terms of species that can be detected. Refractive index detection, on the other hand, is relatively insensitive, but can be used for almost any dissolved ion. This combination of properties makes electrochemical detection valuable in specific cases, but limits the overall usefulness of the method.

The application of "ion chromatography" to the analysis of carbohydrates and alcohols [11], which are ionic only under relatively harsh conditions, is further evidence of the blurring of distinctions between ion chromatography and HPLC.

C. Absorbance Detection

Direct Absorbance

Direct UV absorbance detection can be particularly useful in the analysis of low levels of UV-absorbing ions such as bromide, iodide, or

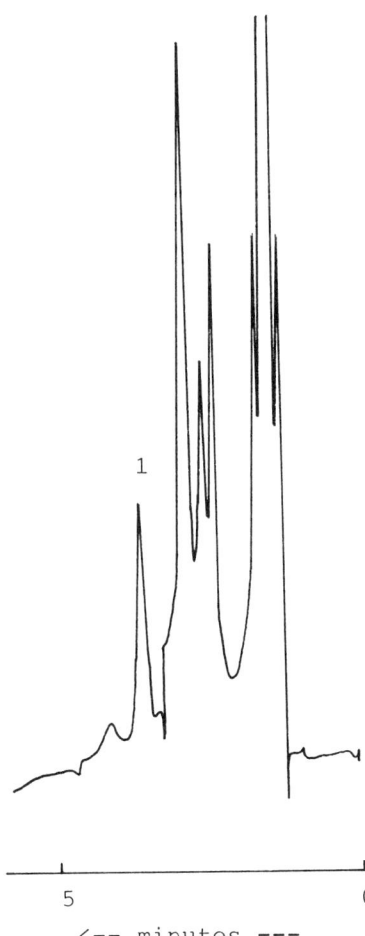

Figure 24. Amperometric detection can be coupled with anion exclusion chromatography to detect phenol in industrial waste. Conditions: column, Anion Exclusion/HS; eluant, sulfuric acid/acetonitrile; flow, 0.8 ml/min; sample, industrial waste; detector, Wescan Model 271 Amperometric. Peaks: (1) phenol, 0.3 ppm.

thiocyanate in matrices that contain high levels of the ubiquitous chloride ion, such as seawater or brine (Figure 25). It is a much less useful technique for the analysis of ions in matrices that are likely to contain significant quantities of UV absorbing organics.

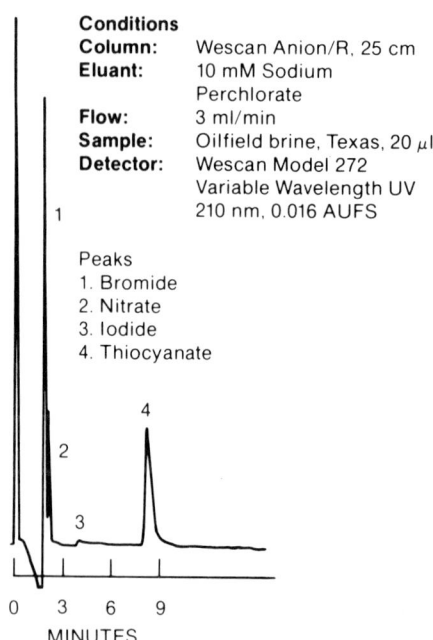

Figure 25. Direct UV detection can be coupled with anion exchange chromatography by use of a UV-transparent buffer for detection of iodide and thiosulfate in oilfield brine.

Indirect Absorbance

The terms *indirect absorbance*, *indirect photometric detection*, or (less correctly) *vacancy chromatography* have been applied to the use of a UV-absorbance detector to measure the decrease in absorbance caused by replacement of UV-absorbing buffer ions by UV-transparent sample ions in an ion exchange system. The arguments for this approach are similar to those used in describing direct electrical conductivity detection: Because an ion exchange system preserves electroneutrality, the concentration of ions in the column effluent must remain constant; only the identity changes as sample ions replace buffer ions. If the molar absorbance of the sample ions is less than that of the buffer ions at a particular wavelength, the result is a decrease in the background absorbance of the eluant as sample ions elute.

In addition to the usual factors such as retention volume and column efficiency, detection sensitivity depends on the molar absorptivity of the eluant buffer and on its concentration. Other things being

Single-Column Ion Chromatography 59

equal, dilute buffers with high molar absorbance provide the best sensitivity, because replacement of a given number of buffer ions provides a large change in signal.

In practice, indirect UV is limited to buffers whose total absorbance is below 1 absorbance unit, because many HPLC UV detectors begin to lose linearity above this level. Working in the linear range of the detector can be accomplished by using extremely low-capacity columns (and hence low concentration of eluants). Under these conditions, eluting sample ions replace a large proportion of the buffer ions. Sensitivity is maximized, but the low-capacity systems are easily overloaded, thus limiting the linear range that can be accommodated.

Granted existing column capacity, the alternative is to "detune" the detector wavelength so that absorbance is measured at a wavelength other than the absorption maximum of the buffer. This approach avoids nonlinearity, but restricts sensitivity because a smaller percentage change in buffer ion concentration (and hence in absorbance signal) is generated for a given concentration of sample ions.

A major advantage of indirect photometric detection is the normalization of detector response [41]. Because what is measured is the disappearance of buffer ion, the equivalent response is the same for all sample ions.

Indirect photometric detection has been coupled to both fixed-site ion exchange and ion-pair separation systems. In general, the same comments made above about eluant selection, system peaks, selectivity, and stability in connection with conductivity detection will also apply to indirect photometric detection.

Comparison of Conductivity and Indirect UV

Detection limits for indirect photometric detection in IC have been quoted at the 1- to 10-ng level [10]. This is comparable to the detection limits routinely obtained with direct conductivity detection [14,33,34]. Direct comparison of sensitivity between two techniques can be a tenuous concept because the same eluant is rarely the optimum for both detection techniques. Thus, comparisons of indirect UV and direct conductivity have tended to find indirect UV generally more sensitive [9,42], although the detection limits obtained for direct conductivity were considerably poorer than the best cited for that technique [14].

IV. ANCILLARY TECHNIQUES AND APPLICATIONS

A. Sample Preconcentration

LC detectors (including the conductivity detectors used in ion chromatography) respond to the *concentration* of sample in the eluant that goes through the detector. If the system is working well, however, a given sample peak will always have about the same volume. In practice, then, the sensitivity of an LC system is well described by the *minimum detectable quantity* of the sample. An SCIC system, for example, can typically detect amounts as low as 1-10 ng chloride in a sample.

If 4 ng chloride (for example) is contained in 100 µl (0.1 ml) of sample, then the minimum detectable concentration is:

$$\frac{40 \times 10^{-9} \text{ g}}{1 \times 10^{-1} \text{ ml}} = 40 \times 10^{-9} \text{ g/ml}$$

or 40 ppb.

One approach to measuring lower concentrations (like 4 ppb) is simply to increase the volume of sample injected. If we inject 1 ml of sample, our concentration is:

$$\frac{4 \times 10^{-9} \text{ g}}{1 \text{ ml}} = 4 \times 10^{-9} \text{ g/ml}$$

or 4 ppb.

Although IC systems tolerate larger injection volumes than we typically associate with HPLC, the direct injection technique does have its limitations. In practice, a volume between 500 µl and 1 ml is the most we can inject directly onto the system without creating problems.

The major problem with direct injection of large volumes is the upset in equilibrium--that is, the huge "water dip" that results from all of that dilute sample passing through the whole column and the detector. What if we could somehow divert this water to drain before it reached most of the column and the detector? We could then let the excess water from the sample drain away without upsetting the equilibrium and then let our sample ions pass through the whole column and detector for analysis.

Sample preconcentration has been applied to both HPLC [43] and to SCIC [44-46] to accomplish exactly this task.

A short cartridge packed with the same type of material used in the separator column is attached to the injection valve in place of the sample loop. When a dilute sample is forced through this cartridge, the ions in the sample are retained by the cartridge while the excess water goes out to the drain. After a sufficiently large volume of sample has been pumped through the cartridge, the injection valve is turned and the trapped sample ions are swept into the separator column by the eluant. The only water that actually gets into the system is the amount physically trapped in the cartridge in the interparticle space (this is typically on the order of 0.5 ml or less).

Single-Column Ion Chromatography

Sample preconcentration is a simple, powerful technique for improving the sensitivity (expressed as minimum detectable concentration) for the analysis of dilute samples. It is not foolproof, however, and the user should be aware of potential problem areas, described in the following two sections.

Problems Inherent in the Sample Preconcentration Concept

Preconcentration cartridges are assumed to strip essentially all ions of a given charge from the sample: These include the ions of interest, as well as other components of the matrix. The warning is obvious: Sample preconcentration is not useful for the analysis of a trace of one component in the presence of a large excess of another component. At best, this will result in an overload of the analytical column. At worst, the interference may displace the sample component of interest to provide unreliable results.

Preconcentration cartridges have a finite capacity; only so much sample can be concentrated before "breakthrough" occurs. In general, the more weakly retained a sample component is on the cartridge packing material, the sooner breakthrough will be a problem. This is because aqueous samples always contain at least some hydronium and hydroxide ions. Although both are weak driving ions, they can be present at significant concentration. To put this into perspective, consider that a solution that is 1 ppb in chloride ion has a molar concentration of:

$$1 \ \mu g/L \times \frac{1}{35} \ \mu mol/\mu g = 3 \times 10^{-8} \ mol/L \ chloride$$

whereas water at neutral pH contains hydroxide at 1×10^{-7} mol/L. Although chloride is more tightly bound than hydroxide (the selectivity coefficient for chloride relative to hydroxide on anion exchange resins is typically on the order of 10-20), hydroxide ion can compete with the sample ion of interest for active sites on the concentrator cartridge.

Sample recovery does not have to be 100%; it merely has to be reproducible. Sample recovery can depend on (among other things) the volume of sample concentrated and the rate at which the sample is pumped through the cartridge. The best rule to follow in sample preconcentration is *always* to base quantitation on a standard that was treated the same way as the sample (i.e., roughly the same concentration, and exactly the same volume and pumping rate).

Problems Resulting from Sample Handling

Because preconcentration is typically used for ultra-low-level (say, the parts-per-billion range and below) analysis, sample handling can become a significant limiting factor. To put it bluntly, it takes a lot of care (and no small degree of luck!) to avoid contaminating aqueous samples with something like chloride.

B. Temperature Control

In the author's experience, the single most critical factor for successful application of SCIC is precise temperature control, especially when direct conductivity detection is used. Because the thermal coefficient of aqueous solutions is typically on the order of 2% per degree Celsius, detection of conductivity changes approaching 1 part in 100,000 implies reduction of short-term temperature fluctuations to below 0.001 degree! In practice, sufficient thermal stability is readily obtained by insulating the column and cell compartment [47]. This approach effectively "damps out" short-term thermal noise, but leaves the system susceptible to longer-term drift as the ambient temperature changes during the course of the day.

Temperature compensation, in which a thermistor is used in the feedback loop of an amplifier to dynamically control detector gain in response to temperature changes, has long been used to minimize thermal drift in conductivity detectors [48]. Even temperature-compensated systems, however, exhibit a certain degree of baseline drift, which generally tracks changes in ambient temperature but is not well correlated with the actual temperature of the eluant. A similar phenomenon has been observed by the author in "indirect"-absorbance UV systems.

Because UV detection is not inherently highly temperature sensitive, the suggestion has been made [49] that such drift is the result of small changes in column equilibration, which effectively change the concentration of eluant buffer ions in the mobile phase as a function of temperature.

C. Applications

Plating Bath Analysis

Pure Water Analysis--Traces of ionic contamination in rinse and process water as well as bath chemicals can dramatically affect yield and performance. Although chloride and sodium are the most frequently measured species, the presence of weak anions such as borate and (especially) silicate can be more insidious, because these species contribute little if anything to the usual bulk conductivity measurement of water purity.

Ion chromatography provides convenient analysis for many anions to below 100 ppb with direct injection, with detection limits below 1 ppb when optional preconcentration techniques are employed. Cation

Single-Column Ion Chromatography

Conditions
Column: Wescan Cation
Eluant: Nitric Acid, pH 2.4
Flow: 2 ml/min
Sample: 10 ml Std. Solution
Detector: Wescan ICM Conductivity

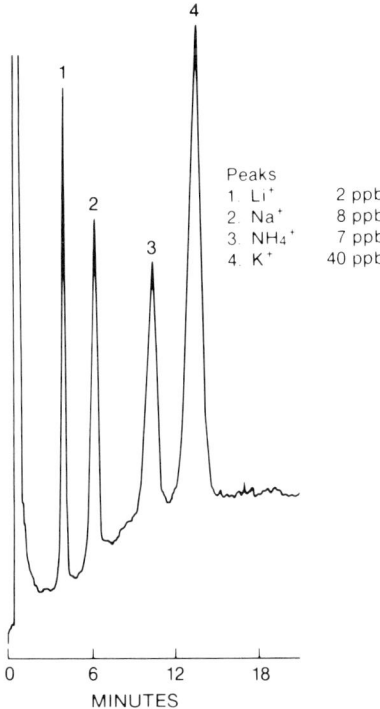

Peaks
1. Li^+ 2 ppb
2. Na^+ 8 ppb
3. NH_4^+ 7 ppb
4. K^+ 40 ppb

Figure 26. Sample preconcentration allows cation analysis with detection limits in the parts-per-trillion range.

analysis is even more sensitive, with detection limits in the low parts-per-trillion range if preconcentration is used (Figure 26).

Process Bath Analysis: Major Components--At the other end of the spectrum, ion chromatography provides a rapid, simple analysis for the bulk composition of process baths. Acid etch baths, for example are easily checked for composition accuracy with minimal sample preparation: Simply dilute the bath into the linear range of the system, and inject (Figure 27). Continual monitoring of bath composition can provide guidance for replenishment cycles on an as-needed basis, as active components are depleted.

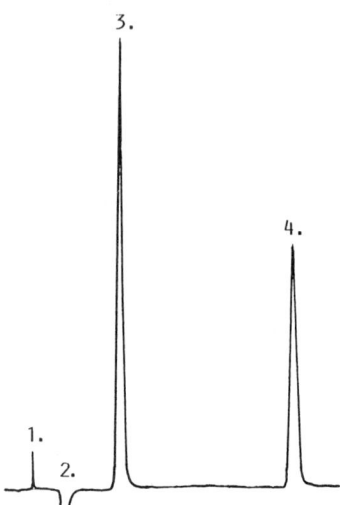

Figure 27. Pan bath analysis. SCIC allows "one-run" analysis of acetic, phosphoric, and nitric acid in an acid etch bath. Conditions: Column, Standard Anion; eluant, 4 mM phthalic acid; flow, 1.6 ml/min; sample, Std., 100 µl; detector, Wescan ICM Conductivity. Peaks: (1) acetate, 23.9 ppm; (2) system; (3) phosphate, 227 ppm; (4) nitrate, 33.7 ppm.

Process Bath Analysis: Trace Components--Ion chromatography provides rapid, simple analysis of the organic acid species used as additives or electrolytes (Figure 28). Ion chromatography also allows trace levels of impurities carried over from previous processing steps to be detected before they accumulate to levels that impair bath performance.

Metals--Ion chromatography allows the direct analysis of metals in a variety of matrices including metal plating baths. Because separation and detection are carried out at the same pH, loss of metal ions before detection is not a problem, and direct electrical conductivity detection is quite feasible. Thus, the same chromatographic system used for anion analysis can be used without modification for the analysis of metal ions (Figure 29).

Complexing Agents--Complexing agents such as EDTA and their metal complexes can also be monitored by ion chromatography. Results can be used to control bath composition and monitor for the degree of depletion (Figure 30).

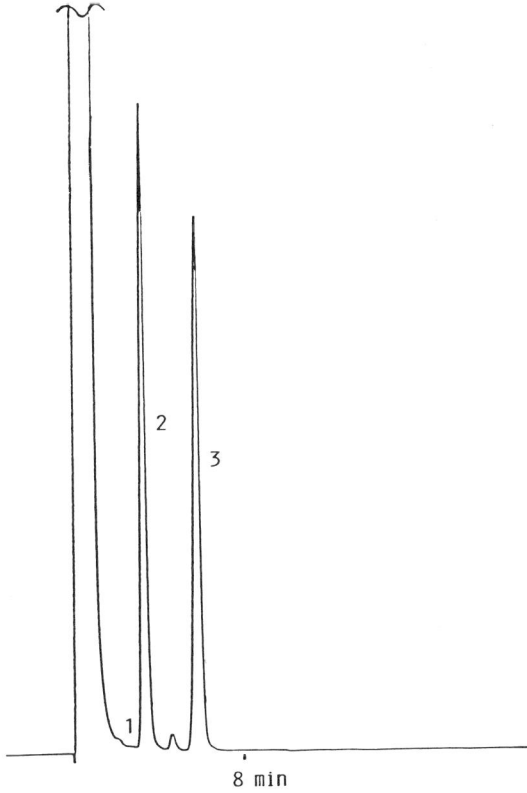

Figure 28. SCIC analysis for organic acids in an electroless nickel plating bath. Conditions: column, Anion Exclusion; eluant, nitric acid; flow, 1.2 ml/min; sample, Electroless Nickel, 1/100, 100 µl; detector, Wescan ICM. Peaks: (1) citric, (2) malic, (3) lactic acids.

Cyanide/Sulfide--Cyanide and sulfide can be analyzed in both plating baths and plating bath waste. Because cyanide forms very strong complexes with metals, the SCIC analysis is used for *free* cyanide (Figure 31). A preliminary distillation is needed in order to break up the cyano complexes prior to ion chromatography.

Pharmaceutical/Biomedical Analysis

Introduction--We typically associate pharmaceutical analysis with organic compounds as analytes and with methods such as gas or high-pressure liquid chromatography. In fact, however, a number of the

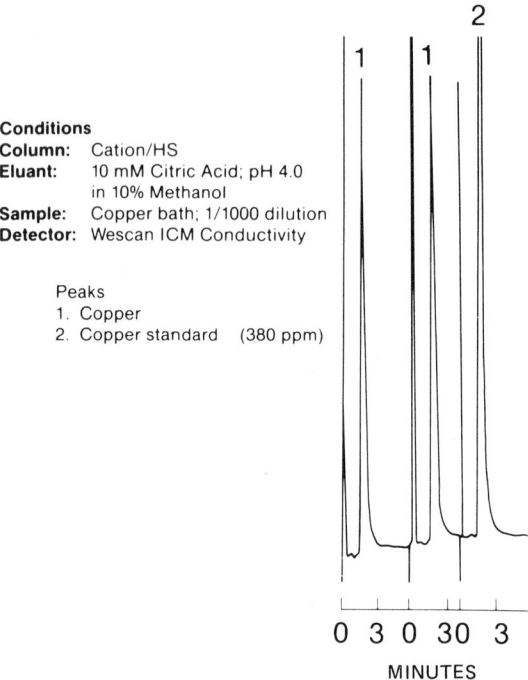

Figure 29. SCIC analysis for copper in a copper plating bath.

most burdensome (expensive and time-consuming) of the analytical techniques used in the pharmaceutical industry deal with ionic (usually inorganic) components.

A large number of ionic species are nonvolatile and, hence, unsuitable for GC analysis. At the same time, HPLC techniques have been developed primarily around organic compounds, with the result that ionic species are often both difficult to separate and difficult to detect when standard HPLC technology is used.

A number of alternative techniques have been developed for the analysis of ionic species, but they all suffer from disadvantages that limit their scope. Wet chemistry, culminating in either gravimetry or colorimetry is tedious and time consuming. Electrode techniques are often subject to inadequate sensitivity, long equilibration times, and interfering substances. Continuous-flow techniques are widely used, but they involve heavy capital expenditure and are usually limited to the determination of one ion at a time.

Single-column ion chromatography, in contrast, is rapid (separation times typically range from less than 1 min to 10 min or so),

Single-Column Ion Chromatography

Conditions
Column: Anion/R-HS
Eluant: 5 mM Sulfuric Acid
Flow: 1.5 ml/min
Sample: A. Electroless Copper Bath (Used)
B. Electroless Copper Bath (New)
Detector: Wescan ICM Conductivity

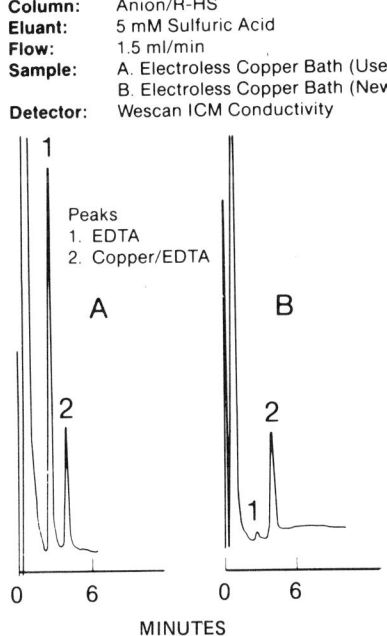

Figure 30. EDTA can be resolved from its copper complex in a plating bath using SCIC.

sensitive (sub-parts-per-million analyses are typical, and sub-parts-per-billion levels can be measured by using preconcentration techniques), economical (capital costs can be under $10,000 for a single-channel system), and easy to use (typical sample preparation involves only dissolution, dilution, filtration, and injection).

Raw Materials Quantitative Analysis--What level of other halide ions is present in sodium chloride USP? What level of chloride is present in boric acid? How pure is reagent-grade potassium iodide? These questions are easily answered by SCIC; simply dilute the sample and inject onto the IC system (Figure 32).

Dosage Form Assay--Many drugs are themselves ionic, and are formulated into dosage forms as an appropriate salt for stability and solubility. Hydrochloride salts of amines, and sodium or ethanolamine salts of acids, for example, are commonly encountered. Frequently, an assay for the counterions can provide important validation of the results of a direct drug assay (Figure 33).

Conditions
Column: Wescan Anion R 25 cm
Eluant: 1.0 mM Boric Acid; 0.3 mM KHP, 10% Acetonitrile adjusted to pH 11.9 with NaOH
Flow: 1.7 ml/min
Sample: Industrial Effluent 100µl
Detector: Wescan Model 215 (ICM)

Peaks
1. Sulfide
2. Chloride
3. Cyanide

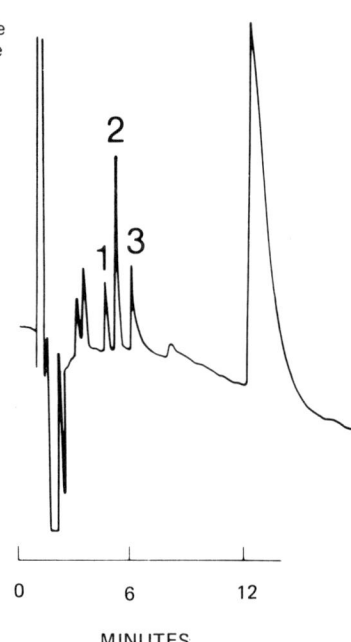

Figure 31. SCIC analysis for free cyanide in industrial waste.

Microchemical Analysis--Information about the equivalent weight of experimental compounds can often be inferred from the result of halide, sulfur, or nitrogen analysis after Schoniger flask combustions (Figure 34). Ion chromatography reproducibility is typically on the order of 2% to 3% relative standard deviation for manual sample injection. Use of an autosampler or extremely careful technique can improve precision to 1% or so relative standard deviation.

Excipients/Preservatives--SCIC technology is readily applicable to the analysis of ionic excipients or preservatives in dosage forms. For this purpose, ion chromatography can be considered to be an extension

Single-Column Ion Chromatography

Conditions
Column: Anion/R-HS
Eluant: 3 mM Nicotinic Acid
Flow: 6 ml/min
Sample: 1.2% Boric Acid
Detector: Wescan ICM Conductivity

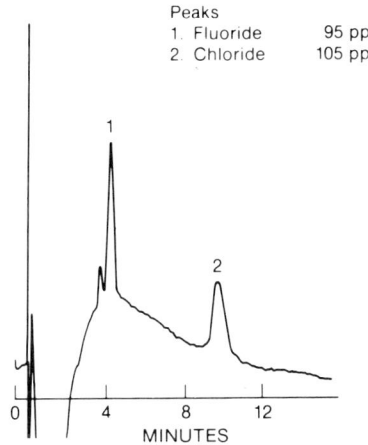

Peaks
1. Fluoride   95 ppb
2. Chloride   105 ppb

Figure 32. Low levels of chloride are readily analyzed in boric acid.

of HPLC with appropriate columns and detectors for the analysis of ionic components (Figure 35).

Cosmetics/Toiletries--How do you analyze for surfactants in shampoo (Figure 23)? For monofluorophosphate in toothpaste, or fluoride in mouthwash (Figure 36)? The answer, of course, is SCIC!

Food/Beverage Analysis

Anions--Relatively small quantities of nitrate and sulfate, for example, can be determined in the presence of a large excess of chloride in samples such as sauerkraut juice (Figure 37a). The same system can be used directly for chloride determination after the sample has been diluted (Figure 37b).

Although the electrical conductivity detector is the workhorse of ion chromatography, the use of an electrochemical detector allows specific analysis for bromide in fruits, nuts, and grain products (an indication of efficacy of fumigation with ethylene dibromide or methyl bromide) (Figure 38). This analysis typifies the simplicity of ion chromatography. A sample of ground nuts, for example, is steeped in water for 1 hr to provide quantitative extraction of bromide

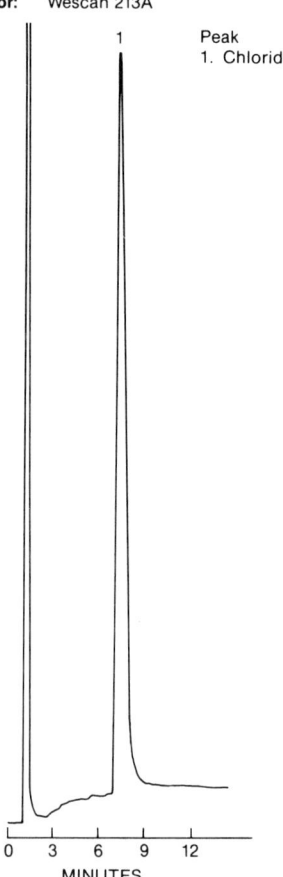

Figure 33. SCIC analysis for chloride in a local anesthetic solution.

(Figure 7). The extract is simply filtered and injected onto the IC column. The linear range exceeds two orders of magnitude (Figure 8), with detection limits well under the parts-per-million level and reproducibilities on the order of 2% to 3% relative standard deviation.

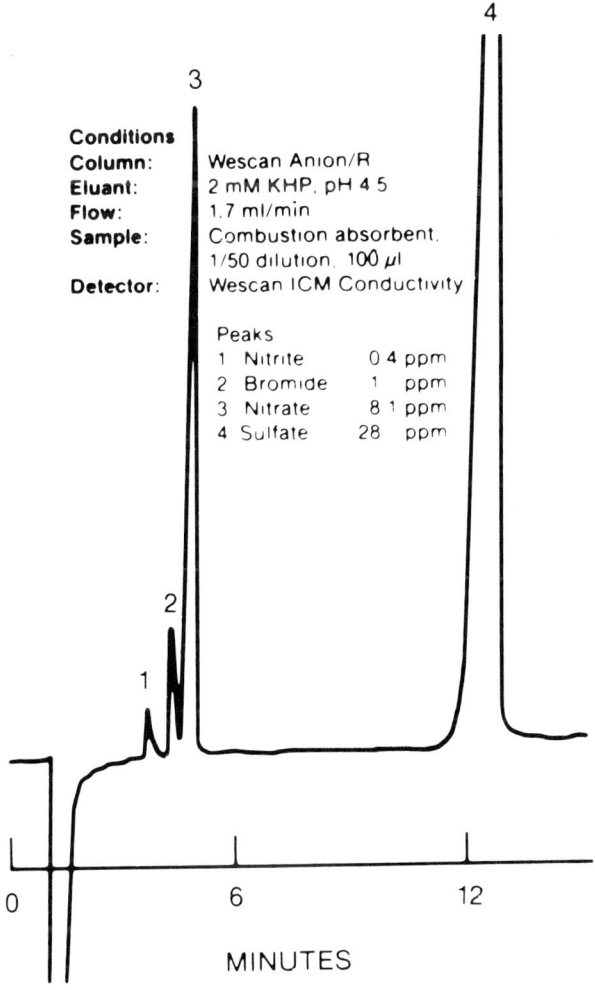

Figure 34. SCIC analysis of anions in the Schoniger flask combustion products of pharmaceutical products.

A similar technique allows the sensitive determination of iodide in a variety of food and beverage products.

Inorganic anions are readily determined in wine (Figure 5). In this case, we recommend the use of a lower- than normal-pH eluant buffer in order to minimize interferences from organic acids. At

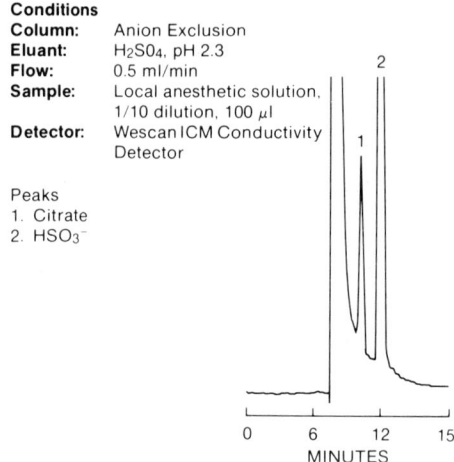

Figure 35. SCIC analysis for citrate and sulfite in a local anesthetic solution.

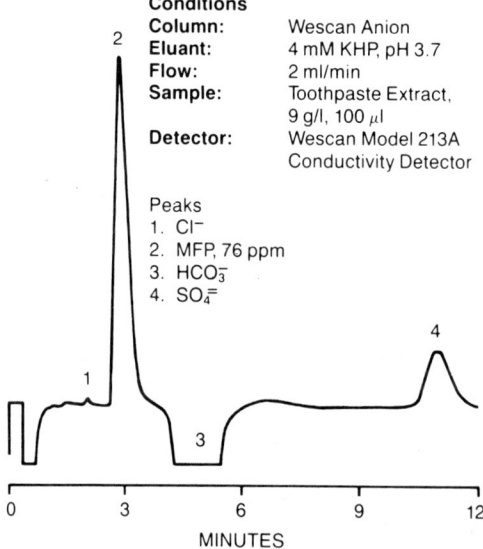

Figure 36. SCIC analysis for monofluorophosphate in toothpaste.

Single-Column Ion Chromatography

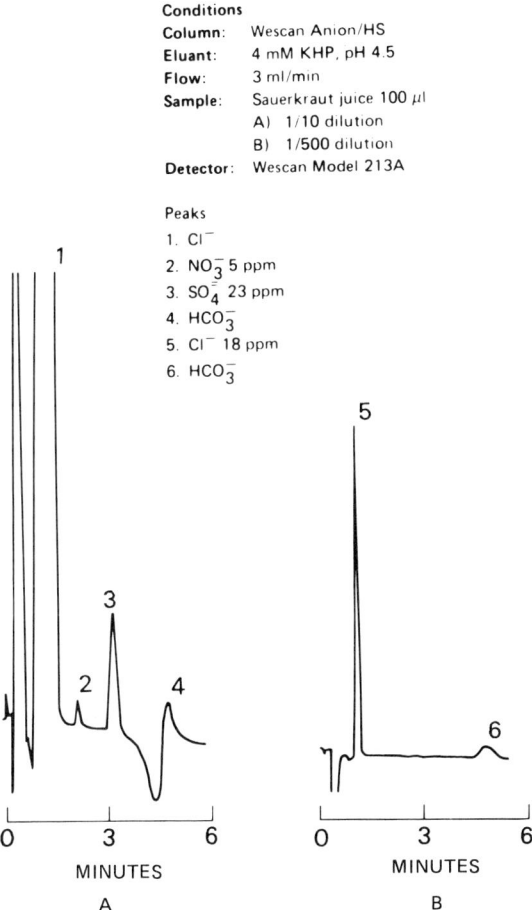

Figure 37. SCIC analysis of inorganic anions in sauerkraut juice. (A) 1/10 dilution; (B) 1/500 dilution.

the relatively low pH (approximately 2.5) used, organic acids are protonated and emerge from the column virtually unretained.

Food Additives--Ion chromatography approaches the technology of HPLC in a number of applications. The analysis of bromate in bread dough additives (Figure 39), for example, illustrates the selectivity that can be achieved when ion chromatography column technology is used in conjunction with the selectivity of a variable-wavelength UV detector.

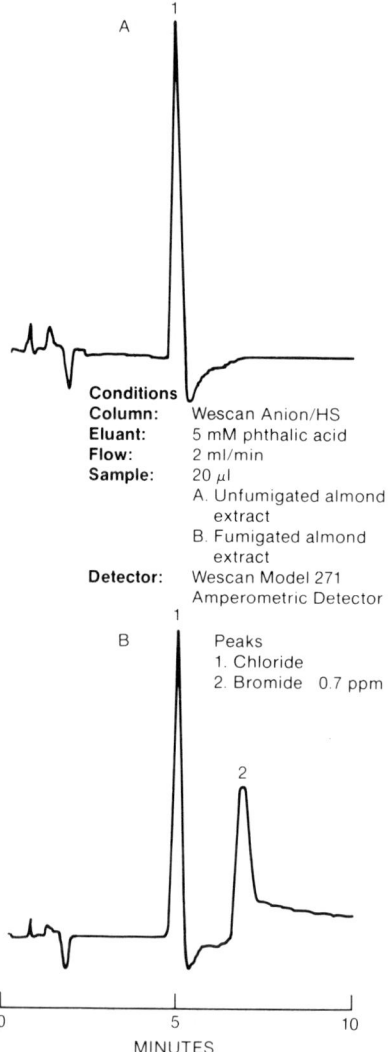

Figure 38. SCIC analysis for bromide in fumigated almonds.

Environmental Analysis

Sulfur Species--Sulfur species can be simply, rapidly, and economically measured in a variety of matrices ranging from stack gas through scrubber baths to acid rain and ground water. Single-column ion chromatography allows the analysis of bisulfite, sulfate, thiosulfate, and thiocyanate in the same run. Similar conditions

Single-Column Ion Chromatography

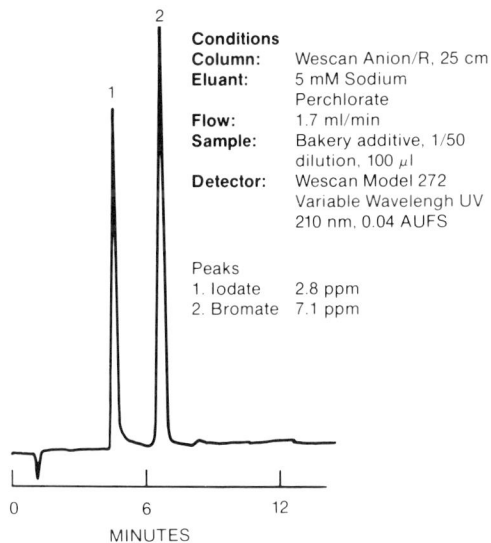

Figure 39. SCIC separation of bromate and iodate.

have been applied to the analysis of thiosulfate in gas scrubber solutions (Figure 40), refinery accumulated waste water, and paper mill pulping liquors.

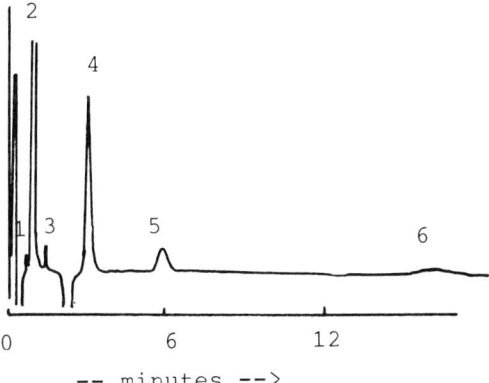

Figure 40. SCIC analysis of sulfur species in a flue gas desulfurization liquor. Conditions: column, Wescan Anion/HS; eluant, 4 mM KHP, pH 3.8; flow, 2.7 ml/min; sample, flue gas desulfurization liquor, 1/125 dilution, 100 µl; detector, Wescan ICM Conductivity. Peaks: (1) bisulfite; (2) chloride; (3) nitrate; (4) sulfate; (5) hydroxylamine disulfonate; (6) dithionate.

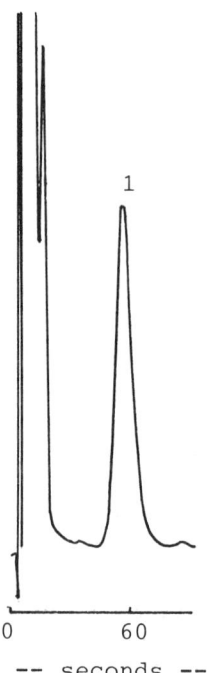

0   60
-- seconds -->

Figure 41. SCIC analysis of sulfate in the presence of a phosphate buffer used for soil extraction. Conditions: column, Wescan Ion-Guard Anion; eluant, 20 mM phthalic acid; flow, 8 ml/min; sample, 0.04 M ammonium dihydrogen phosphate; detector, Wescan ICM Conductivity. Peaks: (1) SO_4^{2-}, 30 ppm.

Modification of the eluant allows the rapid (less than 1 min per sample) analysis of sulfate and nitrate in ground water for monitoring the effects of acid rain. The analysis for sulfate is particularly rugged and can be applied to the analysis of matrices such as soil samples (extracted with 0.04 M phosphate buffer) (Figure 41).

Nitrogen Oxides--Nitrate is readily analyzed under roughly the same conditions as sulfate (see many of the previously referenced chromatograms). Nitrate, sulfate, and chloride, for example, can be determined simultaneously in acid mists from workplace atmospheres after sorption on a silica-gel collector tube.

For analyses of alkaline matrices, such as stack gas collected via alkaline permanganate impingers, we recommend the use of the

Single-Column Ion Chromatography

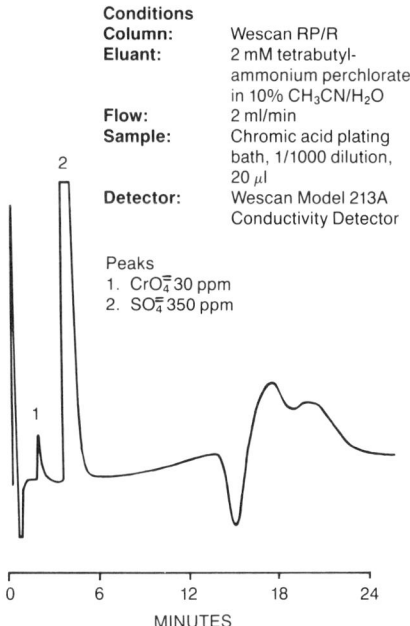

Figure 42. SCIC analysis of sulfate and chromate in a metal plating bath.

resin-based Anion/R column after the sample is passed through a hydrogen-form cation-exchange column to reduce pH.

Other Species--A number of particular analyses are best done with specific eluants. The determination of sulfate and chromate in plating bath wastes provides an example (Figure 42). Specialized conditions have also been developed for the analysis of methylphosphonate from the breakdown of organophosphate pesticides (Figure 43) and for the analysis of fluoride in vegetation resulting from aluminum smelter refinery particulate emissions. Similar conditions are used for the analysis of arsenates in agricultural uses (Figure 44).

ACKNOWLEDGMENTS

The author thanks Wescan Instruments, Inc. for the use of the chromatograms included in this chapter.

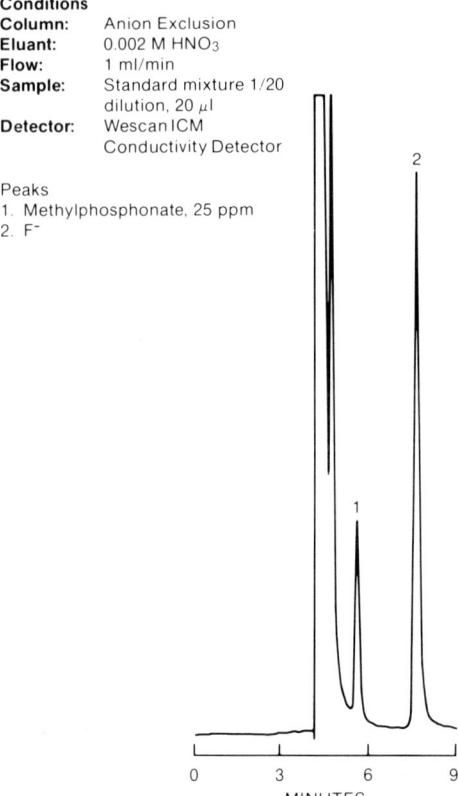

Figure 43. SCIC analysis of organophosphate pesticide breakdown products.

Special thanks are extended to Dave Togami, Pat Barthel, and Doug Gjerde of Wescan, from whose applications and research work many of the chromatograms are taken.

REFERENCES

1. H. Small, T. S. Stevens, and W. D. Baumann, *Anal. Chem.* 47, 1801 (1975).
2. J. S. Fritz, D. T. Gjerde, and C. Pohlandt, *Ion Chromatography*, Huethig, New York, 1982.

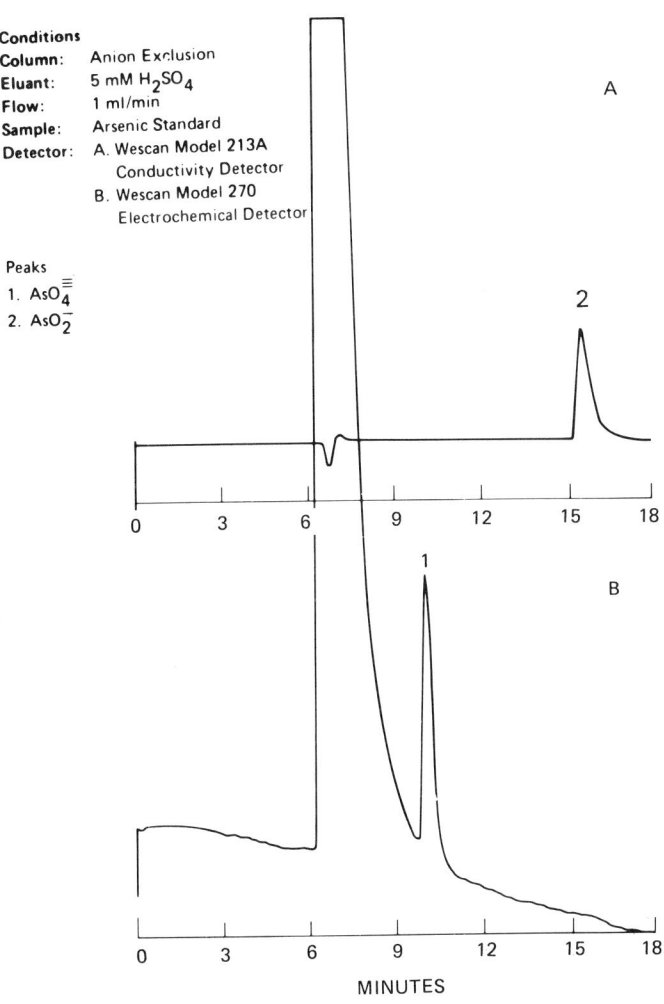

Figure 44. SCIC separation of arsenic species.

3. F. A. Buytenhuys, *J. Chromatogr.* 218, 57 (1981).
4. J. P. de Kleijn, *Analyst* 107, 223 (1982).
5. N. E. Skelly, *Anal. Chem.* 54, 712 (1982).
6. K. J. Stetzenbach and G. M. Thompson, *Ground Water* 21, 36 (1983).

7. R. W. Siergiej and N. D. Danielson, *J. Chromatogr. Sci. 21*, 362 (1983).
8. R. A. Cochrane and D. E. Hillman, *J. Chromatogr. 241*, 392 (1982).
9. M. Denkert, L. Hackzell, G. Schill, and E. Sjogren, *J. Chromatogr. 218*, 31 (1981).
10. H. Small and T. H. Miller, *Anal. Chem. 54*, 462 (1982).
11. S. Hughes, P. L. Meschi, and D. C. Johnson, *Anal. Chim. Acta 132*, 1 (1981).
12. R. D. Rocklin, *LC 2*, 588 (1984).
13. T. H. Jupille, *LC 1*, 24 (1983).
14. T. Okada and T. Kuwamoto, *Anal. Chem. 55*, 1001 (1983).
15. D. T. Gjerde, J. S. Fritz, and G. Schmuckler, *J. Chromatogr. 186*, 509 (1979).
16. D. T. Gjerde, G. Schmuckler, and J. S. Fritz, *J. Chromatogr. 187*, 35 (1980).
17. J. S. Fritz, D. T. Gjerde, and R. M. Becker, *Anal. Chem. 52*, 1519 (1980).
18. Z. Iskandarani and D. J. Pietrzyk, *Anal. Chem. 54*, 1065 (1982).
19. Z. Iskandarani and D. J. Pietrzyk, *Anal. Chem. 54*, 2427 (1982).
20. Z. Iskandarani and D. J. Pietrzyk, *Anal. Chem. 54*, 2601 (1982).
21. T. Jupille, M. Gray, B. Black, and M. Gould, *Am. Lab. 13*(8), 80 (1981).
22. Wescan Ion Analyzer No. 4 (Wescan Instruments, Inc., Santa Clara, CA).
23. V. T. Wee and J. M. Kennedy, *Anal. Chem. 54*, 1631 (1982).
24. J. X. Khym, *Analytical Ion Exchange Procedures in Chemistry and Biology*, Prentice-Hall, Englewood Cliffs, NJ, 1974.
25. J. A. Glatz and J. E. Girard, *J. Chromatogr. Sci. 20*, 266 (1982).
26. J. Hertz and U. Baltensperger, *LC 2*, 600 (1984).
27. T. H. Jupille, D. W. Togami, and D. E. Burge, *Ind. Res. Dev. 25*(2), 151 (1983).
28. T. H. Jupille, D. E. Burge, and D. W. Togami, *Chromatographia 16*, 312 (1982).
29. D. P. Lee, *J. Chromatogr. Sci. 22*, 327 (1984).
30. J. R. Benson, D. J. Woo, and N. Kitagawa, *LC 2*, 398 (1984).
31. J. S. Fritz, D. L. Duvall, and R. E. Barron, *Anal. Chem. 56*, 1177 (1984).
32. T. Jupille, D. Burge, and D. Togami, *Res. Dev. 26*(3), 135 (1984).
33. S. Dogan and W. Haerdi, *Chimia 35*, 339 (1981).
34. Wescan Ion Analyzer No. 7 (Wescan Instruments, Inc., Santa Clara, CA).
35. D. J. Reuter, R. G. Buechele, and T. L. Rudolph, Paper No. 8 presented at the 23rd Rocky Mountain Conference, 1981.

36. G. J. Sevenich and J. S. Fritz, *Anal. Chem. 55*, 12 (1983).
37. V. T. Turkelson and M. Richards, *Anal. Chem. 50*, 1420 (1978).
38. R. M. Cassidy and E. Elchuck, *Anal. Chem. 54*, 1558 (1982).
39. T. Jupille, *J. Chromatogr. Sci. 17*, 160 (1979).
40. P. T. Kissinger, *Anal. Chem. 49*, 447A (1977).
41. D. R. Jenke, *Anal. Chem. 56*, 2468 (1984).
42. P. A. Perrone and J. R. Gant, *Res. Dev. 26*(9), 96 (1984).
43. Rheodyne Technical Notes No. 2 (Rheodyne, Inc., Cotati, CA).
44. K. M. Roberts, D. T. Gjerde, and J. S. Fritz, *Anal. Chem. 53*, 1691 (1981).
45. Wescan Ion Analyzer No. 4 (Wescan Instruments, Inc., Santa Clara, CA).
46. Wescan Ion Analyzer No. 8 (Wescan Instruments, Inc., Santa Clara, CA).
47. D. R. Jenke and G. K. Pagenkopf, *Anal. Chem. 54*, 2603 (1982).
48. K. Harrison and D. Burge, Paper No. 301 presented at the Pittsburgh Conference on Analytical Chemistry and Applied Spectroscopy, 1979.
49. D. T. Gjerde, personal communication.

3
Comparison of Eluant-Suppressed and Single-Column Ion Chromatography

James G. Tarter
Department of Chemistry
North Texas State University
Denton, Texas

Edward L. Johnson
Dionex Corporation
Sunnyvale, California

Thomas Jupille*
Wescan Instruments, Inc.
Santa Clara, California

Ion chromatography can be performed by either of two general procedures, as described in detail in the previous two chapters. Each of these two methods has distinct advantages and disadvantages. Just as for most analytical instrumentation, the potential user must consider which method will be most cost efficient for his/her analytical situation. This chapter is intended to help those unfamiliar with the various ion chromatographic techniques to better understand the capabilities of each system. The purpose of this chapter is not to fully inform the reader of the most up-to-date information and bypass the salesperson, but rather, is to inform the reader so that the reader can ask intelligent and useful questions of the salesperson.

A chapter such as this is, almost by definition, out of date as soon as it is written because of the rapid growth of ion chromatography. An indication of this growth can be easily seen by perusing the scientific literature and observing the number of new products being

---

Current affiliation: Pi Technologies, Orinda, California.

introduced for ion chromatography. Accordingly, this chapter will discuss and compare the fundamentals of the two techniques because the fundamentals will remain constant longer than the exact method in which the fundamentals are applied.

A note is in order concerning the tables which follow. The information in Table 1 is not intended to be complete; it is merely informative and designed for comparative purposes only. For more detailed information and specifics, see the first two chapters of the book and the references contained therein. The second table is provided merely as an initial source of companies that provide equipment used in ion chromatography. The manufacturers listed in Table 2 are those who advertise products with specific emphasis on ion chromatography and, it is hoped, have the ion chromatographic expertise to assist when problems arise.

Table 1. Comparison of Eluant-Suppressed and Single-Column Ion Chromatographic Techniques.

Reference topic	Single-column ion chromatography	Eluant-suppressed ion chromatography
Columns		
Polymer	used for both anion and cation analysis	pellicular, PS-DVB-based, several selectivities available
Silica	anion columns, pH 0-7	not commonly employed
Eluants		
Weakly retained anions	hydroxide, nicotinic acid, or phthalic acid (pH 2.7)	dilute bicarbonate or tetraborate eluants
Moderately retained anions	potassium hydrogen phthalate (pH of 3.8) p-hydroxybenzoic acid	mixtures of bicarbonate and carbonate
Strongly retained anions	phthalate ion, pH 4.5	mixtures of carbonate and bicarbonate
Monovalent cations	dilute mineral acid (HCl, HNO_3); add methanol to get better amine separation	dilute mineral acids (HCl, HNO_3)

Table 1. (Continued)

Reference topic	Single-column ion chromatography	Eluant-suppressed ion chromatography
[Eluants]		
Divalent cations	ethylenediammonium ion	phenylenediamine at pH 2-3 lysine hydrochloride
Transition metals	ethylenediammonium ion plus complexing agent such as citric or tartaric acids	not applicable
Organic anions	ion exclusion; water, dilute nitric acid	ion exclusion; dilute acids
	ion pair; tetrabutylammonium ion	MPIC; ion-pair reagents with organic modifiers
Organic cations	dilute mineral acid plus methanol	MPIC; ion-pair reagents with organic modifiers
		ion exchange; dilute HCl or HNO_3
Detectors		
Conductometric	most common type	bipolar pulse
Electrochemical	amperometric	amperometric and pulsed amperometric
UV-visible	both used, both direct and indirect methods	both used, often with post-column reactors
Other	refractive index	fluorescence for amino acids
Advantages	rapid, sensitive, easy sample preparation, reverse phase, normal phase	easy to use, no significant equilibration time, long column life, no corrosion, wide dynamic range, highest sensitivity, wide range of columns, wide range of eluants
Disadvantages	requires difference in conductance between eluant and analyte ions; temperature stability crucial	$pK_{a,b}$ must be less than 7 for conductometric detection

Table 2. Ion Chromatography Equipment Manufacturers

Total systems	Components
Bio-Rad	Detectors
Dionex	Perkin-Elmer
ESA, Inc.	Columns
LDC	Interaction
Waters	Perkin-Elmer
Wescan	Varian
	Vydac

4
Detection Methods in Ion Chromatography

Paul R. Haddad
Department of Analytical
 Chemistry
University of New South Wales
Sydney, Australia

Petr Jandik
Millipore, Ltd.
Waters Chromatography
 Division
Milford, Massachusetts

I. INTRODUCTION

In this chapter, detection methods are discussed, with emphasis on both the hardware considerations and the chemistry underlying each approach. The discussion will be restricted to separations achieved using ion exchange because this separation mode is the most widespread; moreover, other separation modes such as ion interaction chromatography generally use the same detection procedures after suitable modification to accommodate any special mobile phases employed. It is convenient to subdivide detection methods into three broad categories: (a) electrochemical methods (that is, those using conductivity, amperometry, or potentiometry as the basis of detection); (b) spectroscopic methods [that is, those using ultraviolet/visible (UV/vis) absorbance, refractive index, fluorescence, atomic absorption, or atomic emission for detection]; and (c) those detection methods based on postcolumn reactions. Each of these approaches will be discussed separately.

In most cases, the detection method is inextricably linked to the separation method employed, and the eluant is usually chosen with both the separation and detection methods in mind. Although the separation methods used in all common forms of ion chromatography have been treated in earlier chapters, it is pertinent to review those aspects of ion exchange chromatography that directly affect the nature of the detection process employed.

Typical anion and cation exchange equilibria are given below, and for simplicity only univalent species are considered:

Anion exchange: $\text{Resin-NR}_3^+\text{E}^- + \text{A}^- \rightleftharpoons \text{Resin-NR}_3^+\text{A}^- + \text{E}^-$ (1)

Cation exchange: $\text{Resin-SO}_3^-\text{E}^+ + \text{C}^+ \rightleftharpoons \text{Resin-SO}_3^-\text{C}^+ + \text{E}^+$ (2)

Equations (1) and (2) illustrate the manner in which the solute ions A^- and C^+ are retained on the column. Implicit in the above equations is the fact that when solute ions elute from the ion exchange column, they replace in the eluant an equivalent number of eluant ions so that electroneutrality of the eluant is maintained.

This situation has great importance in the detection of the eluted solute ions. Detection can be achieved by direct measurement of a specific property of the solute ion; however, if this approach is used, it must be remembered that the property measured must enable the solute ion to be differentiated from the relatively concentrated background of eluant ions that must necessarily be present in order to achieve elution. Because of the nature of the ion exchange process itself, it is common for the eluant and solute ions to have similar properties; for this reason, direct detection is generally applied in cases where selective detection of a limited number of solutes is desired.

An alternative and much more widely applicable detection method exists. This method is based on the fact that eluted solute ions replace an equivalent number of eluant ions in the mobile phase, as discussed above. It follows that if a property of the eluant is monitored by the detector, then changes in detector signal will occur upon elution of a solute ion that has a dissimilar value of that same property. This approach can best be described as differential detection and will be successful, provided there is a large enough difference in the values of the measured property between the eluant and solute ions. The majority of ion chromatographic detection methods is based on this differential approach, and for the purposes of the subsequent discussion in this chapter, a further distinction will be made. That is, methods where the eluant ion has a *lower* value of the measured property than the solute ion are described as *direct* detection methods, whereas methods in which the eluant ion has a *higher* value of the measured property than the solute ion are described as *indirect* detection methods.

II. ELECTROCHEMICAL METHODS OF DETECTION

A. Conductivity Detection

Principles

The advantages of conductivity detection for inorganic ion analysis have long been recognized; these advantages are that all ions are electrically conducting, so that conductivity detection should be

Detection Methods in Ion Chromatography

universal in response, and that conductivity detectors are relatively simple to construct and operate. Conductivity detection is very widely employed in ion chromatography, and applications are therefore abundant. The technique will be discussed here in terms of the principles of its operation, the modes of detection employed, the cell designs, and the performance characteristics of conductivity detectors. Applications will not be specifically addressed.

The principle of operation of a conductivity detector in ion exchange chromatography can be illustrated by considering the conductance of a typical eluant prior to and during the elution of a solute ion. For the purposes of illustration, an anion exchange system will be considered; however, the detector response equations developed are equally applicable to cation exchange, provided the obvious amendments are made. The response equations developed below follow the method of Fritz et al. [1-3].

The conductance, G, of a solution (expressed in micro Siemens, μS) is given by the equation:

$$G = \frac{(\lambda_+ + \lambda_-) \, CI}{10^{-3} \, K} \qquad (3)$$

where λ is the limiting equivalent conductance of the cation or anion, C is the normality, I is the fraction of the eluant that is ionized, and K is a constant (called the *cell constant*) with units of reciprocal centimeters, which takes into account the physical dimensions of the cell. This equation permits calculation of the approximate conductances of typical eluants used in ion chromatography, provided the cell constant and the limiting equivalent conductances of the eluant anion and cation are known. The cell constant can be readily determined by measuring the conductance of a solution containing species with known limiting equivalent conductances (e.g., KCl for which $\lambda_+ = 74$ and $\lambda_- = 76$ at 25°C [4]). The values of conductance calculated from Eqn. (3) are only approximate because values of limiting equivalent conductance are used, and, strictly speaking, these apply only to very dilute solution. In addition, the cell constant determined by the abovementioned method is of limited accuracy, and temperature variations can also lead to error.

If it is assumed that the eluant has a total concentration of C_E and contains the cation E^+ and the anion E^-, then the background eluant conductance (G_B) calculated from Eqn. (3) is:

$$G_B = \frac{(\lambda_{E^+} + \lambda_{E^-}) \, C_E \, I_E}{10^{-3} \, K} \qquad (4)$$

where I_E is the degree of ionization of the eluant.

As discussed in the introductory part of this chapter, eluted sample anions (represented as S^-) displace an equivalent number of eluant anions from the mobile phase. If the concentration of solute passing through the detector is given by C_S and the degree of ionization of the solute is I_S, then the eluant concentration in the detector cell during sample elution is $C_E - C_S I_S$. The conductance of the cell at this time originates from the eluant and solute anions, together with the eluant cations that are required to maintain electroneutrality. The solute cations need not be considered because these are unretained on the anion exchange column used in this example. The conductance during solute elution (G_S) is therefore given by:

$$G_S = \frac{(\lambda_{E^+} + \lambda_{E^-})(C_E - C_S I_S) I_E}{10^{-3} K} + \frac{(\lambda_{E^+} + \lambda_{S^-}) C_S I_S}{10^{-3} K} \tag{5}$$

The change in conductance (ΔG) that accompanies the elution of the solute can be obtained by subtracting G_B from G_S to give:

$$\Delta G = G_S - G_B = \left[\frac{(\lambda_{E^+} + \lambda_{S^-}) I_S - (\lambda_{E^+} + \lambda_{E^-}) I_E I_S}{10^{-3} K} \right] C_S \tag{6}$$

The solute concentration during elution will follow an approximately Gaussian profile, and the detector response will therefore change in a similar manner. Equation (6) is applicable to all forms of ion chromatography and shows that the detector response depends on solute concentration, the limiting equivalent conductances of the eluant cation and of the eluant and solute anions, and the degrees of ionization of both the eluant and the solute. These latter parameters are in turn governed by the eluant pH (because the eluant is generally well buffered), and Eqn. (6) suggests that weak acid anions such as acetate, fluoride, formate, phosphate, oxalate, etc. should show appreciable decreases in sensitivity as the eluant pH is lowered; this behavior has been observed in practice [2].

It is interesting to note that the degree of ionization of the eluant (I_E) has a significant effect on detector response. As the degree of dissociation of the eluant decreases, Eqn. (6) predicts that detector response will increase. This prediction has been verified [2] by use of benzoic acid as eluant, with which the detector response was approximately seven times greater than that obtained with a sodium benzoate eluant of equivalent eluting power. This increase in response can be attributed to elution of solute anions by the protonated form of the eluant according to the equation:

$$\text{Resin-}S^- + HBz \rightleftharpoons \text{Resin-}Bz^- + H^+ + S^- \tag{7}$$

where HBz is benzoic acid.

If the solute anion remains completely ionized, the conductance change accompanying solute elution is given by:

$$\Delta G = \left[\frac{(\lambda_{H^+} + \lambda_{S^-}) - (\lambda_{H^+} + \lambda_{E^-})}{10^{-3} K} \right] C_S \qquad (8)$$

and the observed increase in detector response (as compared to when salts are used as eluants) is due to the presence of the highly conducting hydrogen ion. A major disadvantage of partly ionized eluants of the type described above is that such eluants are very weak, and the high concentrations necessary for suitable elution of solutes can lead to baseline disturbances resulting from adsorption of the neutral form of the eluant molecule [5,6].

If the eluant and solute are fully ionized, Eqn. (6) may be simplified to:

$$\Delta G = \frac{(\lambda_{S^-} - \lambda_{E^-}) C_S}{10^{-3} K} \qquad (9)$$

This equation shows that if a conductivity detector is used to monitor the effluent from an ion exchange column, the signal observed when a solute ion elutes is proportional to the solute concentration and to the difference in limiting equivalent conductances between the eluant and solute ions. A similar relationship has been derived for the conductometric detection of cations after ion exchange separation [3]. These relationships are fundamental to the understanding of the function of conductivity detection and can be used as a basis for discussing the various modes of conductivity detection employed in ion chromatography, as outlined in the next section.

Conductivity Detection Modes

From Eqn. (9), it is clear that sensitive detection can result as long as there is a considerable difference in the limiting equivalent conductances of the solute and eluant ions. This difference can be positive or negative, depending on whether the eluant ion is strongly or weakly conducting. It should be remembered here that the response equations developed are applicable to any ion exchange system, provided the conductivity detector is situated immediately after the ion exchange column; that is, the equations relate only to nonsuppressed ion chromatography (alternatively described as single-column ion chromatography). The exception to this is Eqn. (8), which is valid for both forms of ion chromatography. The detection modes used in single-column ion chromatography will be discussed first; further discussion will then be devoted to suppressed ion chromatography, within the context of conductivity detection.

Table 1. Limiting Equivalent Conductances of Some Anions and Cations

Anions	λ_-	Cations	λ_+
OH^-	198	H^+	350
F^-	54	Li^+	39
Cl^-	76	Na^+	50
Br^-	78	K^+	74
I^-	77	NH_4^+	73
NO_3^-	71	Mg^{2+}	53
SO_4^{2-}	80	Ca^{2+}	60
Benzoate$^-$	32	Sr^{2+}	59
Phthalate^{2-}	38	Ba^{2+}	64
Citrate^{3-}	56		

Source: Data from Refs. 1 and 2.

If the limiting equivalent conductance of the eluant is kept low through the use of carefully selected eluants such as phthalate [7], benzoate [8], or other aromatic acid salts, then direct detection (in terms of the definition given earlier) results. That is, an increase in conductance occurs when the solute enters the detection cell. This form of detection is exemplified in the ion chromatographic method developed by Fritz and co-workers [7,9]. Table 1 shows values of the limiting equivalent conductances of some anions and cations, as well as those of typical eluants used in ion chromatography, and Figure 1 illustrates a typical chromatogram obtained with single-column conductivity detection.

An alternative strategy is to use a highly conducting eluant and to monitor the decrease in conductivity that accompanies the elution of a solute ion. This is an indirect detection mode and has been applied to the detection of anions in a potassium hydroxide eluant [11] and cations in a dilute nitric acid eluant. These detection modes are illustrated in Figures 2 and 3, and an example of the application of this same approach to the detection of transition metal and alkaline earth cations with ethylenediamine/tartrate as eluant, is shown in Figure 4. Each of the eluants used in these examples provides excellent detector response because of the very high limiting equivalent conductances of

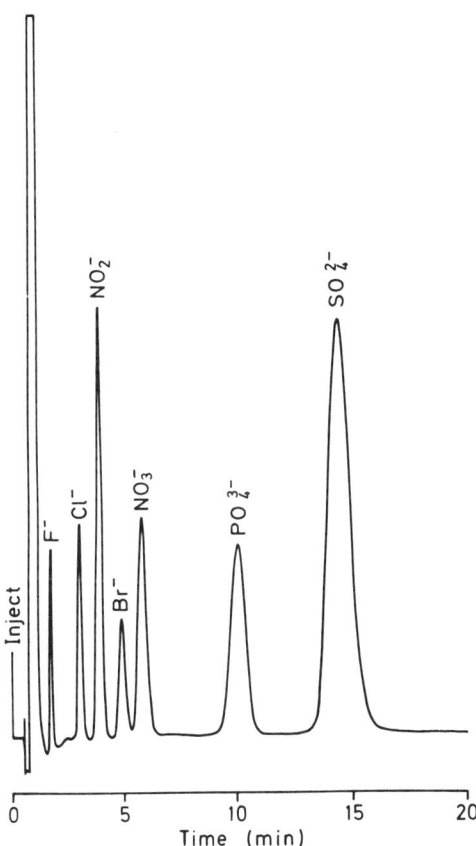

Figure 1. Separation of inorganic anions with conductivity detection. Column: TSK-GEL IEX-620 ion exchanger. Eluant: 1.3 mM $Na_2B_4O_7$ - 5.8 mM H_3BO_3 - 1.4 mM K-gluconate at pH 8.5 in water-acetonitrile (88:12). Solute concentrations: 5-40 ppm. (From Ref. 10.)

the hydroxide, hydrogen, and ethylenediammonium ions; however, the detector noise (and hence the ultimate sensitivity attainable) increases with increasing conductivity. This aspect is further discussed below.

In suppressed ion chromatography, sensitive conductivity detection is achieved by a twofold process. Firstly the eluant is converted by the suppressor column into a poorly conducting form, that is, one with low limiting equivalent conductance. For example, if the case of anion chromatography is considered where a sodium carbonate/sodium bicarbonate buffer is used as eluant, then passage of this buffer through a

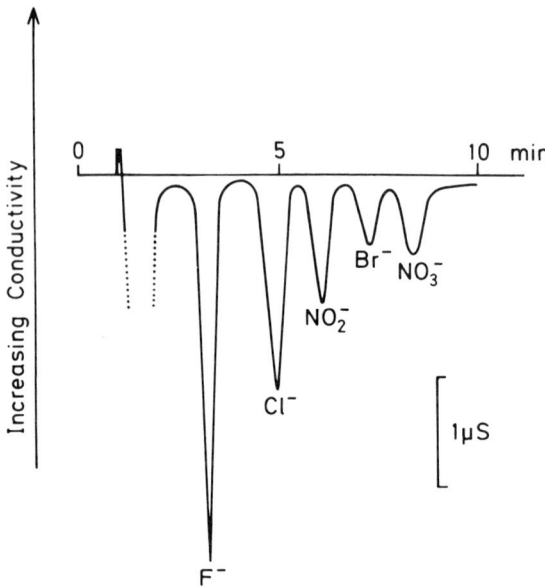

Figure 2. Conductivity detection of anions in single-column ion chromatography, using an eluant with high background conductance. Column: TSK-GEL 620 SA. Eluant: 2 mM KOH. Solute concentrations: 5 ppm. (From Ref. 11.)

suppression device results in exchange of sodium ions for hydrogen ions. Because both eluant anions are the conjugate bases of a weak acid, then the exchanged hydrogen ions are largely consumed in the formation of undissociated eluant acid (in this case, carbonic acid). This step alone causes an increase in the detectability of solute anions, in accordance with Eqn. (6). Detection sensitivity is further enhanced by the fact that sodium ions associated with the solute anions are also exchanged for highly conducting hydrogen ions. Provided that the solute anion is the conjugate of a moderately strong acid, then no significant protonation of the solute will occur and the detector will respond to both the solute anion and the hydrogen ions introduced by the suppressor, resulting in sensitive detection. This situation is identical to that existing in the single-column mode when an acidic eluant (such as benzoic acid) is used. It should be noted that the suppression of eluant conductivity via protonation reactions can be applied to other eluant types, such as zwitterionic buffers [13]; an example of this approach is given in Figure 5.

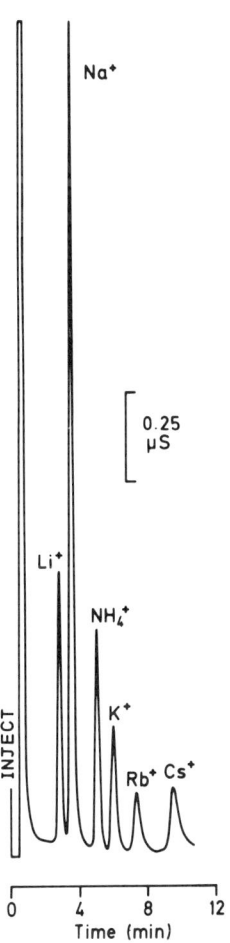

Figure 3. Conductivity detection of cations in single-column ion chromatography, using an eluant with high background conductance. Column: Waters IC PAK C. Eluant: 1.0 mM picolinic acid adjusted to pH 3.0 with nitric acid. Solute concentrations: 10^{-5} M in each species, except for sodium, which has twice this concentration. The peaks are in the direction of decreasing conductance.

In many ways, the distinction commonly made between suppressed and single-column ion chromatography is somewhat arbitrary, especially as far as the detection method is concerned. In both cases, the detector is required to offset a background conductivity which is

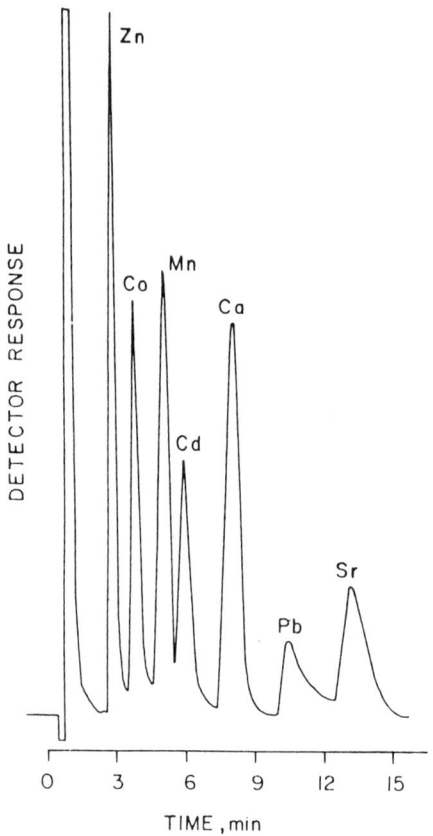

Figure 4. Detection of transition metal and alkaline earth cations, using an eluant with high background conductance. Column: home-packed with Benson Company low-capacity cation exchange resin. Eluant: 1.5 mM ethylenediamine and 2.0 mM tartrate at pH 4.0. Solute concentrations: 10-20 ppm. The peaks are in the direction of decreasing conductance. (From Ref. 12.)

often similar in magnitude for the two methods [2]. In addition, when an acid eluant is used for the single-column approach, the detector response equation developed earlier [i.e., Eqn. (8)] is equally applicable to the suppressed case. Indeed, it has been shown that for a chromatographic system with constant parameters, the conductance change occurring on solute elution in a suppressed ion chromatographic system using 0.003 M $NaHCO_3$ and 0.024 M Na_2CO_3 as eluant is only

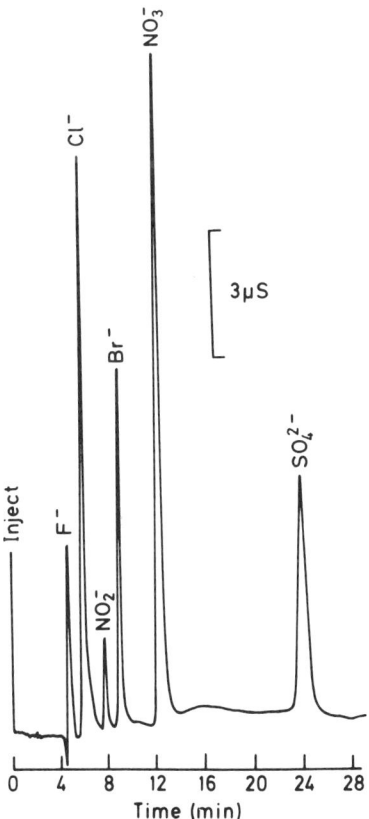

Figure 5. Suppressed ion chromatography involving formation of poorly conducting zwitterions in the suppressor column. Separator column: μBondapak C18. Suppressor column: Dionex anion suppressor. Eluant: 10 mM tetrabutylammonium hydroxide and 11 mM 2-(N-morpholino)ethanesulfonic acid. Solute concentrations: 10-30 ppm. (From Ref. 13.)

27.5% greater than that obtained with single-column ion chromatography using an eluant containing 8.4×10^{-4} M benzoic acid [2].

Cell Design

The design of conductivity detectors suitable for the chromatography of inorganic ions has been the subject of considerable study. The general requirements are small cell volume (to minimize dispersion

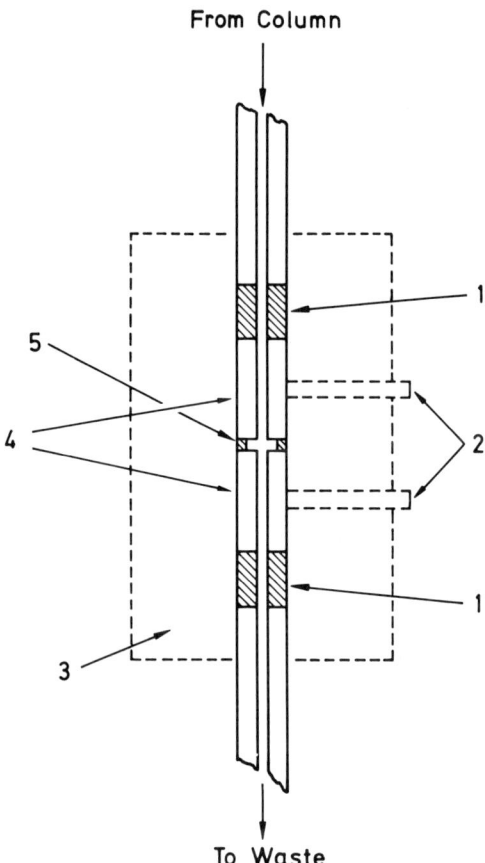

Figure 6. Small-volume flow-cell for conductivity detection. (1) Teflon spacers, (2) electrode contacts, (3) Teflon union, (4) electrodes, (5) spacer. (From Ref. 14.)

effects), high sensitivity, wide range of linearity, rapid response, and acceptable stability. The cell itself generally consists of a small chamber, typically less than 5 μl in volume, fitted with two electrodes (Figure 6); some cell designs, however, have multiple electrodes, and one commercial instrument features a cell with five electrodes [15] (Figure 7). Electrodes are usually constructed of platinum, stainless steel, or gold.

In conductivity detection, the actual parameter measured is the resistance of the solution passing through the cell. This measurement

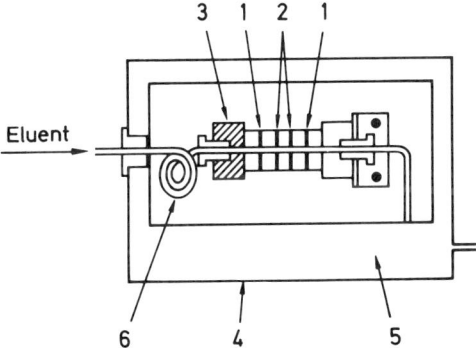

Figure 7. Five-electrode conductivity cell used in Waters M430 conductivity detector. (1) Reference electrode, (2) detection electrode, (3) guard electrode, (4) cell unit, (5) cell oven, (6) heat exchanger. (From Ref. 15.)

may be accomplished by incorporating the cell as one arm of a Wheatstone bridge circuit [16], wherein an alternating voltage is applied to the cell. Cell current is an important aspect because excessive currents lead to heat dissipation in the cell, resulting in a noisy, drifting baseline. A specially designed conductivity cell has been reported [17] that minimizes cell current and provides greater thermal stability than conventional cell configurations, giving a concomitant decrease in baseline noise and hence improved sensitivity.

One problem introduced when a conductivity cell is incorporated into an AC bridge circuit is the appearance of undesirable capacitance effects which limit the working range of the detector. One method by which this problem can be overcome is through the use of the bipolar-pulse technique for fast conductance measurements [18,19]. This involves the application of two successive constant-voltage pulses of opposite polarity to the conductance cell and measurement of the current flowing in the cell at the end of the second pulse. In this way, capacitance effects are eliminated, and electronic nulling of the background conductance of the eluent is also possible so that the bipolar pulse technique is suitable for single-column ion chromatography.

Many modifications to the basic design of conductivity detectors have been suggested in order to improve sensitivity, baseline noise, and drift and to extend the working range of the detector. An attenuation detector with logarithmic response over six decades of conductivity was reported by Svoboda and Marsal [20]. A 3-µl cell with stainless steel electrodes was excited with an alternating symmetrical square wave (1 V, 2.5 kHz), and the detector was used for peaks with steep band edges, which would have required range switching on conventional conductivity detectors. Oscillometric flow cells suitable for use in both conductivity and permittivity (i.e., capacitance measurement) detectors have been developed [21,22]; however, these have not proved to be as sensitive in ion chromatographic applications as other conductivity detectors. A paired-flow-cell differential detector has been designed for use with single-column ion chromatography [23] in an effort to minimize baseline drift resulting from temperature changes and variation in eluant composition. A further specialized cell design for ion chromatography is one in which the electrodes are not in galvanic contact with the solution to be measured [21]. The principal advantage of this innovation is that polarization, corrosion, and other side effects that may alter the condition of the measuring electrode surface, and therefore also the cell constant, are eliminated.

Detector Performance

Thermal stability of conductivity detectors is a problem that has been widely recognized [24-26]. Because the temperature coefficient of conductivity measurements is 0.5-3%/°C [16,27], it is clear that very close temperature control or the use of reference cells is necessary in order to minimize baseline noise, especially when nonsuppressed eluants are used. The extreme temperature sensitivity of conductivity detection can be revealed by the detector disturbance created when a hand is placed on the column or steel tubing [26]. To minimize such temperature effects, it is necessary to temperature-stabilize (either by suitable insulation or active temperature control) all components of the chromatographic system including the injector, precolumn, analytical column, detector cell, and all related tubing. This treatment has been observed to reduce baseline noise by a factor of seven [24].

The parameter that ultimately determines the sensitivity achievable with a particular detector is the level of baseline noise produced by that detector. In the case of conductivity detection, baseline noise is primarily a function of the background conductivity of the eluant, regardless of the type of eluant used [24]. This dependence of baseline noise on background conductance is illustrated in Figure 8, which shows that although the conductance value of the noise increases at high background conductivities, the measured noise (in volts) on the recorder in fact decreases because of the higher range settings used with strongly conducting eluants. These results, when considered together with the detector response equations given earlier, suggest that the sensitivity of conductivity detection (calculated as the concentration of solute in the detector cell required to give a signal-to-noise ratio of two) improves as the background conductance of the eluant is reduced.

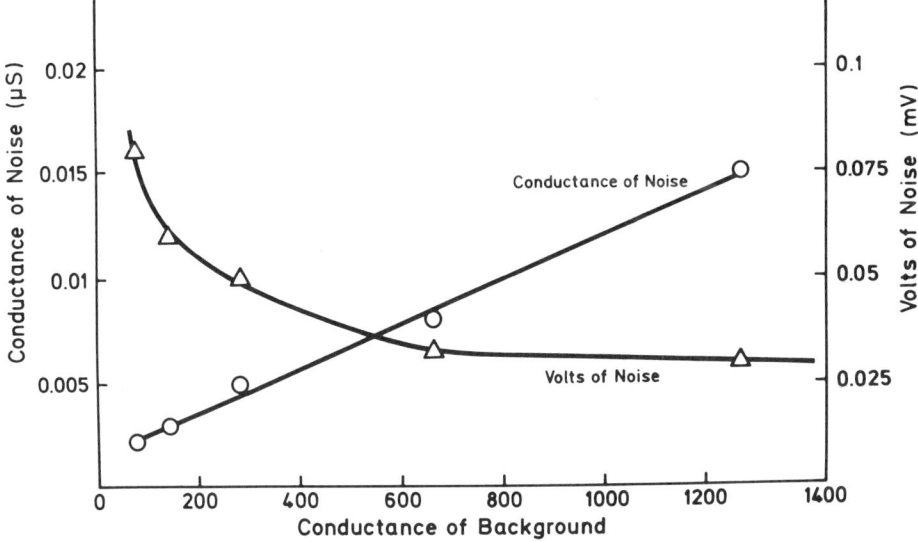

Figure 8. Dependence of baseline noise on background conductance of the eluant. Aqueous KCl solutions of conductance 75, 141, 285, 669, and 1,270 µS were used at appropriate detector range settings. The noise in terms of voltage is the value measured on a 10-mV recorder. (Courtesy of N. Baba, TSK, Japan.)

Calibration graphs in suppressed ion chromatography are generally not linear if a conductivity detector is used. The main reason for this is that the dissociation equilibrium of the eluant (which is present predominantly in the undissociated weak-acid form after passage through the suppression device) is affected by the hydrogen ions which the suppressor also contributes to the solute band. As discussed above, these hydrogen ions strongly influence the measured conductance change that occurs on sample elution, and this leads to an improvement in sensitivity. It must be remembered that the degree of eluant ionization, I_E, (and hence the background conductance) will be dependent on the concentration of hydrogen ions in the detector cell, and the hydrogen ion concentration in turn depends on the solute concentration. The result of these interdependences is that the background conductance of the eluant will vary with the concentration of solute injected, leading to curvature of the calibration graph, in accordance with Eqn. (8). It is, of course, possible to derive a new detector response equation that considers changes in the degree of eluant dissociation with solute concentration, and such an equation has been reported [28]. This equation enables calculation of the solute concentration from the measured conductance signal, provided the limiting equivalent conductances of the eluant and solute ions are known, together with the pK_a of the eluant acid, the eluant concentration, the cell constant, and the capacity factor of the solute ion.

If single-column ion chromatography is used with a fully ionized eluant, it is possible to quantitate solute concentrations without the use of standards [29]. This procedure, based on Eqn. (9) above, involves the measurement of detector response, using first an eluant with a high limiting equivalent conductance and then an eluant with a low limiting equivalent conductance. Two equations of the form of Eqn. (9) are then obtained, which are used to solve for the two unknowns C_S (the solute concentration) and λ_S (the limiting equivalent conductance of the solute ion). Indeed, it is also possible to calibrate the detector response by measuring the peak areas of each of the two eluting ions using the other as eluant at a known concentration; when this is done, it is no longer necessary to determine the cell constant or any physical properties of the eluant. The individual charges of the solute ions can be deduced from a study of adjusted retention times, so that molar concentrations of solutes can be determined.

B. Amperometric Detection

Introduction

The past ten years have seen an increase of interest in the usage of amperometric methods of liquid chromatographic detection [30]. In ion chromatography especially, these methods achieve significantly better sensitivity than the traditional methods described elsewhere in this chapter. This gain in sensitivity is accompanied in most cases by an improvement in selectivity, which means that amperometric methods are less universally applicable than other detection methods.

The discussion provided in this chapter is intended for two different groups of readers. The first group comprises chemists who seek immediate and workable solutions to specific problems in routine analysis and who understandably prefer to use commercially available instrumentation. The second group consists of research workers who desire information about the full scope of the technique, including some noncommercial prototypes developed in laboratories around the world.

The steadily increasing range of amperometric hardware offered by manufacturers for application in ion chromatography suggests that many instruments described here as prototypes may become available for routine analytical tasks in the near future.

Detection Methods in Ion Chromatography

Basic Principles

Amperometric detection for liquid chromatography has emerged from the polarographic and voltammetric methods developed since the 1920s. The first polarographic analyzer was developed by Heyrovsky [31], who was awarded the Nobel Prize in 1959 for his achievements. Polarography today is a subsection of *voltammetry*, which is a more general term used for techniques involving controlled changes of potential between a working electrode and a reference electrode, with measurement of current resulting from the reaction of analyzed species at the surface of the working electrode. The term *polarography* is specific for those techniques that obtain current-potential dependences, using mercury as a working electrode. Amperometry describes the technique in which a fixed potential is applied to a working electrode (which is often situated in a flowing stream), and the resultant current is measured. Recognition of the origins of amperometry will assist the reader in understanding the present status of the method and will provide guidelines for the development of its full potential.

Six decades of voltammetric and polarographic data have been recently processed and compiled into tables by Meites, Zuman, and coworkers [32]. This information is a valuable resource not only to electrochemists but also to anyone involved in the optimization and development of amperometric methods for chromatographic detection.

The interrelations between voltammetric and amperometric detection can be explained with the help of Figure 9. If an oxidizable species is present in the analyzed solution, an S-shaped step (i.e., a wave) appears in the voltammogram. Its height is proportional to the concentration of the electroactive species, and the position of the wave on the potential (E) axis is specific for a given compound. An analogous observation can be made for a reducible species; however, its signal will be located in the right, upper portion of the current-potential plot (i.e., reduction current, negative potentials). In the absence of electroactive species, the current-potential plot does not show any waves, and the recording represented by the dotted line in the left part of the voltammogram and the solid line in the right part is obtained. The increases in current at the extreme positive (E_1), and extreme negative (E_2) potentials are caused by oxidative and reductive decomposition of the supporting electrolytes or by oxidation and reduction of the working electrode.

For successful application of amperometric detection in ion chromatography, the optimal oxidation or reduction potential for a given solute species must be established. The values for such potentials can be found either in the chromatographic literature or in electrochemical tables [32]. Let us assume that we wished to detect an oxidizable anion and that we were able to find the optimal potential (E_3) for its detection. The potential E_3 must be maintained between the working

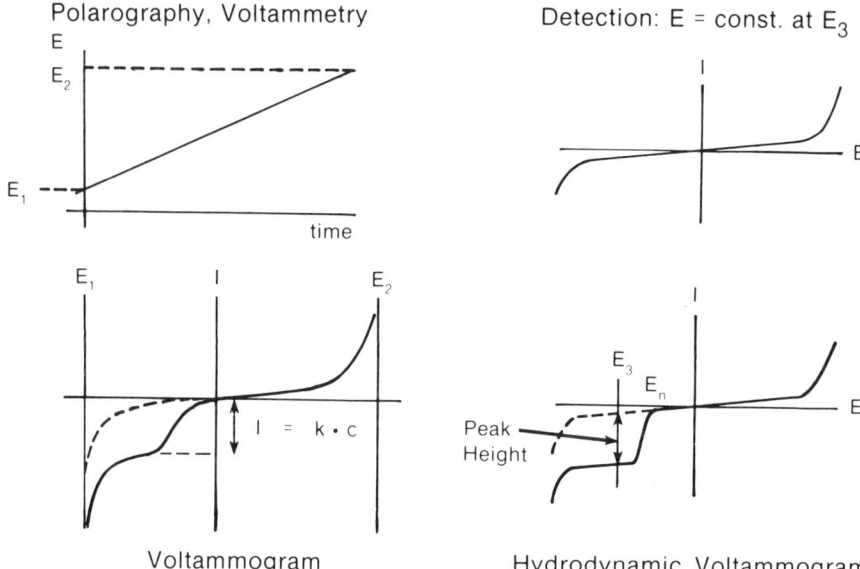

Figure 9. Interrelations between polarographic, voltammetric, and amperometric detection.

and reference electrodes. As long as the solute anion is retained on the column, only a minimal residual current will be recorded, corresponding to the difference between the dotted line in the plot and the abscissa at I = 0. As soon as the solute anion leaves the column and contacts the working electrode, the current will increase. By recording this current change as a function of time (or elution volume), a chromatographic peak is obtained.

In the few cases where no information is available regarding the optimal potential to be used, current-potential plots must be generated, using the ion chromatograph in conjunction with an electrochemical detector. The procedure to be followed involves the injection of the same amount of solute at different applied potentials and evaluation of the peak heights as a function of these potentials. To distinguish the plot obtained by chromatographic procedure from voltammograms measured under static conditions, the designation *hydrodynamic voltammogram* has been suggested [33]. Hydrodynamic voltammograms are useful not only for the determination of unknown detection potentials, but also for the evaluation and comparison of the performance of electrochemical detectors.

The currents generated in detector cells depend on the concentration of the analyte, but are also influenced by a number of other parameters. Exact equations for different amperometric and voltammetric

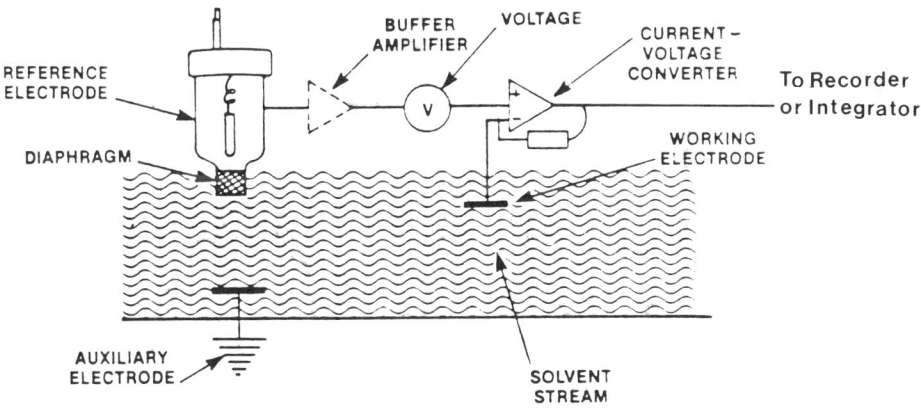

Figure 10. Basic configuration for amperometric detection.

cells can be found in the literature [33-38]. A simplified relationship is given below [33]:

$$i_l \propto nF(C_i - C_f) \tag{10}$$

where i_l is the limiting current (which is the total current obtainable for a given species at the optimal potential), n is the number of electrons transferred in the redox reaction, F is 96,485 C/equivalent, C_i is the concentration (in equivalents per volume) of analyte entering the cell, and C_f is the concentration of analyte leaving the cell. The ratio C_f/C_i is a function of the cell geometry and of the flow rate. In most amperometric cells this ratio is much less than unity (typically between 0.05 and 0.1). If this ratio approaches unity, the detection method is described as being coulometric.

The term *coulometry* implies a measurement of electric charge (coulombs). Because the peak areas obtained in amperometric detection do represent the charge, all chromatographic results based on areas determined with I,t coordinates should in theory be described as coulometric. The results obtained using peak heights, even in cases where C_f/C_i is unity, should be termed *amperometric*. Not wishing to add to the confusion that already exists, we will adhere to the convention that reserves the designation of coulometry for methods involving the total conversion of the electroactive analyte and discuss it as a special case of amperometric detection.

A simplified drawing showing the manner in which the electronic circuitry functions in conjunction with the detector cell is shown in Figure 10. The recorded analog signal is commonly generated by a

Table 2. Basic Principles of Amperometric and Voltammetric Detection

Method	Controlled quantity	Measured quantity
Amperometry	Potential is held constant.	Current
Coulometry	Potential is held constant.	Peak heights, current. Peak areas, charge.
Potentiodynamic voltammetry	Potential is changed as a function of time.	Current is evaluated either continuously or by sampling.
Amperometry with potentiodynamic cleaning of the electrodes	Optimum change of potential is applied during the cleaning interval. Potential is held constant during the measurement.	Current is evaluated between two cleaning intervals.

conversion of amplified oxidation or reduction currents to voltage (this is achieved in the current-voltage converter in Figure 10). More detailed descriptions of the circuitry for voltammetric and amperometric measurements are given elsewhere [33-38].

Table 2 gives a summary of the basic principles of electrochemical detection techniques.

Flow-Through Electrode Cells for Amperometric Detection

Cell Design--Each cell design shown in the following figures can be used with all the modes of amperometric and voltammetric detection

Cell design	Electronic unit	Remarks
Three electrodes	A potentiostat control	Less than 10% conversion of the analyte at the electrode. Electrode materials: carbon, silver, copper, cadmium, gold, mercury. Well-established technique. Large choice of commercial instrumentation.
Working electrodes larger than in amperometry	See Amperometry.	ca. 100% conversion of the analyte. Well-established technique. Few commercial instruments available.
Three electrodes	Voltammetric instruments capable of providing modulated potentials and various sampling modes for the current.	This technique, although still in the developmental stage can be carried out combining amperometric flow through cells with voltammetric instruments, for example, from BAS, EG&G, and Tacussel, etc.
Three electrodes	See potentiodynamic voltammetry.	Reproducibility of amperometric detections on some solid electrodes (for example, platinum) can be improved by this technique.

described in Table 2. A notable exception to this statement is the large-surface-area coulometric cell (see Figure 14), which cannot be employed for any potentiodynamic techniques.

The first commercial amperometric detector, introduced in 1974, contains a thin-layer cell (Figure 11), which has received wide acceptance in comparison with alternative designs. Carbon, silver, gold, and platinum electrodes are available for this cell. A modified thin-layer arrangement that simplifies the handling of the cell and the maintenance of the working electrodes has recently been introduced by Waters Associates and is shown in Figure 12. Glassy carbon, silver, and copper electrodes are offered with this design. Other commercially available

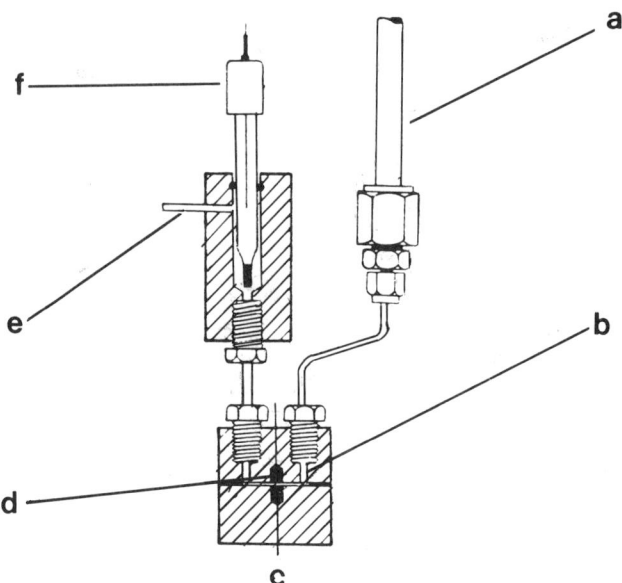

Figure 11. Thin-layer amperometric cell. (a) HPLC column, (b) cell inlet, (c) working electrode, (d) auxiliary electrode, (e) cell outlet, (f) reference electrode. In the original version, the cell outlet was used as the auxiliary electrode. (Courtesy of Bioanalytical Systems.)

detectors utilize the wall-jet configuration (Figure 13) or cylindrical [39] and rod [40] working electrodes.

As explained in the discussion of Eqn. (10), there are two possibilities for the achievement of total (or close to total) electrochemical conversion of an analyte in a detector cell. The first approach is based on the utilization of large areas for the working electrode; the most widely used design of this kind is depicted in Figure 14. This design is the only commercially available large area coulometric cell. The second approach involves the use of smaller electrodes, but with very low flow rates; this approach has been used to achieve 100% coulometric efficiency with a thin-layer cell employed in conjunction with a microbore column and a flow rate of 2 μl/min [38,41].

Polarographic detector cells with either stationary or dropping mercury electrodes have been the subject of some controversy. Doubts have been expressed [33] that the dropping mercury electrode will remain as a competitive analytical device beyond the turn of the century;

however, this view has been opposed [38] on the grounds that the mercury electrode is undoubtedly competitive in comparison to solid electrodes in certain applications. More discussion on this subject will be offered in the section dealing with applications. One of the few commercially available polarographic cells is depicted in Figure 15.

Reference and Auxiliary Electrodes--Early cell designs featured two electrodes--the working and reference electrodes; in such designs the reference electrode served both to complete the circuit in order to allow the charge to flow through the cell, and to maintain a constant potential between itself and the analyzed solution. The latter function can be fulfilled only if the system is kept under zero current conditions [33]. In all recent detector cells, an auxiliary electrode is incorporated into the system in order to carry the current, and the reference electrode acts solely to monitor the potential of the working electrode in a potentiometric (zero current) mode.

The most widely utilized reference electrodes are the silver/silver chloride and calomel electrodes. A palladium-hydrogen electrode and quasi-reference electrodes based on platinum and other materials have also been used [42]. Potentials of reference electrodes are listed in the literature [34] and should be considered when an attempt is made to reproduce a detection method originally developed with a different reference electrode.

Auxiliary electrodes should be inert and should be situated as close as possible to the working electrode. Typical auxiliary electrode materials are platinum and glassy carbon. In some detector cells, the outlet capillary (Figure 11) or the stainless steel body of the cell close to the working electrode (Figure 12) serves as the auxiliary electrode.

Selected Ion Chromatographic Applications on Various Working Electrode Materials

The widespread use of polarography in stationary solutions is due to several unique characteristics of the dropping mercury electrode. These include the constant renewal of the electrode surface with each drop, leading to uniform and reproducible electrochemical behavior, the high hydrogen overvoltage (-0.8 V) of mercury in comparison with platinum, catalysis effects of mercury, and the participation of mercury in redox reactions. The gradual introduction of solid electrodes stemmed in part from the limited applicability of mercury at positive potentials. The voltage "window" for mercury on the positive potential side extends only to about +0.5 V, whereas platinum and carbon can be utilized for oxidations at potentials well in excess of +1.0 V.

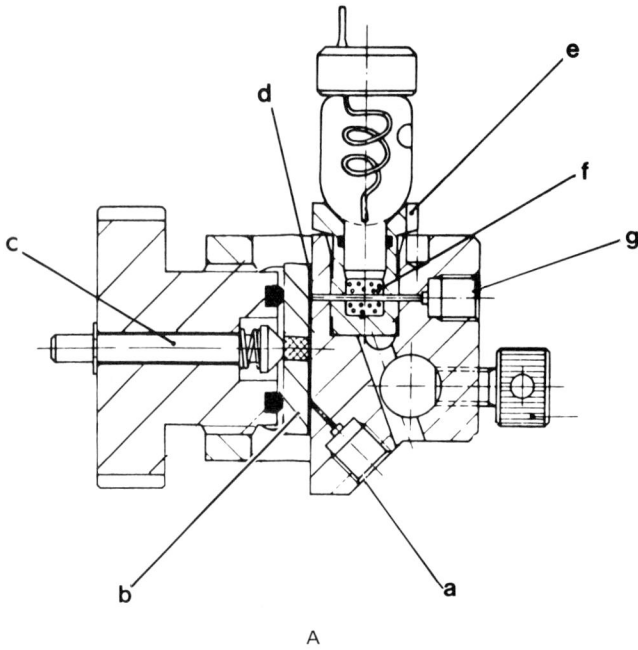

Figure 12. Waters M460 amperometric cell, assembled (A) and disassembled (B). (A) Key: (a) cell inlet, (b) holder for working electrode, (c) contact for working electrode, (d) gasket, (e) Teflon holder for reference electrode, (f) flow-through diaphragm, (g) outlet.

If a solid electrode is used in a flowing solution, or conversely if the electrode itself is moved in a static solution, the performance of the electrode improves because the constant-renewal effect evident with the mercury electrode is imitated to a certain degree. In chromatographic detection cells, hydrodynamic conditions are provided by the flow of mobile phase past the working electrode. On the other hand, inert solid electrodes of platinum or carbon cannot emulate the catalysis and reactive effects of mercury. Mercury is capable of facilitating the detection of many inorganic anions through complexation reactions such as:

$$Hg + 2L^{n-} \rightleftarrows HgL_2^{2(1-n)+} + 2e^- \tag{11}$$

$$2Hg + 2L^{n-} \rightleftarrows Hg_2L_2^{2(1-n)+} + 2e^- \tag{12}$$

Detection Methods in Ion Chromatography

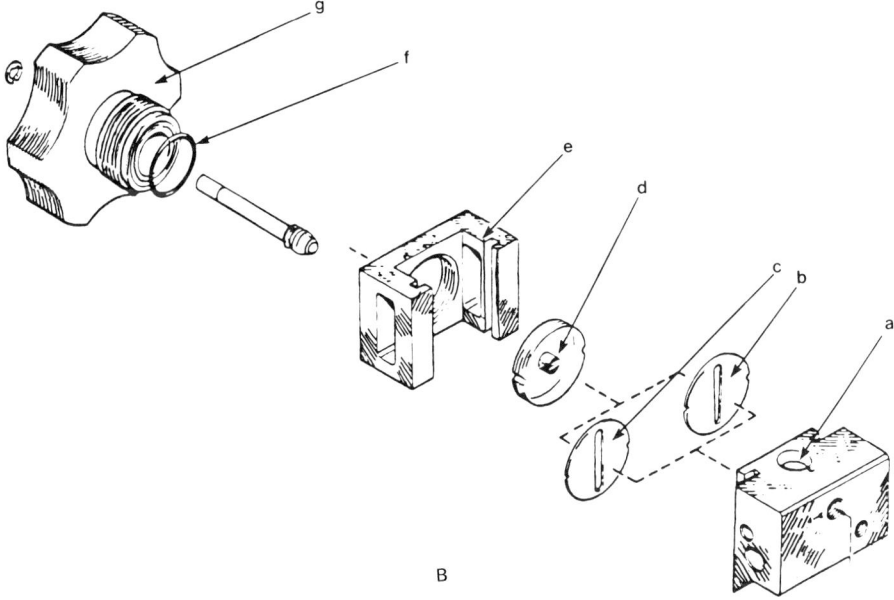

Figure 12. (Continued) (B) Key: (a) opening for reference electrode, (b, c) gaskets which determine the flow path, (d) working electrode, (e) upper part of cell assembly, (f) gasket, (g) Teflon screw. No tools are necessary to dismantle the cell, and the reference electrode is refillable. (Courtesy of Waters Associates.)

At carbon or platinum, direct oxidation of L^{n-} is required, and this necessitates the use of large positive potentials. Successful applications of inert electrodes to the detection of ions are thus relatively rare. One example is the detection of transition metals in the form of coordination complexes on a glassy carbon electrode [43,44].

It has only recently been shown that the amperometric detection of inorganic ions can be improved through the use of electrode materials that mediate or participate directly in the electrode reaction [38,45]. Silver is one such material, for which the following electrode reactions can be postulated:

$$L^- + Ag \rightleftarrows AgL + e^- \tag{13}$$

$$X^- + Ag \rightleftarrows AgX + e^- \tag{14}$$

where either complexes (AgL) or precipitates (AgX) are formed.

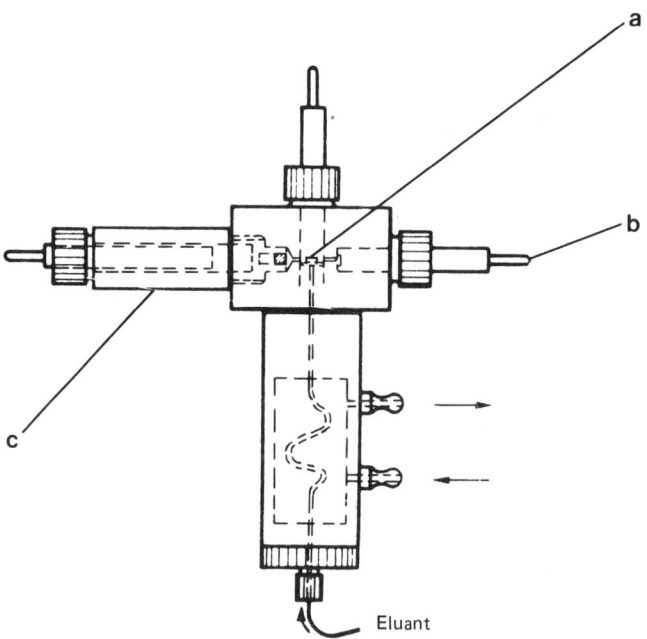

Figure 13. Wall-jet amperometric cell: (a) working electrode, (b) auxiliary electrode, (c) reference electrode. (Courtesy of Metrohm AG.)

A chromatogram of a standard solution of cyanide illustrating the high sensitivity of amperometric detection with a silver electrode is shown in Figure 16a. Injection of the same standard into a mobile phase containing a lower concentration of hydroxide ions results in a significant decrease in peak height (Figure 16b). The role of hydroxide ions in the amperometric detection of cyanide ions has been

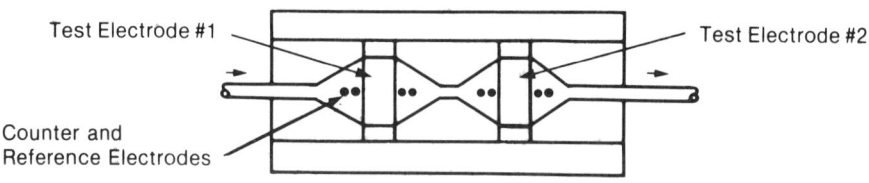

Figure 14. Dual coulometric flow-through cell with large-area working electrodes. (Courtesy of ESA.)

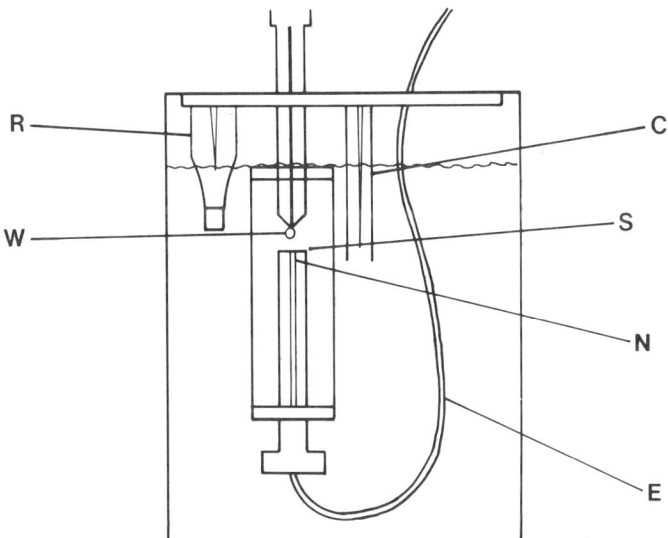

Figure 15. Polarographic detector cell, showing reference electrode (R), mercury drop electrode (W), supporting electrolyte (S), platinum auxiliary electrode (C), nozzle directing the eluant toward the mercury electrode (N), and Teflon tubing connecting the column to the cell (E). (Courtesy of EG&G - PAR.)

investigated [46]. Two other typical examples of the detection of inorganic anions on silver working electrodes are given in Figure 17. The sensitivities of this detection method for the three anions discussed are compared in Table 3, which shows clearly that the complexation of silver by the analyzed ions facilitates detection. Sulfite, which does not exhibit the same complexing ability as cyanide or iodide, has to be oxidized directly at a higher potential, and the sensitivity of detection is diminished in comparison to the complexing anions. When glassy carbon is used as the electrode material, sensitivity for sulfite is further decreased, and a potential in excess of +0.5 V is required [47]. These observations suggest some mediating role of silver, even for the oxidation of sulfite. Further examples of oxidative detection of anions on silver are presented in Figure 18.

An electrochemical detector with a silver electrode has been reported [48] for the detection of nonoxidizable species. The detector cell was placed in the flow stream after a fiber suppressor, and the response to changes in eluant pH was recorded (Figure 19). Silver working electrodes are provided by a variety of manufacturers (e.g.,

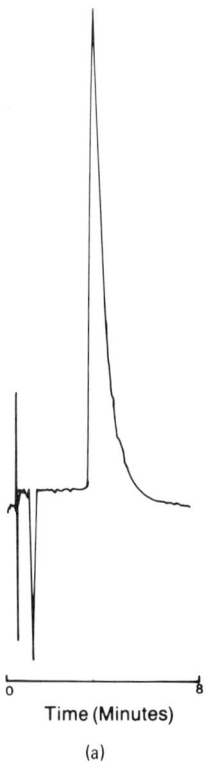
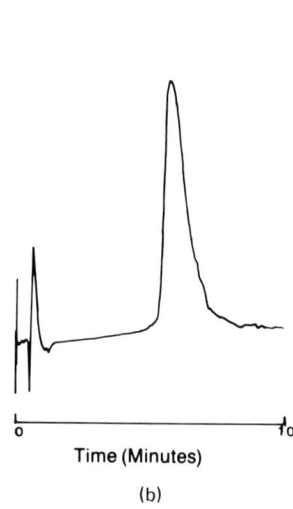

Figure 16. Amperometric detection of cyanide at two different concentrations of hydroxide ions in the mobile phase. Sample: 100 μl of 100 ppb cyanide. Detector: Waters M460 with silver working electrode. Column: Waters IC PAK A. Mobile phase: 5 mM (a) and 2 mM (b) KOH. The same sensitivity setting was used for both chromatograms.

Waters, BAS, and Dionex), and at present, silver represents the most widely used and commercially supported electrode for the detection of inorganic anions.

Another electrode material that facilitates the detection of anions by participation in complexation reactions is metallic copper [49]. This material has been used for the amperometric detection of amino acids (Figure 20). A theoretical model describing the dependence of the current on the rate of the complexation reaction was verified experimentally. Varying detection sensitivity for different amino acids was caused by variation in the rates of the complexation reaction

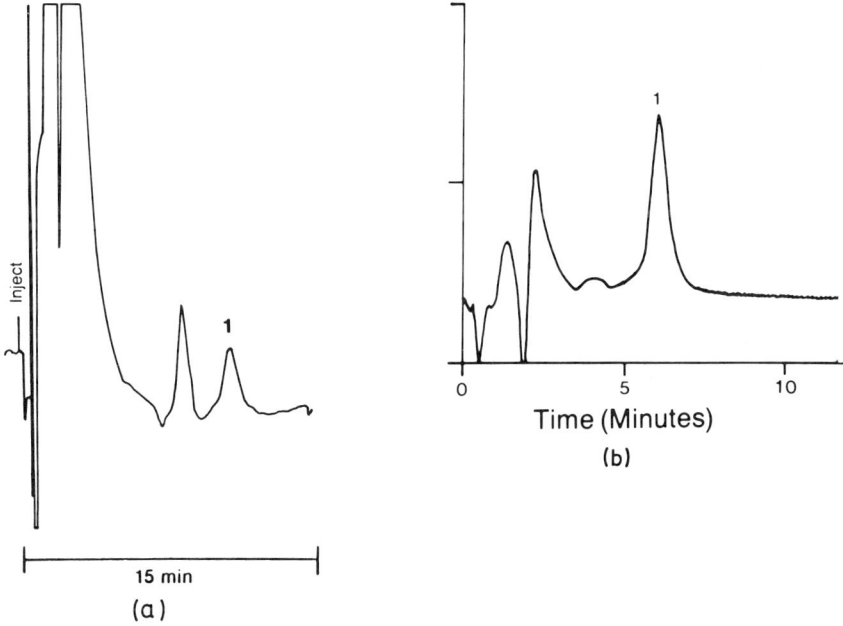

Figure 17. (a) Iodide in infant formula. Sample: 20 µl of the 1:100 diluted formula. Detector: Waters M460 with silver working electrode. Eluant: 2 mM potassium hydrogen phthalate. Column: Waters IC PAK A. Peak 1 is 8.4 ppb iodide in the diluted sample. (b) 4.4 ppm sulfite in beef stew. The sample was prepared according to the procedure given in Ref. 72. Detector: Waters M460 with silver working electrode. Eluant: 1 mM lithium phthalate at pH 10.9. Column: Waters IC PAK A.

Table 3. Data for Amperometric Detection of Iodide, Cyanide, and Sulfite with the Use of a Silver Working Electrode

Anion	Sensitivity (nA/nmol)	Optimum potential (V)	Stability constant (log K_{AgLx})
Iodide	29.7	0.1	approx. 14
Cyanide	16.6	0.21	approx. 21
Sulfite	3.3	0.32	--

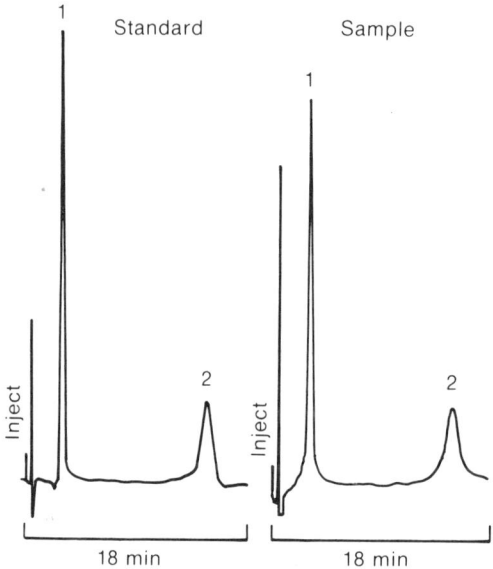

Figure 18. Sulfide and thiosulfate in Kraft White Liquor. Standard: 1 ppm sulfide (1) and 1 ppm thiosulfate (2). Sample: Kraft White Liquor diluted 1:10,000, showing peaks for sulfide (1) and thiosulfate (2). Detector: Waters M460 with silver working electrode. Eluant: 5 mM Na_2HPO_4 at pH 6.5.

between the amino acids and copper. Other preliminary results indicate that the copper electrode may be useful for a wide range of other ligands; in the next section, the use of such an electrode for potentiometric detection in ion chromatography will be discussed.

Appropriate modification of some other metal electrodes opens new possibilities for sensitive and selective detection. Adsorbed iodine on platinum permits interference-free reductive detection of chromate. The reduction of oxygen, which usually complicates the reductive detection of other ions, was significantly inhibited with this type of electrode pretreatment [45]. Other techniques for dealing with oxygen in the reductive mode are given elsewhere [50,51]. Copperized cadmium was successfully used as an electrode material for the selective detection of nitrate [52], and simultaneous determination of nitrite and nitrate in the cathodic mode was achieved with a tubular cadmium electrode [53].

As already mentioned, there are few examples of the use of unmodified inert electrodes for the detection of ions. A carbon cloth

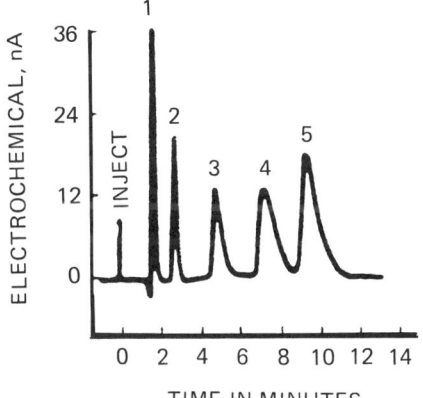

Figure 19. Detection of nonoxidizable anions via amperometric detection: (1) fluoride, (2) chloride, (3) phosphate, (4) nitrate, (5) sulfate. Columns: Dionex HPIC-AS3 anion column and AFS fiber suppressor. Eluant: 0.003 M $NaHCO_3$ and 0.0024 M Na_2CO_3. Detector: Dionex electrochemical detector with silver working electrode. (From Ref. 48.)

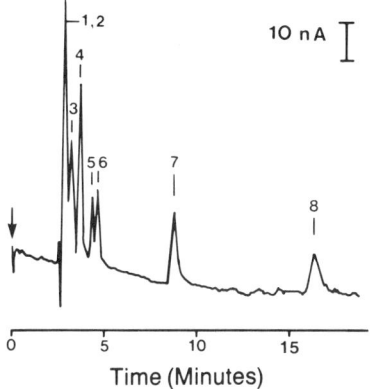

Figure 20. Detection of amino acids by means of a copper working electrode: (1) lysine, (2) threonine, (3) valine, (4) methionine, (5) isoleucine, (6) leucine, (7) phenylalanine, (8) tryptophan. Sample: 25-µl injection containing 10^{-5} M of each amino acid. Eluant: 0.025 M phosphate buffer (pH 7.2) containing 10% methanol. Detector: Metrohm EA 1096. (From Ref. 49.)

electrode has been utilized for the reductive detection of uranyl ions [54], and glassy carbon in a wall-jet type cell has been employed for the detection of several anions at an applied potential of +1.0 V [55].

From the material presented thus far, the conclusion can be drawn that at least some of the advantageous properties of mercury as an electrode material can be simulated by optimally selected solid electrodes employed under hydrodynamic conditions such as those encountered in ion chromatography. However, the mercury electrode remains unequaled for some applications, such as the simultaneous determination of tetraalkyl lead additives in gasoline [38]. Performance of this same analysis on solid electrodes constructed of platinum, gold, or glassy carbon required the use of potentials in excess of +1.0 V, and under these conditions, many other components of the gasoline were observed to be electroactive [56]. The detection potential used with mercury was only about +0.5 V, and very few gasoline constituents were oxidized, leading to a simple and straightforward analysis. Figure 21 illustrates the use of a mercury electrode for this application.

Potentiodynamic voltammetric methods on mercury-containing electrodes have been reported for the determination of organotin [57,58] and organomercury [59] cations. Two examples of this work are presented to conclude this discussion on the current status of amperometric and voltammetric detection methods for ion chromatography (Figures 22, 23). Table 4 lists literature references for further selected applications of amperometric and voltammetric detection in ion chromatography.

C. Potentiometric Detection

Principles

Potentiometry is the technique whereby the potential of an indicating electrode is measured with respect to a reference electrode, under conditions of zero or insignificant current flow. The potential of the indicator electrode varies with the concentration of a particular ion in the solution in contact with the indicator electrode, in accordance with the Nernst equation, giving a logarithmic relationship between electrode potential and solute concentration. The most familiar potentiometric devices are glass electrodes for the measurement of pH, and ion-selective electrodes for the determination of inorganic anions and cations. Although these devices are usually employed for measurements in static solutions, they are also adaptable to measurements in flowing solutions such as the effluent from an ion chromatograph. In such cases, a suitable flow cell must be designed to accommodate the indicator and reference electrodes while keeping the cell volume to a level where solute dispersion in the cell is not excessive. Many

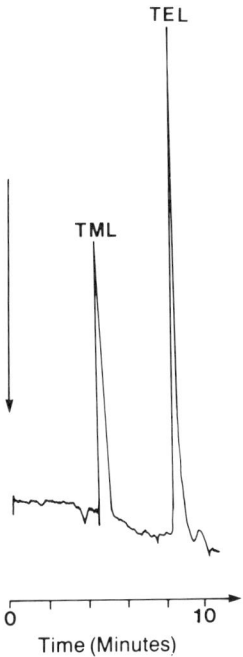

Figure 21. Determination of tetraethyl lead (TEL) and tetramethyl lead (TML) in gasoline. Column: Waters Rad Pak C18. Eluant: 0.05 M tetraethylammonium perchlorate in acetonitrile. Detector: EG&G - PAR model 310 used in the hanging mercury drop mode (see Fig. 15). (From Ref. 38.)

micro flow-cells have been reported, and these have been discussed elsewhere [e.g., 73,74].

The main drawbacks of potentiometric detectors (particularly ion-selective electrodes) in ion chromatography are the slow response of many electrodes and the fact that they respond at best to only a few different species. This inherent selectivity arises both from the nature of the indicator electrode and also from the presence of an ion-selective barrier which separates the indicator electrode from the surrounding solution. This barrier is deliberately chosen to be as selective as possible and therefore naturally restricts the ions that are detected by the indicator electrode. Such selectivity is advantageous when it is required that the electrode respond to one ion in the presence of others, but is particularly disadvantageous in chromatographic applications where a broader range of electrode response is desired.

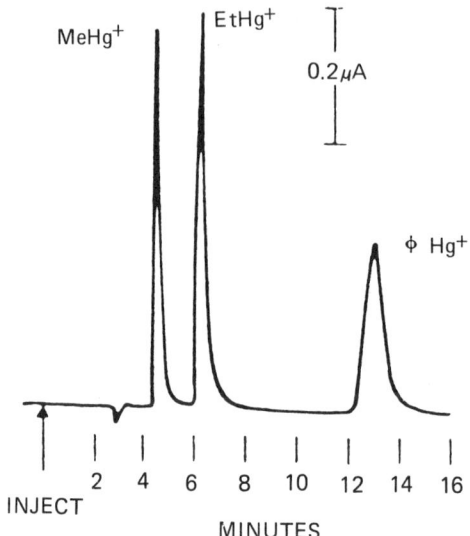

Figure 22. Detection of some organomercury cations. Column: Altex Spherisorb ODS. Eluant: 40% methanol/water containing 0.06 M ammonium acetate. Detector: Modified BAS detector cell (Figure 11) in conjunction with PAR 174 Polarographic Analyzer. The working electrode was a mercury film on gold. (From Ref. 59.)

Because of the above considerations, it is not surprising to note that potentiometry has not been widely applied to ion chromatography, and that those applications that have been reported are generally concerned with the detection of an ion in the presence of an excess of other species that either elute close to or even coelute with the analyte ion, that is, where selective detection is essential.

Applications

Some electrodes show a relatively general potentiometric response and are applicable to the detection of more than one species, for example, a silver/silver chloride electrode for the detection of halides [74], a silver/silver salicylate electrode for the detection of halides and thiocyanate [75], and a silver sulfide membrane electrode for the detection of sulfide and cyanide [76]. In the case of the silver/silver salicylate indicator electrode, it was found that the electrode response was enhanced and the baseline stabilized when salicylate was also used as the eluant in the ion chromatographic system. Applications of electrodes with liquid and plastic membranes have also been reported; for example, a liquid-state nitrate electrode has been used in

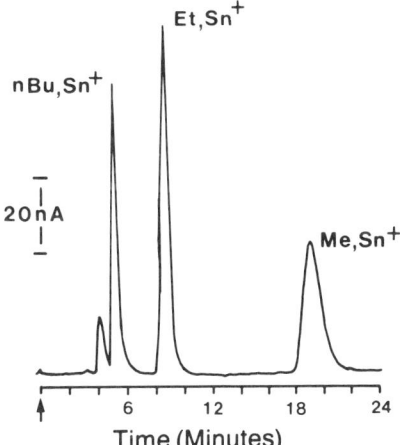

Figure 23. Detection of some organotin cations. Column: Whatman PXS cation exchange. Eluant: 60% methanol/water containing 0.042 M ammonium acetate. Detector: Prototype thin-layer cell with a mercury film on gold working electrode, used with a PAR 174 Polarographic Analyzer. (From Ref. 57.)

the determination of nitrite and nitrate [77], whereas electrodes with plastic membranes containing neutral ligands have been successfully applied to the ion chromatography of monovalent cations [78].

Potentiometric detection may also be used in an indirect mode, and several anions have been detected in this manner with a chloride ion-selective electrode [79], and amino acids detected with a copper-sensitive membrane electrode, using postcolumn reaction of eluted species with cupric ions [80]. On the basis of results reported for flow-injection measurements, a possible application of indirect detection of transition and rare earth metal ions via a copper-sensitive membrane electrode has been suggested; however, no chromatographic data have yet been reported [81]. Another application of indirect potentiometric detection is the determination of carboxylic acids with a glass pH electrode [82].

Some published applications of potentiometric detection in ion chromatography are summarized in Table 5.

Potentiometric Detection with a Metallic Copper Electrode

The only general-purpose potentiometric detector capable of sensitive detection of a wide range of solute species is based on the use of a

Table 4. Selected Applications of Amperometric and Voltammetric Detection in Ion Chromatography

Species	References
Amino acids	49
Antimony	69
Arsenite	60
Bromide	54, 61, 62, 64
Cadmium	44, 54, 63
Chloride	48, 54, 61
Chromate	45, 61
Cyanide	54, 62, 66, 67, 68
EDTA	61
Fluoride	48, 54
Hydrogen peroxide	70
Iodide	61, 62, 64, 65, 71
Mercury, methyl-, ethyl, phenyl-	59
Nitrate	48, 52, 54, 61
Nitrite	52, 61, 66
Phosphate	54
Sulfate	54
Sulfide	62, 68
Sulfite	47, 61, 66
Thiocyanate	47, 61, 66
Thiosulfate	47, 61, 66
Tin, butyl-, ethyl-	57, 58
Transition metals	43, 44, 54, 63
Uranyl	54
Zinc	54, 63

metallic copper wire or tube as the indicator electrode [86-96]. Detector response arises from changes in the concentration of cuprous or cupric ions at the electrode surface which occur when a solute species comes into contact with the electrode. The main strength of the metallic copper electrode detector is the fact that it provides a certain degree of selectivity yet can respond in a variety of different modes to a wide range of species. It is this latter capability that differentiates the copper electrode detector from previously reported potentiometric sensors, and this aspect merits more detailed discussion.

The cell design reported [88] for the copper electrode detector is shown in Figure 24. The eluant passes over a copper wire housed in

Table 5. Selected Applications of Potentiometric Detection in Ion Chromatography

Sensor[a]	Separation mode	Species separated	Reference
Ag/AgCl	Ion exchange	halides, cyanide, thiocyanate	74
Nitrate-ISE	Ion exchange	nitrite, nitrate	79
H^+ glass	Ion exchange	carboxylic acids	82
Ag_2S	Ion exchange	cyanide, sulfide	76
Nonselective membrane cell	Ion exchange	Na^+, Li^+, F^-, $H_2PO_4^-$ acetate, sulfate	83
Cu-ISE	Reverse phase	amino acids, diamines	80
Cu-ISE	Reverse phase	transition metal and rare earth ions	81
Iodide-ISE	Ion exchange	iodide	85
Ag coated with Ag-salicylate	Ion exchange	halides, thiocyanate	75
Fluoride-ISE, chloride-ISE	Ion exchange	fluoride, chloride	84
Metallic Cu	Ion exchange	organic acids	91
Metallic Cu	Reverse phase	amino acids	87
Metallic Cu	Ion exchange	transition metal ions	88
Metallic Cu	Ion exchange	alkaline earth ions	90
Metallic Cu	Ion exchange	inorganic anions, iodate, bromate, chlorate	92

[a]ISE: Ion-selective electrode.

a flow cell of 2-μl volume, which also contains a silver/silver chloride reference electrode. An auxiliary electrode constructed of platinum is also included to enable the cell to be used in an amperometric mode. Considering only the copper indicator and reference electrodes to be the functional components of the cell for potentiometric detection, the potential of the copper electrode in a flowing stream will be dependent

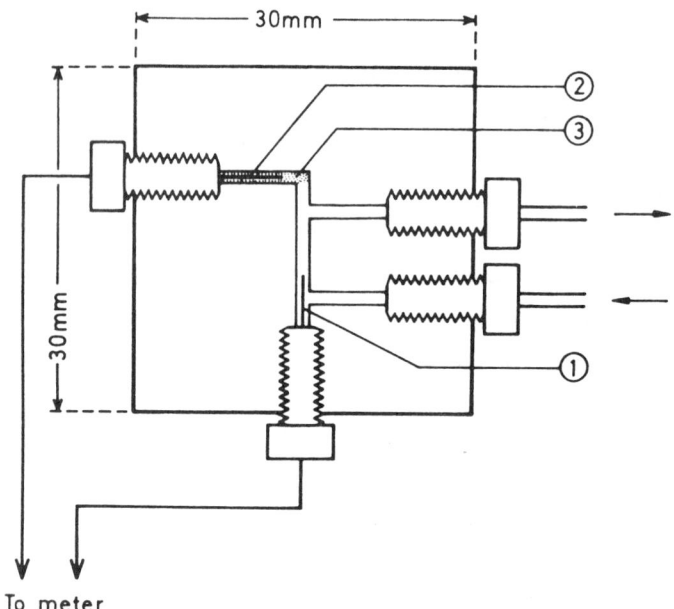

Figure 24. Flow-cell for potentiometric detector based on a metallic copper electrode. (1) Copper wire indicating electrode, (2) Ag/AgCl reference electrode, (3) agar gel. (From Ref. 88.)

on the concentration of copper ion (i.e., cuprous and cupric) at the electrode surface, in accordance with the Nernst equation.

$$E = E^0_{Cu^{2+},Cu^0} + \frac{RT}{NF} \ln \left[\frac{[Cu(II)]}{\alpha_{Cu^{II}(E,OH)}} \right] \quad (15)$$

where all terms have their usual meanings, and $\alpha_{Cu^{II}(E,OH)}$ refers to the side-reaction coefficient for the complexation of copper (II) by the eluant (E) and hydroxyl ions. For simplicity, Eqn. (15) refers only to the cupric ions in the solution.

The concentration of copper ions depends on a number of factors which are either constant over the period of the analysis or show variation. Factors of the former type include the oxygen content of the eluant and the concentration of the eluant ion, which may be copper complexing. A stable electrode voltage therefore results, and this background signal may be easily zeroed. When a solute ion elutes, the detector voltage may change, depending on the characteristics

Detection Methods in Ion Chromatography

of that particular solute. The possible detection modes that can thereby result are summarized as follows:

1. If the eluted solute forms a more stable complex with copper ions than does the eluant ion, a local decrease in copper ion concentration will occur at the electrode surface, leading to a decrease in the electrode potential. This process assumes that an ion exchange separation is being employed, so that solute ions replace an equivalent number of eluant ions in the mobile phase at the time of elution.
2. If the eluted solute forms a less stable complex with copper ions than does the eluant ion, a local increase in the copper ion concentration will occur at the electrode surface, leading to an increase in the electrode potential.
3. If the eluted solutes are strong oxidants and are able to oxidize the surface of the metallic copper electrode, a local increase in copper ion concentration will occur, and the electrode potential will increase. Reducing solutes can also be detected by reducing cupric ions to cuprous; this reduction will also result in an increase of the electrode potential, due to the higher standard electrode potential of the Cu^+/Cu^0 couple (+0.520 V) compared to the Cu^{2+}/Cu^0 couple (+0.337 V).

The first and last detection modes can be classified as direct, whereas the second mode is indirect. Each of these detection mechanisms has been realized in practice, and Table 5 gives a summary of the applications of the metallic copper electrode detector to ion chromatography. Some chromatograms obtained with the device are given in Figures 25-27. Provided that the potential change measured at the copper electrode is restricted to less than a few millivolts (the exact potential range depends on the sample type and the response mode employed), then detector response is observed to be linear with respect to solute concentration. This behavior is illustrated in Figure 28, which also shows that the expected Nernstian response is observed at higher concentrations of the solute. Linear detector response can be explained by appropriate mathematical manipulation of the response equation to give [92]:

$$H = \text{constant} + \frac{RT\beta_1 N}{2FD\alpha_{L(H)}} \tag{16}$$

where H is the peak height (millivolts), N is the number of moles of injected solute, D is the dispersion factor for the chromatographic system, β_1 is the formation constant for complexation of copper by the injected solute, $\alpha_{L(H)}$ is the side-reaction coefficient for protonation of the solute, and the remaining terms have their usual meanings.

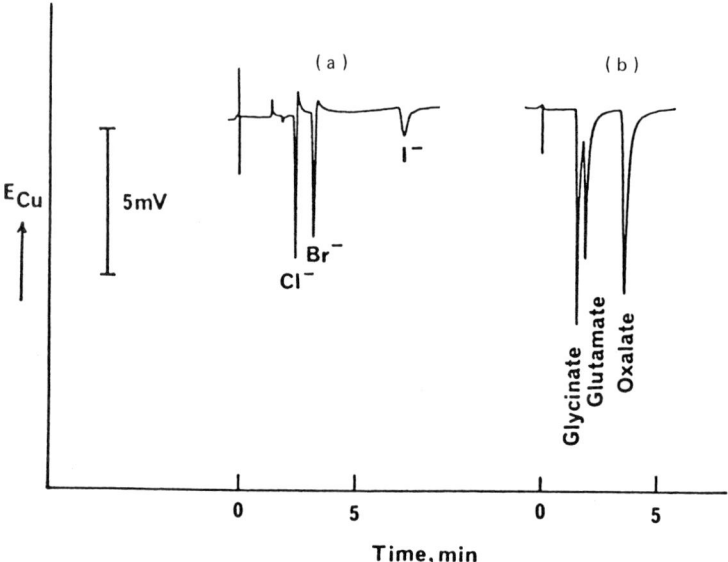

Figure 25. Chromatograms of species exhibiting direct response with a metallic copper electrode. Column: Vydac 302 IC 4.6. Eluant (a) 1 mM sodium tartrate at pH 3.2, (b) 1 mM potassium orthophosphate at pH 7.0. Injected amounts: (a) 0.5-50 nmol, (b) 5 nmol of each species. (From Ref. 94.)

III. SPECTROSCOPIC METHODS OF DETECTION

A. Introduction

Spectroscopic methods of detection will, for the purposes of the present discussion, be interpreted to include UV/visible spectrophotometry, refractive index measurements, fluorometry, atomic emission spectroscopy (using various excitation sources), and atomic absorption spectroscopy. To aid discussion, it is convenient to divide these methods into molecular spectroscopic techniques and atomic spectroscopic techniques.

B. Molecular Spectroscopic Techniques

Direct Detection Methods

Perhaps the most straightforward method for the detection of solutes in ion chromatography is to monitor their UV absorbance. A considerable

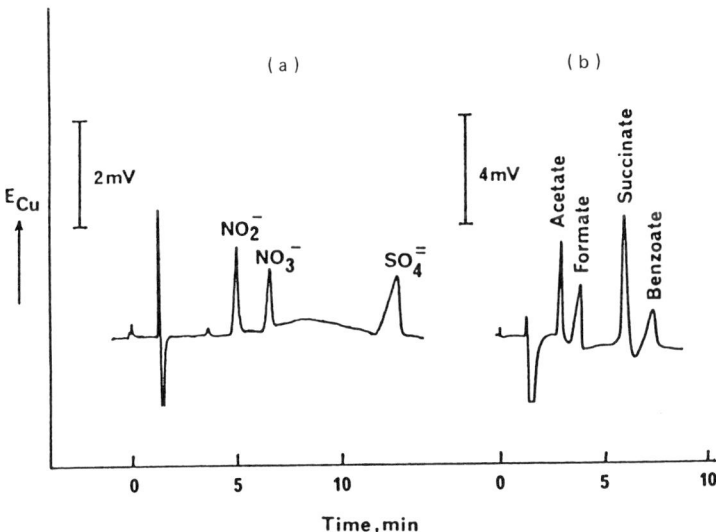

Figure 26. Chromatograms of inorganic and organic anions exhibiting indirect potentiometric response with a metallic copper electrode. Column: Vydac 302 IC 4.6. Eluant: 2 mM potassium phthalate at pH 4.6 (a) or 4.0 (b). Injected amounts: (a) 60-120 nmol, (b) 250 nmol of each species. (From Ref. 94.)

number of ions shows appreciable absorbance in the range 190-220 nm [97,98]; these include many inorganic anions and cations, as well as some organic acids. However, the use of direct UV absorption for detection has been confined mainly to inorganic anions, particularly nitrate, nitrate, bromide, bromate, iodide, iodate, periodate, thiocyanate, and thiosulfate. Table 6 lists the detection limits and suitable wavelengths for some of these UV-absorbing inorganic anions. Some anions, such as chloride, fluoride, sulfate, phosphate, perchlorate, and cyanide do not show appreciable absorption of UV radiation except at very low wavelengths. Therefore, some degree of detection selectivity exists, which enables common interferences such as high levels of chloride or sulfate in samples to be minimized or even eliminated. Alternatively, choice of a low wavelength such as 190 nm permits direct UV absorption to be used as a more general detection mode; this approach has been applied to the detection of common anions in atmospheric precipitation [99].

Direct UV absorption detection has found considerable usage in the determination of nitrite and nitrate [100-102], for which it is

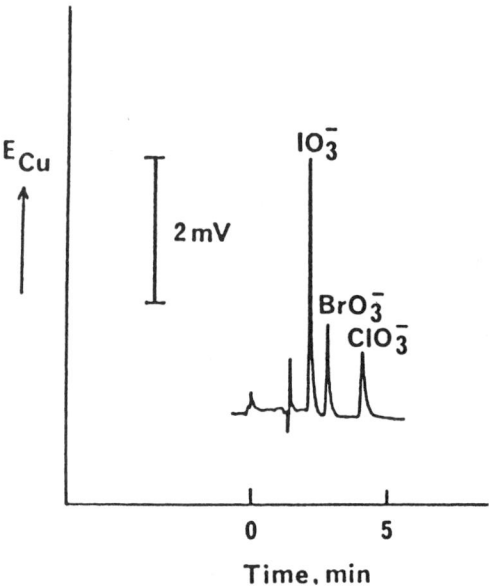

Figure 27. Chromatogram of oxidizing anions detected at a metallic copper electrode. Column: Vydac 302 IC 4.6. Eluant: 20 mM sodium tartrate at pH 3.2. Injected amounts: 1-100 nmol. (From Ref. 94.)

particularly suited in view of the relatively high molar absorptivities of these ions (e.g., 9000 L mol^{-1} cm^{-1} at 210 nm for nitrate). An obvious restriction on the eluant composition is that all components must be transparent or at least only weakly absorbing at the detection wavelength employed. For this reason, eluants such as phosphate buffer [103], chloride [104], sulfate [105], carbonate/bicarbonate buffer [106], citrate [55], and alkylsulfonates [107] have been used. Figure 29 shows a chromatogram obtained for the determination of nitrite and nitrate in cured meats, with the use of methane sulfonate as eluant; this figure shows that no interference from the large amounts of chloride present in the sample was observed [102]. Detection of UV-absorbing metal-chloro complexes at 215 nm is illustrated in Figure 30 [108].

UV-absorbing anions have been determined simultaneously with metal cations by means of direct UV-absorption detection at 210 nm and anion exchange chromatography with ethylenediaminetetraacetic acid (H$_4$EDTA) as eluant [109]. Under these conditions, the metal ions (M^{2+}) form anionic MEDTA^{2-} complexes, which are separated by

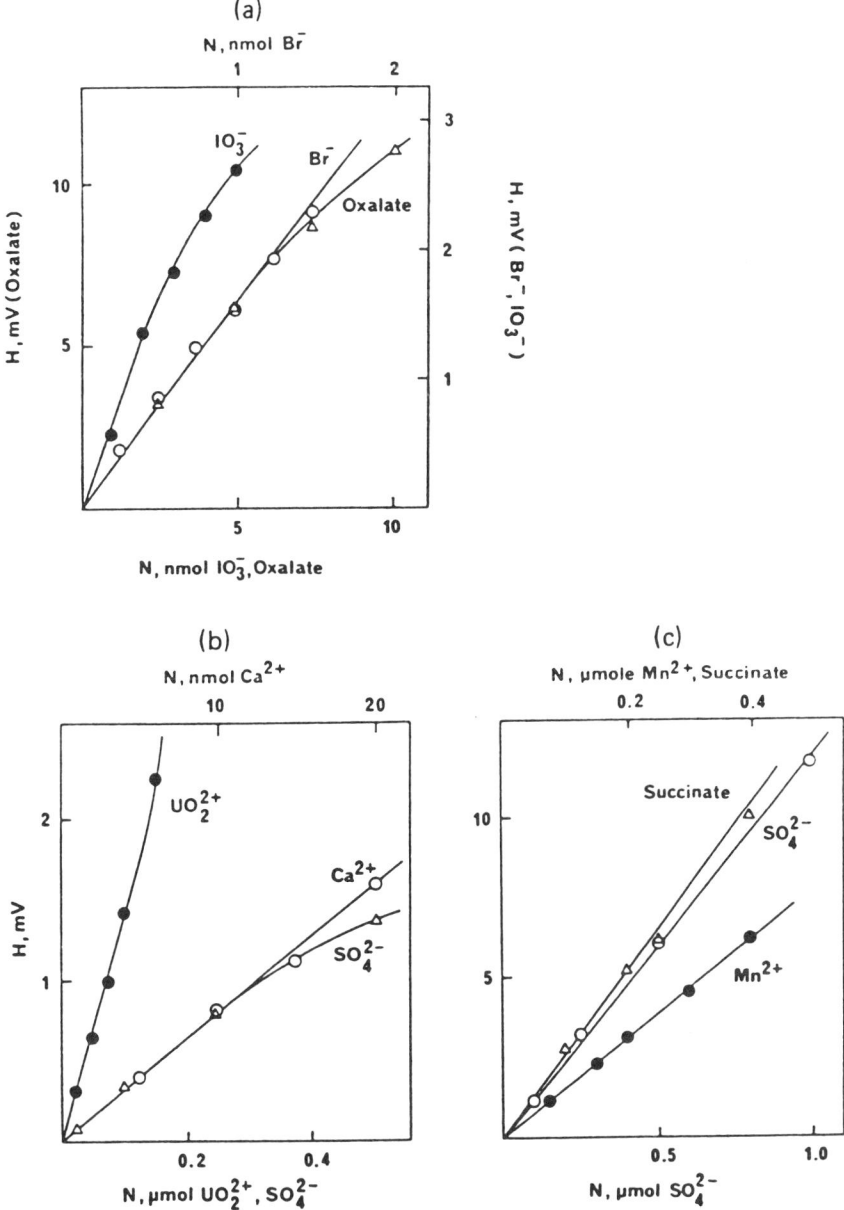

Figure 28. Examples of calibration plots for direct (a) and indirect (b, c) detection with a metallic copper electrode. Conditions as given in Figures 25-27, except for Mn^{2+} and UO_2^{2+} (Nucleosil 10SA column with 5 mM tartaric acid and 7.5 mM ethylenediamine at pH 4.5 as eluant), and Ca^{2+} (Wescan cation column with 1 mM tartaric acid and 1 mM diethylenetriamine at pH 4.6 as eluant). (From Ref. 94.)

Table 6. UV Absorption Detection Limits for Some Inorganic Anions, Obtained Using Latex Agglomerate Columns with Various Eluants[a]

Anion	Detection limit (ppm)	Wavelength (nm)	Eluant
Arsenate	1.5	200	A
Arsenite	1.2	200	A
Azide	0.3	195	B
Bromate	0.16	195	B
Bromide	0.1	195	B
Chlorate	4	195	B
Chloride	2	192	B
Iodate	0.08	195	B
Iodide	0.15	195	C
Nitrate	0.1	195	B
Nitrite	0.1	195	B
Selenate	15	195	B
Selenite	0.5	195	B
Sulfide	0.4	200	A
Thiocyanate	0.2	195	C

[a]Eluants: (A) 10 mM HCl; (B) 3 mM $NaHCO_3$ - 2.4 mM Na_2CO_3; (C) 6 mM Na_2CO_3.
Source: Data from Ref. 106.

ion exchange with $HEDTA^{3-}$ acting as the eluting anion. The metal complexes show strong absorbance at 210 nm, whereas the eluant is only weakly absorbing at this wavelength; thus sensitive detection is achieved. Detection limits for metal ions ranged from 5.0 to 70.0 ng in a 100-µl injection.

It has recently been demonstrated that direct refractive index detection can also be applied in ion chromatography [110-112]. For example, phosphonates of the Dequest series separated on an anion exchange column can be detected with a differential refractive index detector. The chromatogram obtained is shown in Figure 31.

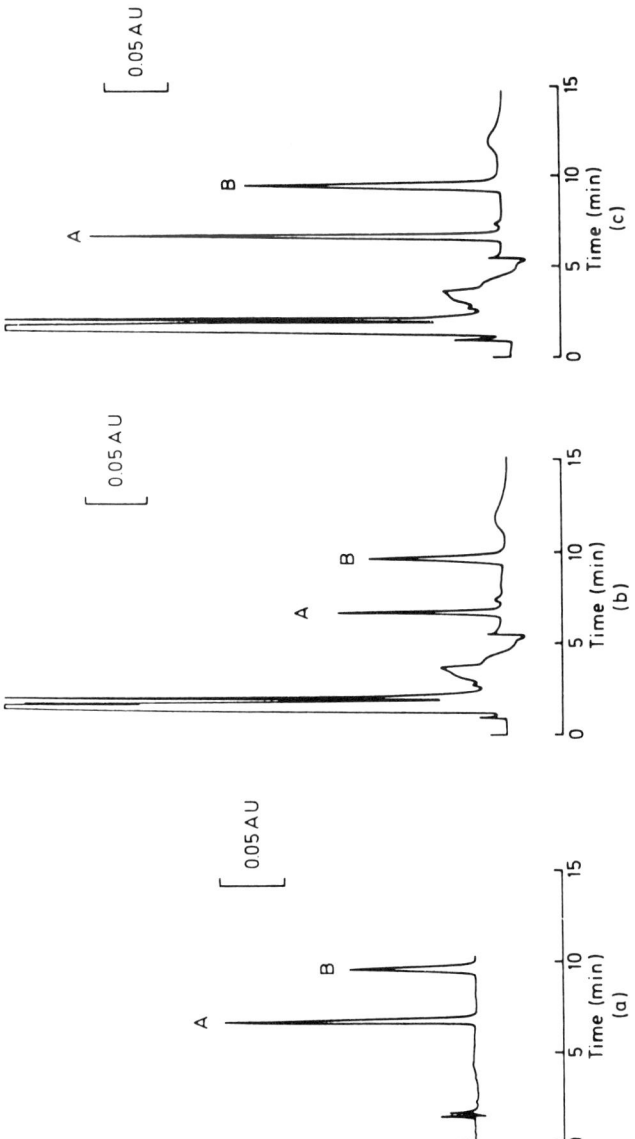

Figure 29. Separation of nitrite (A) and nitrate (B) in standards (a), bacon sample (b), and spiked bacon sample (c). The sample for (c) was spiked with 0.5 μg of nitrite and nitrate in a 25-μl injection. Column: Vydac 302 IC 4.6. Eluant: 11.0 mM methanesulfonic acid at pH 5.0. Detection at 210 nm. For chromatogram (a), 10 μl of a 50-ppm solution of nitrite and nitrate were injected. (From Ref. 102.)

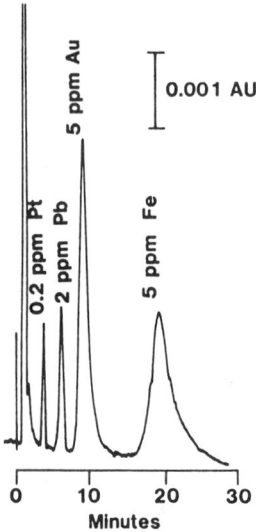

Figure 30. UV detection of metal-chloro complexes at 215 nm. Column: Dionex HPIC-AS5 separator. Eluant: 0.30 M $NaClO_4$ and 0.05 M HCl. (AU = Absorbance units.) (From Ref. 108.)

Indirect Detection Methods

One of the most widely used detection methods in ion chromatography is indirect UV absorption detection, also called *indirect photometric chromatography* [113] and *vacancy detection*. This detection mode arose from the use of eluants containing aromatic acid anions as competing ions in anion exchange methods that employed conductivity detection. As discussed above, the rationale for this was that these species have low limiting equivalent conductances and so enable sensitive detection to be achieved in the single-column mode. These ions also have high molar absorptivities in the UV region, and if the background absorbance of the eluant is monitored at a suitable wavelength, a change in absorbance will occur on elution of a solute ion that has a molar absorptivity different from that of the eluant. This results directly from the fact that solute ions replace eluant ions on elution from an ion exchange column.

If an ion exchange system is considered (as was done earlier in the discussion of conductivity detection), it is straightforward to develop a detector response equation. The background absorbance of the eluant (A_B) is given by:

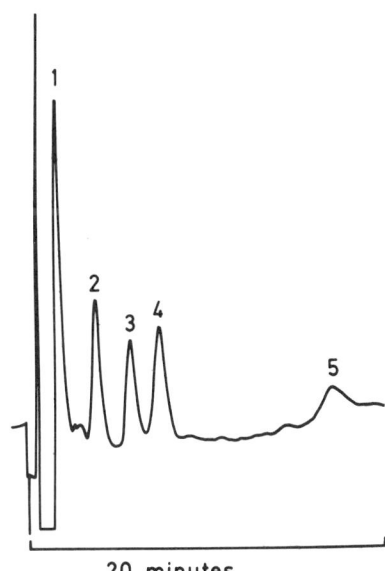

Figure 31. Direct refractive index detection of phosphonates of the Dequest series. Column: Waters IC PAK A. Eluant: 15 mM HNO$_3$. Detector: Waters Model 410 RI. Peak identities: (1) 2010, (2) 2054, (3) 2006, (4) 2041, (5) 2060 Dequest phosphonates. (Courtesy Dora Wong, Waters Associates, Milford.)

$$A_B = [\varepsilon_E C_E I_E + \varepsilon_{HE} C_E (1 - I_E)] \times l \tag{17}$$

where the eluant E is assumed to exist as the neutral form HE and the ionized form E$^-$, with molar absorptivities ε_{HE} and ε_E, respectively. C_E is the total eluant concentration, I_E is the degree of ionization of the eluant, and l is the path length of the detector cell. It is assumed in Eqn. (17) that the eluant cation does not absorb radiation at the detection wavelength used.

When a fully ionized solute (S$^-$) is eluted at a concentration C_S, the concentration of the eluant anion in the cell is given by $C_E I_E - C_S$, and the absorbance signal measured by the detector during sample elution (A_S) is given by:

$$A_S = [\varepsilon_E (C_E I_E - C_S) + \varepsilon_{HE} C_E (1 - I_E) + \varepsilon_S C_S] \times l \tag{18}$$

where ε_S is the molar absorptivity of the solute.

The change in absorbance resulting from elution of the sample is obtained by subtracting A_B from A_S to give:

$$\Delta A = A_S - A_B = (\varepsilon_S - \varepsilon_E) C_S l \qquad (19)$$

It is assumed here that the eluant is well buffered, and I_E therefore remains constant. Equation (19) shows that the detector absorbance depends on the solute concentration, the detector cell path length, and the difference in molar absorptivities between the solute and eluant anions. In most cases, the eluant is deliberately chosen to be more absorbing than the solute, so a net decrease in absorbance results when the solute elutes. Conventionally, the polarity of the detector output is reversed to give positive peaks on the recorder. Provided Beer's law is followed, then linear calibration plots will result, thereby permitting sample quantitation. The similarity of the form of Eqn. (19) to that of Eqn. (9) suggests that the method of quantitating unknown solutions without the use of standards (as discussed in detail above for conductivity detection) should also be applicable to indirect UV detection. This suggestion has been shown to be valid [114].

The sensitivity attainable with this method will depend on a number of factors. In the first place, it is clear from Eqn. (19) that if the molar absorptivity of the eluant is maximized by selection of the wavelength of maximum absorption, the absorbance change accompanying solute elution is also maximized. However, problems can thus be created due to the high eluant background absorbance such that the detector is incapable of being zeroed. Furthermore, photometric error is reduced if the detector is operated in the approximate absorbance range 0.2-0.8; thus it is desirable that the background absorbance be maintained within this range. A further factor is that the detector baseline noise increases with the eluant concentration, so that the signal-to-noise ratio (and hence the sensitivity of the method) improves as the eluant concentration is reduced [113]. The above considerations show that the results obtained with indirect UV absorption detection will be strongly dependent on the performance characteristics of the particular detector used.

For anion chromatography, eluants such as phthalate [115,116], nitrate [117], sulfobenzoate [118], and benzenetricarboxylate [113] have been used. A typical chromatogram obtained with phthalate as eluant is shown in Figure 32. For cation chromatography, copper (II) [113] and anilinium [115] have been used as eluants; a chromatogram of inorganic cations is shown in Figure 33. It might appear at first sight that aromatic bases would serve as ideal eluants for indirect UV detection of cations because these are the cationic equivalents of the aromatic acid eluants which have proved so successful in anion

Detection Methods in Ion Chromatography

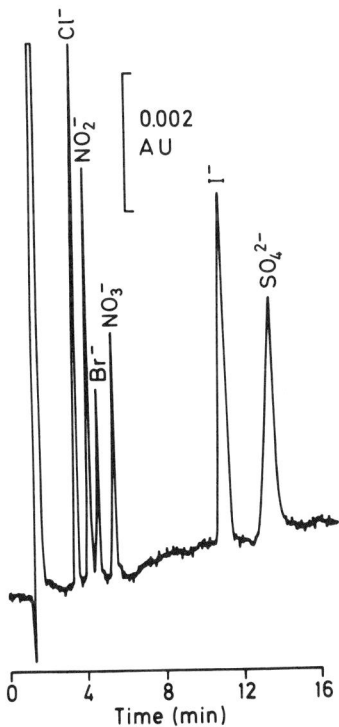

Figure 32. Indirect UV absorption detection of inorganic cations. Column: Vydac 302 IC 4.6. Eluant: 6.0 mM potassium hydrogen phthalate at pH 4.0. Detection by UV absorbance at 285 nm. Solute concentrations: 2.5-7.5 ppm. The peaks are in the direction of decreasing absorbance. (From Ref. 10.)

exchange chromatography. When such eluants are used in practice--for example, picolinic acid at pH 3.5 [119], only very small changes in absorbance are observed when a solute cation elutes. The reason is that elution of solute cations is being accomplished jointly by the protonated aromatic base cations and also by hydrogen ions. Indeed, hydrogen ions are a very effective eluant for monovalent cations, and nitric acid is the preferred eluant for these species when conductivity detection is used. A change in background absorbance will occur only when a protonated base cation is involved in the ion exchange process, so it is clear that detection sensitivity will be reduced as long as hydrogen ions continue to act as competing ions.

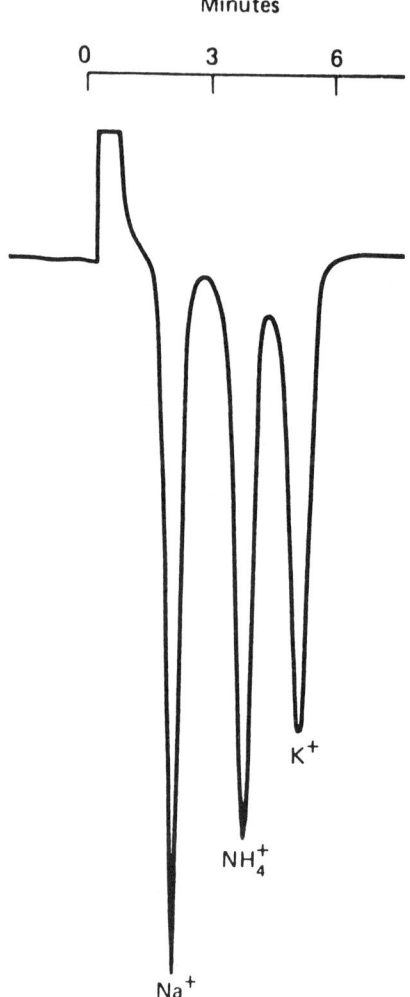

Figure 33. Indirect UV absorption detection of inorganic cations. Column: Dowex 50 resin. Eluant: 0.01 N copper sulfate. Detection at 252 nm, 0.05 AUFS (absorbance units full-scale). Solute concentrations: 0.01 N for each species. (From Ref. 113.)

Indirect UV absorption detection has been successfully applied to the simultaneous independent determination of anions and cations by means of a single detector [120]. A dual-column system comprising an anion exchange column and a cation exchange column connected in series was used, and the eluant was 2.5×10^{-4} M copper o-sulfobenzoate. This eluant was selected because both the anion and cation components could exhibit appreciable elution power in an ion exchange mode, and both also showed significant UV absorbance. A desirable feature of this technique is that the eluant anion and cation should have absorption maxima at different wavelengths and that each species should be transparent at the absorption wavelength of the other species. This is not possible with the copper o-sulfobenzoate eluant, so it is necessary to monitor the absorbance of the eluant at two wavelengths and to apply ratioing and scaling techniques to obtain independent chromatograms of solute anions and cations. A dual variable-wavelength detector with ratioing capability is therefore required.

The principle of the method can be described briefly as follows: If the two wavelengths monitored by the detector are λ_1 and λ_2, then the ratio of the peak heights or areas obtained at these wavelengths is determined by the ratio of the molar absorptivities of the eluant species at the same two wavelengths. Because different components of the eluant are responsible for elution of solute cations and anions (i.e., copper and o-sulfobenzoate, respectively), then the ratio referred to above will be different for cations and anions. If the ratio for cations is given by R_C and that for anions is given by R_A, and the detector is operated so that its output signal (S) is the difference in the absorbance readings at the two wavelengths, that is,

$$S = A_2 - A_1 \tag{20}$$

then application of the ratio factors R_C and R_A can enable deconvolution of the anion and cation chromatograms. That is, the cation chromatogram is obtained by plotting A_C versus time, whereas the anion chromatogram is obtained by plotting A_A, where A_C and A_A are given by:

$$A_C = R_A \cdot A_2 - A_1 \tag{21}$$

and

$$A_A = R_C \cdot A_2 - A_1 \tag{22}$$

This procedure is illustrated in Figure 34, for which the two detection wavelengths were 270 nm and 240 nm, giving values of 0.44 and 0.74 for the ratio factors R_C and R_A, respectively. It is worthy of note that the approach described above is applicable only to those absorbance detectors that enable simultaneous monitoring of two wavelengths, together with the facilities for applying a scaling factor to

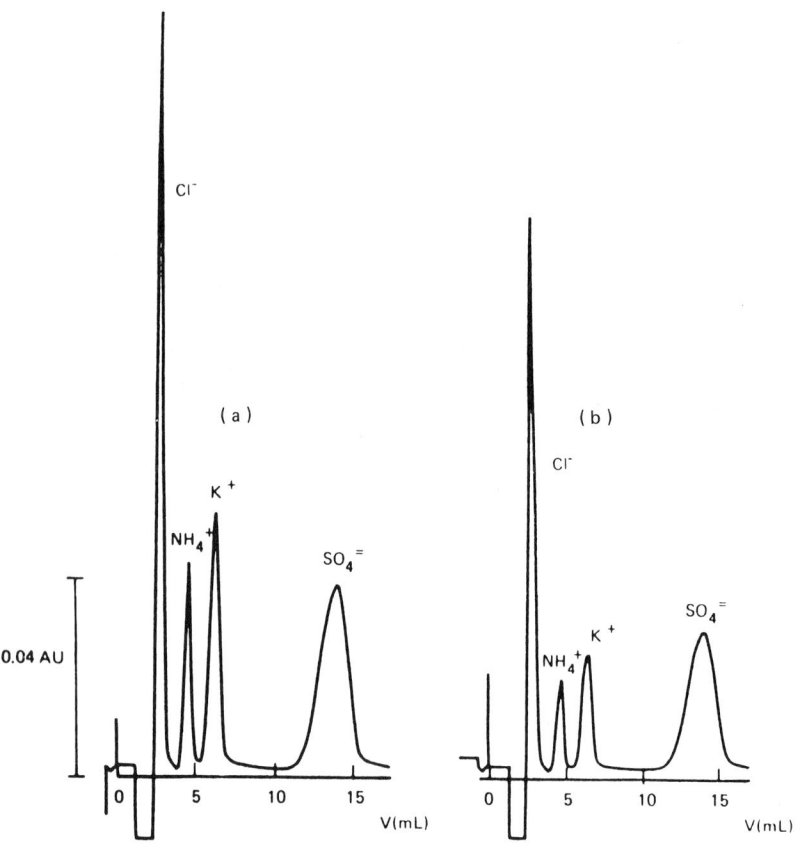

Figure 34. Joint and independent separations of anions and cations. Columns: Zipax SAX and SCX columns in tandem. Eluant: 0.5 mM copper o-sulfobenzoate at pH 4.3. Detection: UV absorbance at 240 nm (a) and 270 nm (b).

the absorbance signal at one wavelength and subtracting from this the absorbance signal measured at the second wavelength. Clearly, a high level of instrumental sophistication is necessary.

The flexibility provided by variation of the wavelength in indirect UV absorption detection can be used to advantage in the elimination of matrix effects. For example, if one considers a sample matrix that contains high levels of nitrate (such as that produced when a sample is digested in nitric acid), then it would be advantageous if nitrate ion were not detectable and a chromatogram could be obtained that

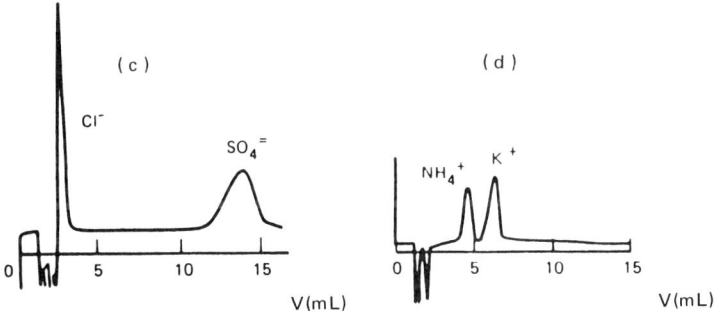

Figure 34. (Continued) Chromatograms (c) and (d) were obtained by applying the scaling factors 0.44 and 0.74 to Eqns. (22) and (21), respectively, to obtain deconvoluted chromatograms for anions (c) and cations (d). Note that the peaks are in the direction of decreasing absorbance, except for (c), where the peaks have opposite polarity (i.e., increasing absorbance). (From Ref. 120.)

showed the other anions present in the sample. Equation (19) indicates that no detector signal will result for a particular solute when the molar absorptivities of the eluant and solute ions are equal. That is,

$$\varepsilon_S = \varepsilon_E \qquad (23)$$

Because nitrate ion is a strongly absorbing species, it should be possible to find an eluant for which Eqn. (23) is satisfied, while at the same time achieving satisfactory ion exchange elution. These conditions can be met by using benzenesulfonate as eluant and by monitoring the eluant absorbance at 239 nm [121]. A chromatogram obtained for a mixture of chloride, nitrite, and nitrate, monitored at two different wavelengths, is shown in Figure 35. Note that the detection sensitivities for chloride and nitrite change with wavelength and that the peak direction for nitrite reverses at 239 nm because of a change in sign for Eqn. (19) at this wavelength.

An unusual feature of some chromatograms obtained with indirect UV detection is the appearance of a slowly eluting peak which is not directly attributable to any injected ion [115,122]. This peak, which is generally called a "system" peak, is most prevalent when partly dissociated aromatic acid eluants are used in the separation of anions. The system peak can be very large and often represents a major interference to the quantitation of later-eluting solute ions. The size

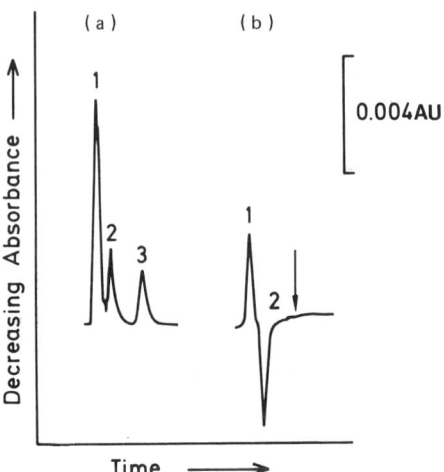

Figure 35. Elimination of the peak of a UV-absorbing anion (nitrate) by wavelength manipulation to satisfy Eqn. (23). Column: TSK-GEL IC-Anion-PW. Eluant: 1 mM benzenesulfonate. Detection: UV absorbance at 225 nm (a) and 239 nm (b). Peak identities: (1) chloride, (2) nitrite, (3) nitrate. The arrow in (b) shows the elution position of nitrate. (From Ref. 121.)

and direction of the system peak have been shown to be dependent on a variety of experimental parameters, including the injection volume, the nature of the solute ion, and the disparity between the pH values of the sample solution and that of the eluant [6]. Figure 36 shows a typical system peak observed when a large injection volume is used. It has been proposed [6] that the system peak results from reverse-phase elution of the neutral, undissociated form of the eluant which can be adsorbed onto the unfunctionalized parts of the ion exchange resin. Evidence in favor of this proposal is that the system peak disappears when a fully ionized eluant is used and that the system peak shows predictable reverse-phase behavior when organic modifiers are added to the eluant (Figure 37). System peaks also occur with conductivity detection, but to a lesser extent than that observed with indirect UV detection.

To conclude the discussion of indirect detection by means of molecular spectroscopic detectors, it should be noted that the technique is also applicable to refractive index and fluorescence detectors. In the former case, anions can be detected by monitoring the refractive index of an eluant containing an aromatic acid [115,123]. This is

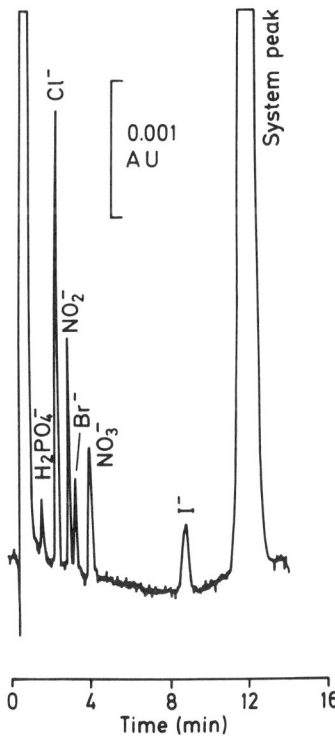

Figure 36. Illustration of the system peak observed with indirect UV absorption detection when a large injection volume is used. Column: Vydac 302 IC 4.6. Eluant: 2.5 mM potassium hydrogen phthalate at pH 4.0. Injection volume: 250 µl. Detection: UV absorption at 285 nm. Solute concentrations: 100-400 ppb. (From Ref. 10.)

illustrated in Figure 38, which shows indirect refractive index detection of several inorganic anions after separation on an anion exchange column with phthalate as eluant. In this detection mode, the background refractive index of the eluant is not restricted (unlike the indirect UV detection mode) because measurements are conventionally made by comparison of the column effluent with pure eluant contained in a reference cell. Sensitivity is therefore limited only by the performance of the detector used, which does, however, impose a practical limitation in that most commercially available refractive index detectors are designed to operate with large solute concentrations and are generally not optimized for high-sensitivity applications. A

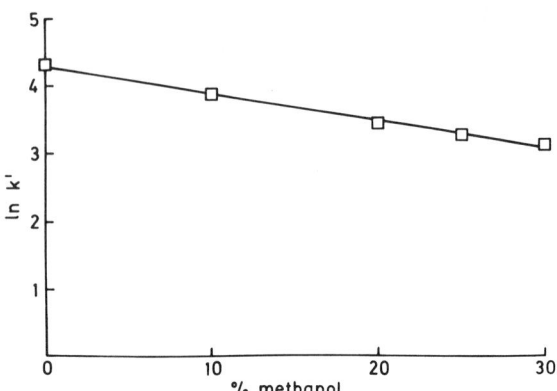

Figure 37. Variation of the capacity factor of the system peak with the percentage of methanol in the mobile phase. Column: Hamilton PRP X100. Eluant: potassium hydrogen phthalate at pH 4.2 containing the indicated percentages of methanol. The eluant concentrations were adjusted so that chloride ion gave the same retention time for each eluant.

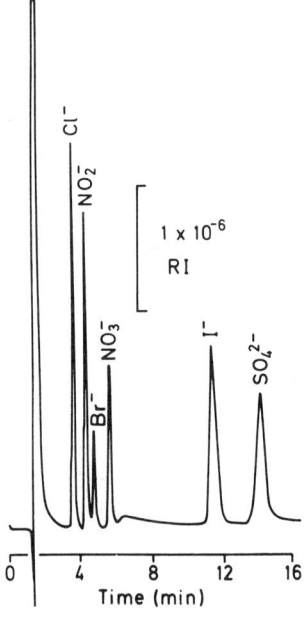

Figure 38. Indirect refractive index detection of inorganic anions. Conditions as for Figure 32, except that an Erma Optical Works refractive index detector was used. (From Ref. 10.)

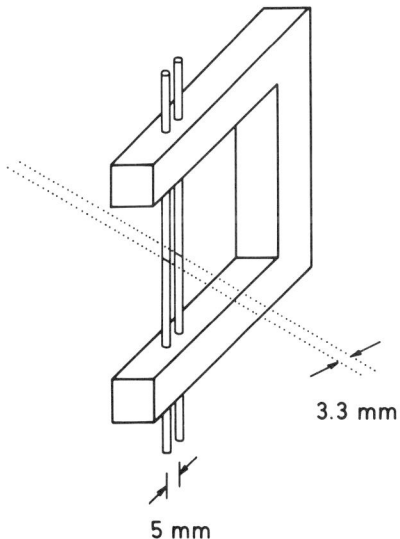

Figure 39. Flow-cells for double-beam fluorometry: two 1.0 mm i.d. × 2.0 mm o.d. quartz tubes are mounted on an aluminum block. The dotted lines show the path of the laser excitation beams. (From Ref. 124.)

further distinction between the indirect UV and refractive index detection modes is that the latter can be used with quite concentrated eluants and is therefore applicable to separations performed on columns with high ion-exchange capacities [123].

An indirect fluorescence detection procedure for ion chromatography has been devised [124] in an attempt to utilize the high sensitivity of fluorometric measurements. Here, sodium salicylate was used as a fluorescent eluting ion for anion exchange chromatography, and decreases in the background fluorescence were used to detect eluted solute anions. A modulated, double-beam, laser-excited fluorimetric detector was designed in an attempt to overcome the inherent flicker noise of the laser source. The cell design used is shown in Figure 39. The detection sensitivity achieved with this system was comparable to those of the indirect UV and conductivity detection methods; however, this was attributed chiefly to the fact that the eluant concentration necessary for solute elution was inappropriately high for the indirect fluorescence detection procedure. The availability of very-low-capacity ion exchange materials would enable the use of more dilute eluants and hence realize more of the potential of this detection mode.

C. Atomic Spectroscopic Techniques

Some detection methods based on atomic spectroscopic techniques have been reported for ion chromatography. These incorporate selective detection methods such as atomic absorption spectroscopy and atomic emission spectroscopy, with a variety of excitation sources being applied to the latter technique.

An atomic absorption (AAS) instrument can be conveniently coupled to an ion chromatograph by direct connection of the column effluent line to the nebulizer intake of the AAS [125]. The chief requirements are that the nebulization rate is controlled [126] and that the mobile phase used in the ion chromatograph does not lead to high or unstable backgrounds in the AAS signal, and that excessive carbon buildup on the burner is not permitted to occur. The advantage of coupling ion chromatography with AAS is that speciation of solutes is possible; for example, a range of arsenic species has been separated and detected in this manner [127,128], and free chromate has been determined in the presence of chromium (III) [125]. Cold-vapor AAS techniques have been applied to the determination of mercury compounds, with stannous chloride or sodium borohydride as the reductant [129].

Atomic emission techniques have also proved relatively successful for selective detection in ion chromatography. The simplest case is the use of a flame as the excitation source (i.e., flame photometry); this approach has been used for the detection of alkaline earth and lanthanide elements after separation on a pellicular ion-exchange column with citrate as eluant [130]. Phosphorus- and sulfur-containing solutes can be selectively detected by flame photometry via the emission of the molecular species HPO and S_2 [131]; the detection of 5'-monophosphate nucleotides at 526 nm is illustrated in Figure 40. Flame photometry has also formed the basis of a new detection scheme called *replacement ion chromatography* [132], which is reported to have the potential for very low detection limits and universal applicability. Here the counter cation associated with an eluted anion is stoichiometrically replaced in a cation exchange column by a cation that shows good flame photometric response (e.g., lithium). The replacement column is situated after the separation column, and the eluant from the replacement column is directed into a flame photometer. The resulting signal observed for lithium emission at 670 nm provides an indirect but quantitative measure of the eluting anion concentration. Similarly, the concentrations of eluted cations can be determined after their replacement in the eluant with lithium. The detection limits attainable with this technique are in the micromolar region and are comparable or superior to those of conventional conductivity detection. Figure 41 shows the detector output obtained in the separation of monovalent cations with nitric acid being used as eluant.

More exotic emission sources such as direct-current, microwave-induced, or inductively coupled plasma systems have been applied

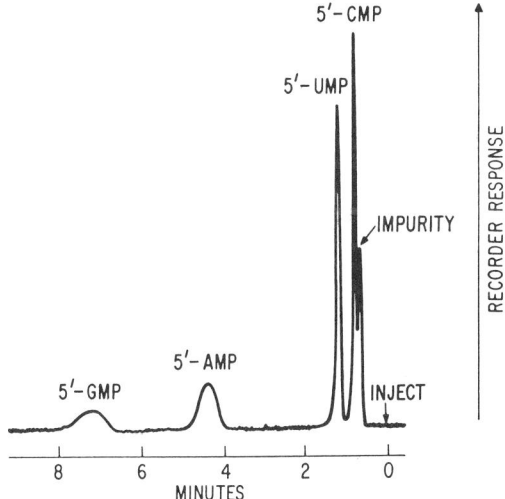

Figure 40. Separation of 5'-monophosphate nucleotides by means of a phosphorus flame photometric detector. Column: Zipax SAX. Eluant: 0.04 M formic acid at pH 3.2. Detection: emission at 526 nm. Solute concentrations: 0.4 mg of each compound per milliliter. Peak identities: 5'-GMP, guanosine 5'-monophosphate; 5'-AMP, adenosine 5'-monophosphate; 5'-UMP, uridine 5'-monophosphate; 5'-CMP, cytidine 5'-monophosphate. (From Ref. 131.)

in a number of cases, but the detection limits achievable with these devices when coupled to chromatographic systems fall well short of those attained when they are used as stand-alone instruments. Reductions in detection limits ranging from twenty to several orders of magnitude have been observed when an inductively coupled plasma atomic emission spectrometer (ICP) is coupled to a HPLC [133,134]. This poor performance stems directly from the inherent inability of the pneumatic nebulizer used in such ICP systems to equilibrate to the rapidly changing concentration of analyte that elutes from the HPLC in a very small volume (e.g., 100 μl), because such nebulizers typically have large internal volumes. An added factor is the solute dilution effect necessarily present in any liquid chromatographic method. Direct-current plasma atomic emission spectrometers (DCP) have more suitable spray chambers for sample introduction, and these devices give greatly improved sensitivity in comparison to ICP systems, when interfaced to a liquid chromatograph [135]. The output of the ICP can be presented as a chromatogram through the use of pulsing techniques or by reconstructing tabular ICP data at the conclusion of a run.

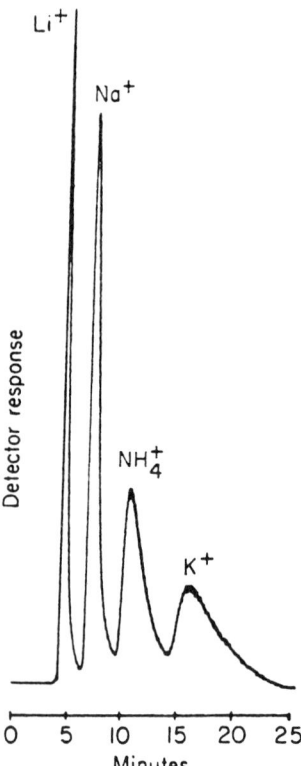

Figure 41. Detection of monovalent cations by replacement ion chromatography. Separator column: Wescan cation column. Suppressor column: 100- to 200-mesh anion exchange resin. Replacement column: 20- to 50-mesh cation exchange column in the lithium form. Detection: emission at 670 nm. Solute concentrations: 1 mM of each species. (From Ref. 132.)

IV. DETECTION VIA POSTCOLUMN REACTIONS

Some postcolumn reaction detection procedures have already been discussed. The most significant of these is the use of a suppressor column to facilitate conductometric detection when highly conducting eluants are employed. Other possibilities also exist, particularly for the detection of cations; these are discussed below in terms of the hardware requirements and reported applications.

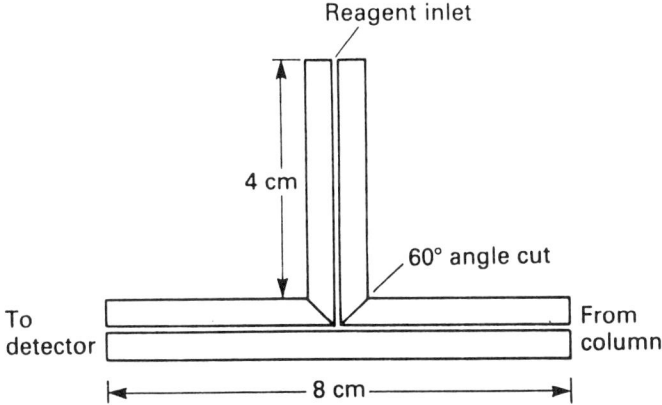

Figure 42. Simple mixing tee for postcolumn reactions. (From Ref. 137.)

A. Hardware Requirements

In most cases, a color-forming reaction is used, involving the addition of a reagent (or reagents) to the column effluent with provision for mixing and reaction with the eluted solutes. The ideal characteristics of color-forming reactions and the hardware suitable for their use have been studied [136], and results have shown that very fast reactions are preferable to eliminate the requirement for inclusion of a reaction coil into the system because of the band broadening that necessarily results from the use of such devices.

The mixing apparatus in which the reactant stream is merged with the column effluent is crucial to the sensitivity and resolution achieved with the postcolumn reactor. If the color-forming reaction is fast, a simple mixing tee can be satisfactory, and a suitable design for such a mixing tee giving 90° impingement of the flowing streams is shown in Figure 42 [137]. The ideal angle of impingement is a matter of some contention, but it has been suggested that a Y-type device yields superior results [76]; in our experience, the angle of impingement is not critical for the applications typically used in ion chromatography. The degree of mixing provided by postcolumn mixing devices can be enhanced, if necessary, by exploitation of the whirlpool effect introduced by tangential entry of the color-forming reagent [76,136,138]. Packed-bed reactors [139] can also be of great value for some postcolumn reactions; these devices eliminate the need for both a pump to deliver the postcolumn reactant and a mixing chamber

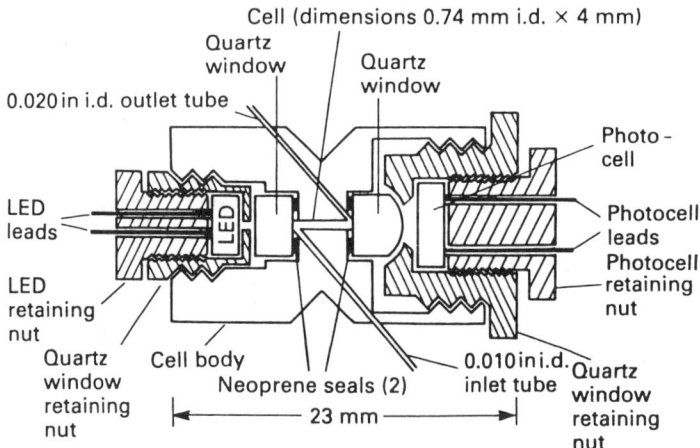

Figure 43. Miniature spectrophotometric detector cell for ion chromatography. (From Ref. 137.)

for reaction to occur. The band broadening introduced when a packed-bed reactor is used has been shown to be minimal [140,141].

A simple detector designed specifically for the detection of cations after postcolumn reaction with PAR has recently been described [137]. This detector is shown in Figure 43, which shows that the device is miniature and incorporates a light-emitting diode (LED) light source and a photocell, mounted directly onto the ends of the detector flow-cell. The detector was found to give minimal contribution to band broadening and provided sensitivities of about 10 ng/ml for transition metal ions.

B. Applications

A limited number of postcolumn reactions have been reported for the detection of anions. Phosphates and polyphosphates are particularly amenable to this mode of detection [142-144] through the formation of the heteropoly blue complex resulting from the reaction of the phosphates with molybdenum reagents, via air-segmented or unsegmented flow systems [145,146]. This reaction is essential to enhance the detectability of the phosphate species because the routine use of a complexing agent (such as EDTA) in the eluant to prevent hydrolysis of polyphosphates means that conductivity detection is not suitable.

An interesting postcolumn reaction method has been described for the detection of common inorganic anions, based on the formation of

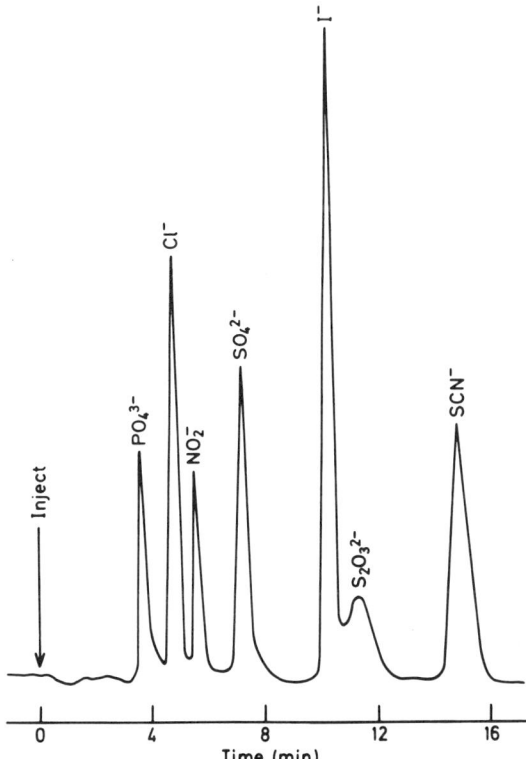

Figure 44. Detection of inorganic anions by UV absorption following postcolumn reaction with iron (III) perchlorate. Column: TSK-GEL IEX-520 QAE. Eluant: 0.5 M acetate buffer at pH 5.48 containing 0.05 M $NaNO_3$. Detection: absorbance at 340 nm, following postcolumn reaction with 0.8 M $HClO_4$ containing 0.05 M $Fe(ClO_4)_3$. (From Ref. 147.)

colored complexes by reaction of these anions with iron (III) in perchlorate media [147]. Figure 44 shows the chromatogram obtained after ion exchange separation of the anions and postcolumn reaction with iron (III) perchlorate. This approach is applicable to the detection of thiocyanate, cyanide, sulfate, chloride, iodide, nitrite, thiosulfate, and phosphate, with detection at 340 nm giving limits of detection in the low parts-per-million range. Sulfur species (sulfide, sulfite, sulfate, thiosulfate, trithionate, tetrathionate, pentathionate, and hexathionate) can be detected after ion exchange separation by postcolumn oxidation with bromine to produce sulfate, which is then reacted with

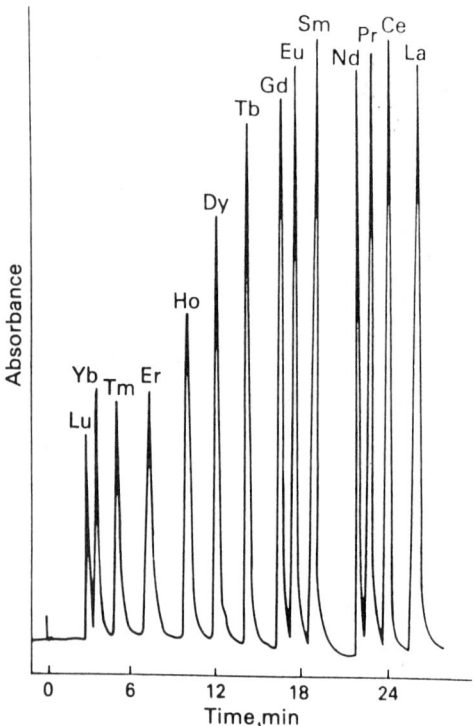

Figure 45. Detection of lanthanides by postcolumn reaction with Arsenazo I. Column: 10-μm Nucleosil SA. Eluant: linear program from 0.01 to 0.04 M 2-methyllactic acid over a 30-min period. Detection: absorbance at 600 nm after postcolumn reaction with 1 mM Arsenazo I. Solute concentrations: 10 ppm of each species. (From Ref. 76.)

iron (III) perchlorate and detected spectrophotometrically at 335 nm [138]. A packed-bed reactor for the detection of sulfate in industrial wastewaters has been described [148], in which the indirect colorimetric method based on reaction of sulfate with barium chloroanilate is employed. Sulfate reacts with barium chloroanilate to produce insoluble barium sulfate (which remains in the postcolumn reactor) and highly colored chloroanilate ions.

Postcolumn reactions have been much more widely applied to the detection of cations than for anions; indeed, it is fair to say that there is no better method for the detection of transition metals and lanthanides. The color-forming reagent should give a low background absorbance, together with high molar absorptivities for as large a range of cations as possible. Many reagents have been evaluated for their suitability as color-forming reagents in postcolumn reaction detection; these include broad-spectrum reagents such as dithizone, Eriochrome Black T, 4-(2-pyridylazo)resorcinol (PAR), xylenol orange, and Arsenazo dyes. The most widely applicable reagents are PAR and Arsenazo I or Arsenazo III, which together provide rapid color reactions with most elements of interest [149]. Figure 45 illustrates the detection of lanthanides after separation by ion exchange using methyllactic acid as eluant, and postcolumn reaction with Arsenazo I, and detection at 600 nm [76]. Ion-interaction methods have been extensively employed for the separation of metal ions [150,151]; these methods are also ideally suited to detection using postcolumn reaction with color-forming species such as PAR or Arsenazo III.

A postcolumn fluorescence detection method for nitrite, thiosulfate, and iodide has been reported [152] in which these anions are reacted with cerium (IV) to produce the fluorescent species cerium (II) in a packed-bed reactor. Nitrate and nitrite may be determined simultaneously by including into the system a postcolumn copperized reductor which reduces nitrate to nitrite, prior to further reaction with cerium (IV) in the packed-bed reactor. The detection limits attainable with this method were in the low parts-per-billion range.

REFERENCES

1. J. S. Fritz, D. T. Gjerde, and C. Pohlandt, *Ion Chromatography*, Huthig, Heidelberg, 1982, p. 119.
2. D. T. Gjerde and J. S. Fritz, *Anal. Chem.* 53, 2324 (1981).
3. J. S. Fritz, D. T. Gjerde, and R. M. Becker, *Anal. Chem.* 52, 1519 (1980).
4. J. A. Dean (Ed.), *Lange's Handbook of Chemistry*, 11th ed., McGraw-Hill, New York, 1973, pp. 6-30.
5. J. S. Fritz, D. L. DuVal, and R. E. Barron, *Anal. Chem.* 56, 1177 (1984).
6. P. E. Jackson and P. R. Haddad, *J. Chromatogr.* 346, 125 (1985).
7. D. T. Gjerde, J. S. Fritz, and G. Schmuckler, *J. Chromatogr.* 186, 509 (1979).
8. J. E. Girard and J. E. Glatz, *Am. Lab.* 13, 26 (1981).
9. D. T. Gjerde, G. Schmuckler, and J. S. Fritz, *J. Chromatogr.* 187, 35 (1980).
10. P. R. Haddad and A. L. Heckenberg, *J. Chromatogr.* 300, 357 (1984).
11. T. Okada and T. Kuwamoto, *Anal. Chem.* 55, 1001 (1983).

12. G. J. Sevenich and J. S. Fritz, *Anal. Chem. 55*, 12 (1983).
13. J. P. Ivey, *J. Chromatogr. 287*, 128 (1984).
14. R. P. W. Scott, *Liquid Chromatography Detectors*, Elsevier, Amsterdam, 1977, p. 88.
15. Operating Manual, Waters M 430 Conductivity Detector.
16. R. P. W. Scott, *Liquid Chromatography Detectors*, Elsevier, Amsterdam, 1977, p. 82.
17. I. Molnar, H. Knauer, and D. Wilk, *J. Chromatogr. 201*, 225 (1980).
18. D. E. Johnson and C. G. Enke, *Anal. Chem. 42*, 329 (1970).
19. J. M. Keller, *Anal. Chem. 53*, 344 (1981).
20. V. Svoboda and J. Marsal, *J. Chromatogr. 148*, 111 (1978).
21. E. Pungor, F. Pal, and K. Toth, *Anal. Chem. 55*, 1728 (1983).
22. J. F. Alder, P. R. Fielden, and A. J. Clark, *Anal. Chem. 56*, 985 (1984).
23. T. Jupille, D. Togami, and D. Burge, Abstract No. 242 presented at Pittsburgh Conference and Exposition on Analytical Chemistry and Applied Spectroscopy, Atlantic City, March 8-13, 1982.
24. R. M. Cassidy and S. Elchuk, *J. Chromatogr. Sci. 21*, 454 (1983).
25. R. M. Cassidy and S. Elchuk, *J. Chromatogr. 262*, 311 (1983).
26. D. R. Jenke and G. K. Pagenkopf, *Anal. Chem. 54*, 2603 (1982).
27. C. A. Pohl and E. L. Johnson, *J. Chromatogr. Sci. 18*, 442 (1980).
28. M. J. van Os, J. Slanina, C. L. de Ligny, and J. Agterdenbos, *Anal. Chim. Acta 156*, 169 (1984).
29. S. A. Wilson, E. S. Yeung, and D. R. Bobbitt, *Anal. Chem. 56*, 1457 (1984).
30. R. E. Majors, H. G. Barth, and Ch. H. Lochmuller, *Anal. Chem. 56*, 300R (1984).
31. J. Heyrovsky, *Chem. Listy 16*, 256 (1922).
32. L. Meites, P. Zuman, and co-workers, *CRC Handbook Series in Inorganic Electrochemistry*, Vols. I-V, CRC Press Inc., Boca Raton, FL, 1980-1985.
33. P. T. Kissinger and W. R. Heineman, eds., *Laboratory Techniques in Electroanalytical Chemistry*, Marcel Dekker Inc., New York, 1984.
34. D. T. Sawyer and J. L. Roberts, *Experimental Electrochemistry for Chemists*, John Wiley, New York, 1974.
35. R. Rveki, *Talanta 27*, 147 (1980).
36. K. Stulik and V. Pacakova, *J. Electroanal. Chem. 129*, 1 (1981).
37. F. Palmisano and P. G. Zambonin, *Ann. di Chim. 74*, 633 (1984).
38. D. C. Johnson, S. E. Weber, A. M. Bond, R. M. Wrightman, R. E. Shoup, and I. S. Krull, *Anal. Chim. Acta*, in press.
39. LDC/Milton Roy, Riveria Beach, Florida.
40. Dionex, Sunnyvale, California
41. E. J. Caliguri and I. N. Mefford, *Brain Res. 296*, 156 (1984).
42. LKB, Bromma, Sweden.

43. A. M. Bond and G. G. Wallace, *Anal. Chem.* 54, 1706 (1982).
44. A. M. Bond and G. G. Wallace, *J. Liq. Chromatogr.* 6, 1799 (1983).
45. J. H. Larochelle and D. C. Johnson, *Anal. Chem.* 50, 240 (1978).
46. P. Jandik and D. Cox, *J. Chromatogr.*, in preparation.
47. T. Imanari, K. Ogata, and S. Tanabe, *Chem. Pharm. Bull.* 30, 374 (1982).
48. J. G. Tarter, *J. Liq. Chromatogr.* 7, 1559 (1984).
49. W. T. Kok, U. A. Brinkman, and R. W. Frei, *J. Chromatogr.* 256, 17 (1983).
50. W. A. MacCrehan and W. E. May, *Anal. Chem.* 56, 625 (1984).
51. K. Bratin and P. T. Kissinger, *Talanta* 29, 365 (1982).
52. G. A. Sherwood and D. C. Johnson, *Anal. Chim. Acta* 129, 101 (1981).
53. P. J. Davenport and D. C. Johnson, *Anal. Chem.* 46, 1971 (1974).
54. J. E. Girard, *Anal. Chem.* 51, 836 (1979).
55. B. B. Wheals, *J. Chromatogr.* 262, 61 (1983).
56. J. K. Kochi and R. J. Klinger, *J. Am. Chem. Soc.* 102, 4790 (1980).
57. W. A. MacCrehan, *Anal. Chem.* 53, 74 (1981).
58. W. A. MacCrehan and R. A. Durst, *Anal. Chem.* 53, 1700 (1981).
59. W. A. MacCrehan and R. A. Durst, *Anal. Chem.* 50, 2108 (1978).
60. J. A. Cox and P. J. Kuleszq, *Anal. Chem.* 56, 1021 (1984).
61. E. A. Ostrovidov, *Zh. Anal. Khim.* 35, 1677 (1980).
62. R. D. Rocklin and E. L. Johnson, *Anal. Chem.* 55, 4 (1983).
63. E. B. Buchanan and J. R. Bacon, *Anal. Chem.* 39, 615 (1967).
64. Ch. Wang, S. D. Bunday, and J. G. Tarter, *Anal. Chem.* 55, 1617 (1983).
65. P. Jandik, D. Cox, and D. Wong, *Am. Lab.* 114-125 (March 1986).
66. T. Imanari, S. Tanabe, and T. Toida, *Chem. Pharm. Bull.* 30, 3800 (1982).
67. B. Philar, L. Kosta, and B. Hristovski, *Talanta* 26, 805 (1979).
68. A. M. Bond, I. D. Heritage, G. G. Wallace, and M. J. McCormick, *Anal. Chem.* 54, 582 (1982).
69. L. R. Taylor and D. C. Johnson, *Anal. Chem.* 46, 262 (1974).
70. G. Sittampalam and G. S. Wilson, *Anal. Chem.* 55, 1608 (1983).
71. W. J. Hurst, J. W. Stefovic, and W. J. White, *J. Liq. Chromatogr.* 7, 2021 (1984).
72. S. Williams, ed., *Official Methods of AOAC*, 14th ed., AOAC Washington DC, 1984.
73. J. Slanina, W. Lingerak, and F. Bakker, *Anal. Chim. Acta* 117, 91 (1980).
74. M. C. Franks and D. L. Pullen, *Analyst (London)* 99, 503 (1974).
75. H. Hershcovitz, C. Yarnitzky, and G. Schmuckler, *J. Chromatogr.* 252, 113 (1982).
76. Wang-nang Wang, Yeong-jgi Chen, and Mon-tai Wu, *Analyst (London)* 109, 281 (1984).

77. F. A. Schultz and D. E. Mathis, *Anal. Chem.* 46, 2253 (1974).
78. K. Suzuki, H. Aruga, and T. Shirai, *Anal. Chem.* 55, 2011 (1983).
79. T. Deguchi, T. Kuma, and H. Nagai, *J. Chromatogr.* 152, 349 (1978).
80. C. R. Loscombe, G. B. Cox, and J. A. W. Dalziel, *J. Chromatogr.* 166, 403 (1978).
81. R. C. Dorey, Report 1981, FJSRL-TR-81-0005. *Chem. Abstr.* 96, 14676u (1982).
82. S. Egashira, *J. Chromatogr.* 202, 37 (1981).
83. R. S. Deelder, H. A. J. Linssen, J. G. Koen, and A. J. B. Beeren, *J. Chromatogr.* 203, 153 (1981).
84. J. Slanina, F. P. Bakker, P. A. C. Jongejan, L. van Lamoen, and J. J. Mols, *Anal. Chim. Acta* 130, 1 (1981).
85. E. C. V. Butler and R. M. Gershey, *Anal. Chim. Acta* 164, 153 (1984).
86. P. W. Alexander and C. Maitra, *Anal. Chem.* 53, 1590 (1981).
87. P. W. Alexander, P. R. Haddad, G. K-C. Low, and C. Maitra, *J. Chromatogr.* 209, 29 (1981).
88. P. W. Alexander, M. Trojanowicz, and P. R. Haddad, *Anal. Lett.* 17(A4), 309 (1984).
89. P. W. Alexander, P. R. Haddad, and M. Trojanowicz, *Anal. Chem.* 56, 2417 (1984).
90. P. R. Haddad, P. W. Alexander, and M. Trojanowicz, *J. Chromatogr.* 294, 397 (1984).
91. P. R. Haddad, P. W. Alexander, and M. Trojanowicz, *J. Chromatogr.* 315, 261 (1984).
92. P. R. Haddad, P. W. Alexander, and M. Trojanowicz, *J. Chromatogr.* 321, 363 (1985).
93. P. R. Haddad, P. W. Alexander, and M. Trojanowicz, *J. Chromatogr.* 324, 319 (1985).
94. P. W. Alexander, P. R. Haddad, and M. Trojanowicz, *Chromatographia* 20, 179 (1985).
95. P. W. Alexander, P. R. Haddad, and M. Trojanowicz, *Anal. Chim. Acta*, in press.
96. P. R. Haddad, P. W. Alexander, and M. Trojanowicz, *J. Liq. Chromatogr.*, in press.
97. R. P. Buck, S. Singhadeja, and L. B. Rogers, *Anal. Chem.* 26, 1240 (1954).
98. L. Goodkin, M. D. Seymour, and J. S. Fritz, *Talanta* 22, 245 (1975).
99. G. P. Ayers and R. W. Gillett, *J. Chromatogr.* 284, 510 (1984).
100. R. G. Gerritse, *J. Chromatogr.* 171, 527 (1979).
101. S. H. Kok, K. A. Buckle, and M. Wooton, *J. Chromatogr.* 260, 189 (1983).

102. P. E. Jackson, P. R. Haddad, and S. Dilli, *J. Chromatogr. 295*, 471 (1984).
103. R. N. Reeve, *J. Chromatogr. 177*, 393 (1979).
104. N. E. Skelly, *Anal. Chem. 54*, 712 (1982).
105. T. Kamiura and M. Tanaka, *Anal. Chim. Acta 110*, 117 (1979).
106. R. J. Williams, *Anal. Chem. 55*, 851 (1983).
107. J. P. Ivey, *J. Chromatogr. 267*, 218 (1983).
108. R. D. Rocklin, *Anal. Chem. 56*, 1959 (1984).
109. S. Matsushita, *J. Chromatogr. 312*, 327 (1984).
110. P. Jandik and D. Wong, unpublished results.
111. D. Bushee, I. S. Krull, R. N. Savage, and S. B. Smith, Jr., *J. Liq. Chromatogr. 5*, 463 (1982).
112. D. Bushee, D. Young, I. S. Krull, R. N. Savage, and S. B. Smith, Jr., *J. Liq. Chromatogr. 5*, 693 (1982).
113. H. Small and T. E. Miller, Jr., *Anal. Chem. 54*, 462 (1982).
114. S. A. Wilson and E. S. Yeung, *Anal. Chim. Acta 157*, 53 (1984).
115. P. R. Haddad and A. L. Heckenberg, *J. Chromatogr. 252*, 177 (1982).
116. R. A. Cochrane and D. E. Hillman, *J. Chromatogr. 241*, 392 (1982).
117. A. Laurent and R. Bourdon, *Ann. Pharm. Fr. 36*, 453 (1978).
118. C. A. Hordijk, C. P. C. M. Hagenaars, and Th. E. Cappenberg, *J. Microbiol. Methods 2*, 49 (1984).
119. R. C. L. Foley and P. R. Haddad, *J. Chromatogr.*, in preparation.
120. Z. Iskandarani and T. E. Miller, Jr., *Anal. Chem. 57*, 1591 (1985).
121. T. Okada and T. Kuwamoto, *J. Chromatogr. 325*, 327 (1985).
122. M. Dreux, M. Lafosse, and M. Pequignot, *Chromatographia 15*, 653 (1982).
123. F. A. Buytenhuys, *J. Chromatogr. 218*, 57 (1981).
124. S-I. Mho and E. S. Yeung, *Anal. Chem.*, in press.
125. J. M. Pettersen, *Anal. Chim. Acta 160*, 263 (1984).
126. J. A. Koropchak and G. N. Coleman, *Anal. Chem. 52*, 1252 (1980).
127. E. A. Woolson and N. Aharonson, *J. Assoc. Off. Anal. Chem. 63*, 523 (1980).
128. G. R. Ricci, L. S. Shepard, G. Colovos, and N. E. Hester, *Anal. Chem. 53*, 610 (1981).
129. M. Fujita and E. Takabatake, *Anal. Chem. 55*, 454 (1983).
130. D. J. Freed, *Anal. Chem. 47*, 186 (1975).
131. B. G. Julin, H. W. Vandenborn, and J. J. Kirkland, *J. Chromatogr. 112*, 443 (1975).
132. S. W. Downey and G. M. Heiftje, *Anal. Chim. Acta 153*, 1 (1983).
133. I. S. Krull, *Trends Anal. Chem. 3*, 76 (1984).

134. I. S. Krull, D. Bushee, R. N. Savage, R. G. Scheecher, and S. B. Smith, Jr., *Anal. Lett. 15*, 267 (1982).
135. G. Nickless, *J. Chromatogr. 313*, 129 (1985).
136. J. P. Sickafoose, Ph.D. dissertation, Iowa State University, Ames, Iowa (1971).
137. G. J. Schmidt and R. P. W. Scott, *Analyst (London) 109*, 997 (1984).
138. J. N. Story, *J. Chromatogr. Sci. 21*, 272 (1983).
139. S. van der Wal, *J. Liq. Chromatogr. 6*, 37 (1983).
140. J. F. K. Huber, K. M. Jonker, and H. Poppe, *Anal. Chem. 52*, 2 (1980).
141. L. Nondek, U. A. Th. Brinkman, and R. W. Frei, *Anal. Chem. 55*, 1466 (1983).
142. N. Yoza, K. Ito, Y. Hirai, and S. Ohashi, *J. Chromatogr. 196*, 471 (1980).
143. H. Yamaguchi, T. Nakamura, Y. Hirai, and S. Ohashi, *J. Chromatogr. 172*, 131 (1979).
144. T. Nakamura, T. Yano, A. Fujita, and S. Ohashi, *J. Chromatogr. 130*, 384 (1977).
145. Y. Hirai, N. Yoza, and S. Ohashi, *J. Chromatogr. 206*, 501 (1981).
146. Y. Hirai, N. Yoza, and S. Ohashi, *Anal. Chim. Acta 115*, 269 (1980).
147. T. Imanari, S. Tanabe, T. Toida, and T. Kawanishi, *J. Chromatogr. 250*, 55 (1982).
148. K. Brunt, *Anal. Chem. 57*, 1338 (1985).
149. J. S. Fritz and J. N. Story, *Anal. Chem. 46*, 825 (1974).
150. S. Elchuk and R. M. Cassidy, *Anal. Chem. 51*, 1434 (1979).
151. R. M. Cassidy and S. Elchuk, *J. Liq. Chromatogr. 4*, 379 (1981).
152. S. H. Lee and L. R. Field, *Anal. Chem. 56*, 2647 (1984).

5
Ion Chromatography Exclusion

Phyllis E. Buell and James E. Girard
Department of Chemistry
The American University
Washington, D.C.

I. INTRODUCTION

Ion chromatography exclusion (ICE) is a high-pressure liquid chromatography (HPLC) technique using ion exchange resin columns, in which ionic substances are rejected by the resin whereas nonionic (or weakly ionic) materials are retained and separated by partition between the liquid inside the resin particles and the liquid outside the particles. Although ion exchange resins are used in this technique, true ion-exchange reactions are not involved. Anions are separated on cation exchangers, and cations are separated on anion exchangers.

The term *ion chromatography exclusion* (ICE) was introduced by Rich et al. of the Dionex Corporation [1]. Other terms that have been used include ion exclusion chromatography (IEC) [2], ion exclusion partition (IEP) chromatography [3,4], Donnan exclusion chromatography (DEC) [5], and ion-moderated partition (IMP) chromatography [6].

In this chapter, ICE will be used to include separations of polar compounds, and of weakly ionic acids and bases on cation and anion exchangers, respectively, regardless of the exact nature of the separation mechanism which may involve other separation modes besides ion exclusion, including normal and reverse phase partition, adsorption, steric exclusion, and ligand formation.

ICE is particularly useful for the analysis of organic acids, and can be used very effectively for the separation of sugars, alcohols, phenols, amino acids, organic bases, and certain weakly ionized inorganic compounds. It is also of value for the group separation of ionic species from nonionic species.

The basic ICE technique is very simple. The analysis of organic acids requires only a standard liquid chromatograph, a column packed with a strongly acidic cation exchange resin in the hydrogen form, dilute mineral acid as eluent, and a UV detector. Regeneration of the column is not required, and even for complex matrices no pretreatment, other than filtration, or centrifugation, is usually necessary.

As will be described later in this chapter, coupling ICE with ion chromatography (IC) adds to the range of anions that can be separated, and allows simultaneous determinations of weak and strong acids.

II. THE THEORY OF ION CHROMATOGRAPHY EXCLUSION

The ion exchange resins used in ICE are synthesized by polymerizing styrene with divinylbenzene followed by introduction of an active ionic group, the degree of crosslinking being determined by the proportion of divinylbenzene present as copolymer. The network structure of the resin accounts for the ion exclusion, adsorption, and partition behavior that has been observed.

The theory of ion chromatography exclusion has been discussed by a number of authors [2,3,7-9]. When an ion exchange resin is placed in aqueous solution, the resin bed consists of three parts: the solid resin network, the occluded liquid inside the resin beads, and the liquid between the resin beads. The resin network acts as a semipermeable membrane between the stationary liquid phase within the resin and the mobile liquid phase between the resin particles. The resin's own ionic groups are fixed, and movement of ions across the membrane occurs according to Donnan theory. Electrostatic forces arising from the high ionic concentration inside the resin prevent strong electrolytes from entering the resin, and, as a result, they are not retarded and pass rapidly through the column. For example, if hydrochloric acid is added to a cation exchange resin in the hydrogen form, the repulsive forces between the chloride ion and the fixed sulfonate groups of the resin prevent the chloride from entering the interior of the resin. The chloride is excluded and elutes with the void volume.

Nonionic or weakly ionic materials, however, are not excluded, or are only partially excluded, and can partition between the two liquid phases. This slows their progress through the column. The degree of retardation increases with decreasing degree of ionization and also depends on polar attractions between the solute and the fixed functional group, and on differing Van der Waals forces between the solute and the hydrocarbon part of the resin. In general, elution order can be predicted from pK_a values, but for aromatic and long-chain aliphatic compounds Van der Waals forces become dominant. The exact mechanisms involved vary from compound to compound and are not fully understood, but the result of the complex interactions is enhanced selectivity [6,10]. For organic acids, the mechanism is predominantly ion exclusion with some reverse phase partitioning. Alcohols are separated primarily by reverse phase partitioning; for carbohydrates, size exclusion becomes a factor.

III. BACKGROUND

A. The Development of ICE

From the time that chemically homogeneous ion exchange resins were available, it became evident that processes other than true ion exchange occurred between resin and analyte.

Early investigators [2,9], using the then newly synthesized sulfonated crosslinked polystyrene cation exchange resin Dowex 50, noted that when the hydrogen form of the resin was equilibrated with hydrochloric acid, the concentration of hydrochloric acid within the aqueous portion of the resin phase was lower than the concentration in the surrounding solution. This distribution pattern was found true for electrolytes in general. Nonpolar or slightly ionized compounds, on the other hand, were generally distributed approximately evenly between aqueous and resin phases, although in many cases additional adsorption on to the resin occurred. Adsorption on ion exchange materials is well known [7,11,12], and recently the mechanisms involved in the retention of organic acids, and other organic species, on functionalized and unfunctionalized styrene-divinylbenzene copolymer resins have been investigated [13,14].

Wheaton and Bauman [2] suggested the use of ion exchange resins for the separation, in aqueous solution, of nonionic from ionic material and were the first to use the term *ion exclusion* for this fractionation. They demonstrated the successful separation of acetic acid from hydrochloric acid on the strongly acidic cation exchange resin Dowex 50-X8 in the hydrogen form (Figure 1). Using the sodium form of the resin, they also separated sodium chloride from a number of organic compounds including formaldehyde, glycols, and amines. Several authors [15-17], employing this technique, described separations of nonionic compounds not only from ionic compounds but also from each other. However, at this time the technique was primarily used as a unit operation for the group separation of electrolytes and nonelectrolytes.

Harlow and Morman [3], in a method described as ion-exclusion partition chromatography, studied the elution of a large number of acids, both organic and inorganic, from a hydrogen form of cation

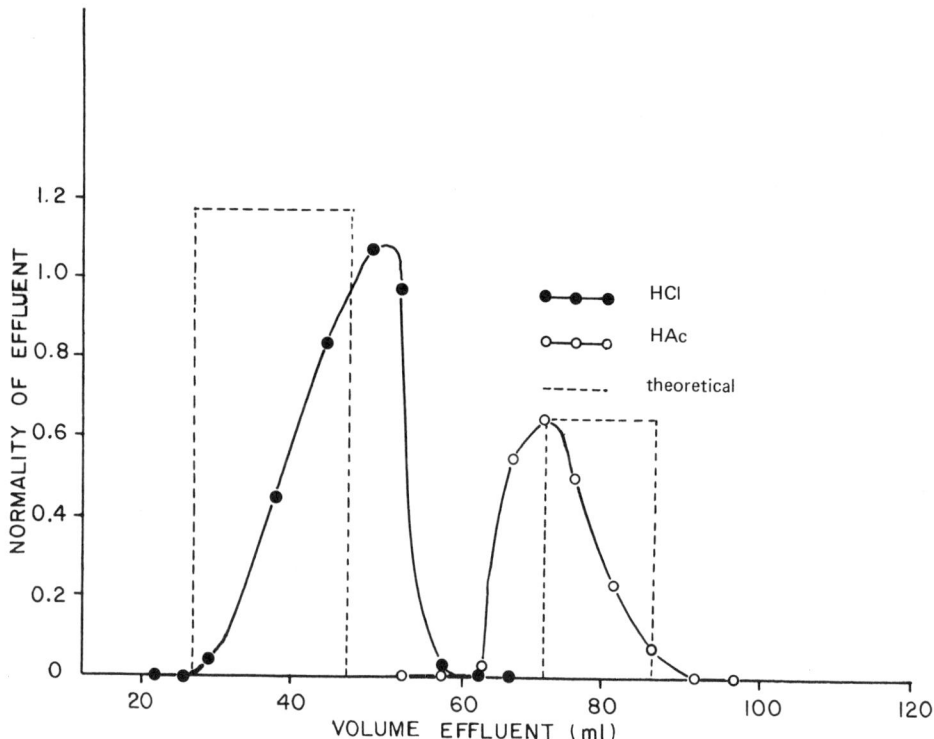

Figure 1. Ion exclusion: complete separation. Resin: Dowex 50 × 8% 50- to 100-mesh. Feed: 15 ml of 1.7 N HCl and 0.66 N acetic acid. (From Ref. 2.)

exchange resin column, using water as eluant. The time of emergence of each acid from the column was compared with that obtained for acetic acid under identical conditions (Table 1). On the basis of elution behavior, Harlow and Morman came to the following conclusions: Strong acids, as expected, are not retained (i.e., they are excluded) and elute with the void volume. Members of a homologous series, such as formic, acetic, and propionic acids, elute in order of decreasing acid strength (increasing pK_a). Dibasic acids emerge sooner than monobasic acids of the same carbon number. For example, oxalic and succinic acids emerge ahead of acetic and propionic acids, respectively. Isoacids elute ahead of normal acids, as shown for isobutyric and butyric acids. A double bond retards elution; thus acrylic acid emerges after the weaker propionic acid. A keto group, on the

Table 1. Emergence Data for Acids Relative to Acetic Acid by Ion Exclusion Partition Chromatography

Acid	Ratio acid/acetic acid	Acid	Ratio acid/acetic acid
Sulfuric	0.57	Fumaric	1.0
Toluenesulfonic	0.57	Glutaric	1.0
Sulfurous	0.58	Chloroacetic	1.0
5-Sulfosalicylic	0.58	Acetic	1.00
Sulfamic	0.58	Levulinic	1.0
Hydrochloric	0.59	Nadic	1.0
Acetylenedicarboxylic	0.59	L-Pyroglutamic	1.13
Trichloroacetic	0.60	Methylenebis-[mercap-toacetic]	1.15
Mucic	0.60		
L-Cysteic	0.61	Propionic	1.17
Maleic	0.61-0.71	Tetrahydrophthalic	1.21
Oxalic	0.62	Acrylic	1.23
Phosphoric	0.63	Carbonic	1.26
Citric	0.64	Isobutyric	1.32
Nitroform	0.67	Butyric	1.45
Itaconic	0.70	Mandelic	1.49
Pyruvic	0.71	Pivalic	1.49
Malonic	0.72	α-Hydroxybutyric	1.57
α-Ketobutyric	0.74	Methacrylic	1.63
Glyceric	0.75	Isovaleric	1.66
Boric	0.75-0.79	t-Butylacetic	1.67
α-Ketovaleric	0.79	Crotonic	1.95
Cyanuric	0.80	Valeric	2.09
Dichloroacetic	0.81	Furoic	2.09
Mercaptosuccinic	0.82	Cyclohexanecarboxylic	3.26
Succinic	0.82	2,4-Dihydroxybenzoic	3.80
Glycolic	0.82	p-Hydroxybenzoic	4.46
Lactic	0.84	Hydrocinnamic	5.40
Formic	0.91	Benzohydroxamic	5.95
Adipic	1.0		

Source: Ref. 3.

other hand, increases the elution rate; therefore pyruvic acid elutes before propionic acid. Substituted benzoic acids are strongly retained.

During the 1960s, reports of separations of organic acids on cation exchange resins continued to appear and were reviewed by Jandera

Table 2. Distribution Coefficients (K_d) of Acids

Acid	K_d	Acid	K_d
HI	0	HF	0.36
HBr	0	HCOOH	0.43
$HClO_4$	0	CH_3COOH	0.65
HCl	0	C_2H_5COOH	0.81
H_2SO_4	0	C_3H_7COOH	1.10
HNO_3	0	H_2CO_3	1.00
$(COOH)_2$	0.01	HCN	1.00
H_3PO_3	0.06	H_3BO_3	1.00
H_3PO_2	0.08	C_6H_5OH	0.98
H_3PO_4	0.09	CH_3OH	1.02
H_2SO_3	0.11	H_2S	1.40

Source: Ref. 19.

and Churacek [18], but there was no widespread interest in the ion exclusion method. The technique still employed gravity flow, separations required several hours, and the polymeric resins available at this time were not sufficiently rigid for high-pressure systems.

Some ten years later, Tanaka et al. [19], using a cation exchange resin with water as eluant and an HPLC technique rather than gravity flow, made a further study of the elution behavior of a large number of strong and weak acids and calculated their distribution coefficients (Table 2). They observed that strong acids were completely ion-excluded and eluted with the void volume. Weaker acids were not completely ion-excluded and permeated selectively into the resin, their retention volumes increasing proportionally with the increase in their pK_a values. Very weak acids ($pK_a > 6.4$), were not excluded at all and permeated totally into the resin, and their retention volumes were independent of pK_a.

Hyakutake and Hanai [20] reported an HPLC separation of a number of the citric acid cycle acids on a weakly acidic cation exchange column (LS 140, polystyrene gel in the carboxymethyl form), using a n-hexane/tetrahydrofuran/tertiary butanol eluant, and UV detection. However, it was the report of Turkelson and Richards [21] in 1978, describing the separation of these same acids on a strongly acidic cation exchange resin, that renewed interest in ICE as a practical technique. Turkelson and Richards separated the citric acid cycle acids on the sulfonated polystyrene resin Aminex 50W-X4 (30-35 μm), using 0.001 N hydrochloric acid as eluant [22], and continuous

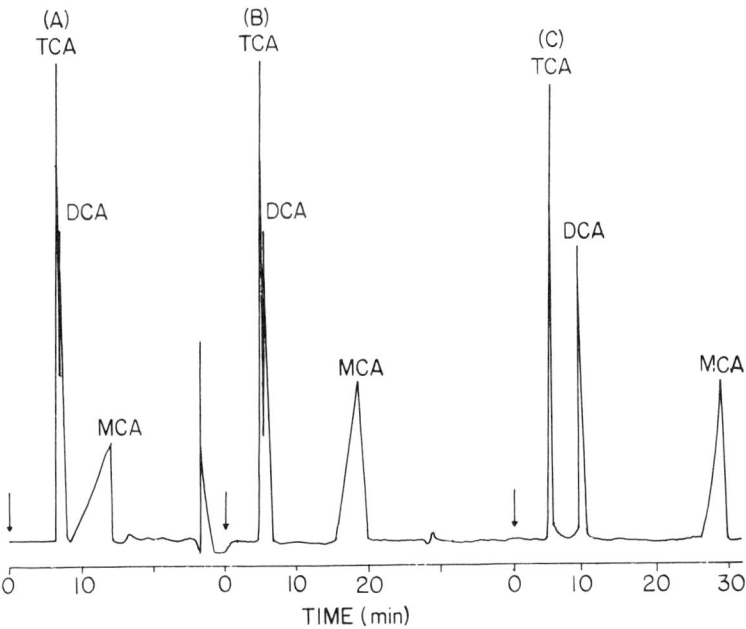

Figure 2. Separation of mono-, di-, and trichloroacetic acid on cation exchange resin. (a) Eluant: deionized water. (b) Eluant: 0.001 N hydrochloric acid. (c) Eluant: 0.01 N hydrochloric acid. Detector: UV at 210 nm and 0.2 aufs. Flow rate: 1 ml/min. Column temperature: ambient. (From Ref. 21.)

UV monitoring. They demonstrated the advantage of using dilute mineral acid, in place of water, as eluant. The addition of mineral acid suppressed the ionization of strong and moderately strong organic acids allowing them to partition into the resin phase which, in turn, increased their retention time on the resin, and improved separation. The degree of retention depended primarily on the pK_a of the acid and the pH of the eluant, as was demonstrated for a mixture of mono-, di-, and trichloroacetic acids (Figure 2). Turkelson and Richards used their technique to analyze human and rat urines, and citrus juices, for the citric acid cycle acids by simply injecting diluted filtered samples onto the column. Sensitivity below 5 ppm was achieved for some acids.

With continuing improvement in the suitability of polymeric packing materials for HPLC [10,23,24], the ICE technique has found increasing applications for organic acid analysis, and it now provides

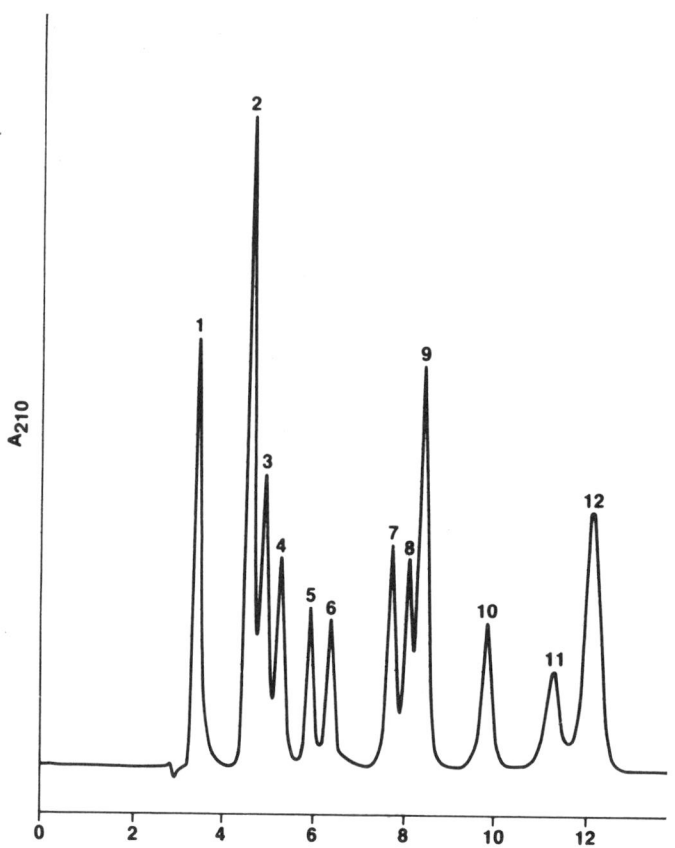

Figure 3. Separation of organic acids. Analysis conditions: column, Interaction ORH-801; temperature, 35°C; eluant, 0.01 N H_2SO_4; flow rate, 0.8 ml/min; pressure, 140 atm; pump, Waters 6000A; detector, Waters model 450 UV. The sample contains the following acids: (1) oxalic; (2) maleic; (3) α-ketoglutaric; (4) citric; (5) glyoxylic; (6) malic; (7) glycolic; (8) formic; (9) acetic; (10) propionic; (11) uric; (12) butyric. (From Ref. 25.)

a much needed practical and efficient method for an analysis that had long been a problem, particularly in the case of low-molecular-weight polar organic acids. The potential of ICE for the analysis of organic acids is demonstrated in Figure 3, which was obtained by Woo and Benson [25] for a mixture of organic acids.

B. Alternate Methods for Analysis of Organic Acids and Weakly Ionized Inorganic Compounds

Alternate chromatographic methods for the determination of organic acids have not been entirely successful for a number of reasons, including the complex matrices in which organic acids are frequently present, their interaction with many column support materials, and their lack of UV chromophores which makes detection difficult. These alternate methods will be discussed briefly.

Until the development of ICE, gas chromatography/mass spectrometry (GC/MS) was the method of choice for accurate, quantitative analysis of organic acids. This elegant technique is very sensitive and selective, but the extraction and derivatization steps that are required make it extremely time consuming [26]. The expense of the equipment is a further disadvantage.

Many HPLC approaches to the analysis of organic acids have been investigated and were reviewed by Schwarzenbach [27] in 1982. Normal phase separations, and reverse phase separations using an acid eluant to suppress the ionization of the weak acids, have been described [27,28]. Derivatization, which occurs readily with weak acids, has been used to obtain products that are both nonionized and strongly UV absorbing [29]. Both normal and reverse phase ion-pair chromatography systems have been used, the introduction of a UV-absorbing counterion allowing for ready detection of the ion pairs [27,30,31].

Ion exchange on bonded phase packings, which permit rapid flow rates, have the disadvantage that the silica support material is unstable at the high pH levels required to increase ionization of weak acids [27,32]. In the newer IC methods that separate anions on polymeric anion exchange materials [33], organic acids such as citric, tartaric, and oxalic tend to elute with excessively long retention times; interference between organic acids and inorganic acids is common; many organic acids cannot be well resolved; and their low conductance remains a problem [27,34].

As well as organic acids, weakly ionized inorganic acids and bases, and their salts, have also been difficult to analyze by standard IC techniques because of their lack of conductance. There is a further problem with the analysis of carbonate in that the conventional Dionex suppressed-conductivity IC system uses a carbonate-bicarbonate eluant, making it impossible to detect these ions. The ICE technique meets a very real need in this case. Ammonia also presents a problem. If determined by conventional IC a nonlinear detector response is obtained because of incomplete dissociation [35]. ICE offers a promising alternative [36].

IV. ICE/IC

Rich et al. [1,34,37] introduced the concept of coupling ICE with IC to improve the chromatographic resolution of inorganic and organic acids in complex matrices. The coupled system is illustrated in Figure 4. The solution to be analyzed is injected onto an ion-exclusion strong cation-exchange separator column, and dilute hydrochloric acid is used as eluant. Strongly ionized acids and their salts are separated, as a group, from weakly ionized acids and their salts, which are separated from each other on the column, the degree of resolution depending on the nature of the weak acids and the chromatographic conditions. At the same time, most cations are removed from the solution by classical ion exchange. Eluant from the separator column then passes to a suppressor column packed with cation exchange resin in the silver (Ag^+) form, for removal of the highly conducting hydrogen and chloride ions. Hydrogen ions are removed by exchange with silver ions, and chloride ions are removed by precipitation as AgCl, allowing other anions eluting from the ICE system to be monitored by conductivity detection. The eluant at this stage is water, and it must be noted that acids with pK_a greater than seven are not appreciably dissociated at pH 7, and so cannot be detected. The ICE peak containing the strong acids is further resolved in the IC system. The peak is collected on a low-capacity anion-exchange resin concentrator column, which is connected to a standard IC system consisting of anion separator column with suppressor, and $NaHCO_3/Na_2CO_3$ as eluant. This high-resolution IC step separates the components in the ICE peak. Ions that have similar k' values in ICE have large differences in k' values for IC, and so can be resolved by the coupled ICE/IC technique. The capability of ICE/IC is demonstrated in Figure 5 [37,38], which shows the determination of both strong and weak acids in coffee extract. The dual system can also be reversed (IC/ICE); this was found advantageous for the separation of certain aromatic acids in urine [39].

The suppressor column required for the Dionex ICE separation with conductivity detection has a number of disadvantages. As more and more AgCl precipitates out, back pressure builds up, resulting in lack of reproducibility and loss of signal sensitivity. This can be prevented by periodically cutting off the expended portion of the column; however, there still remains the drawback that the column cannot be regenerated and must be replaced.

To avoid the need for the silver form suppressor, with its attendant problems, the Dionex Corporation recently introduced an anion fiber suppressor which uses tetrabutylammonium hydroxide (TBAOH) as regenerant [40]. The usual hydrochloric acid eluant is replaced by either octanesulfonic acid or tridecafluoroheptanoic acid, both of

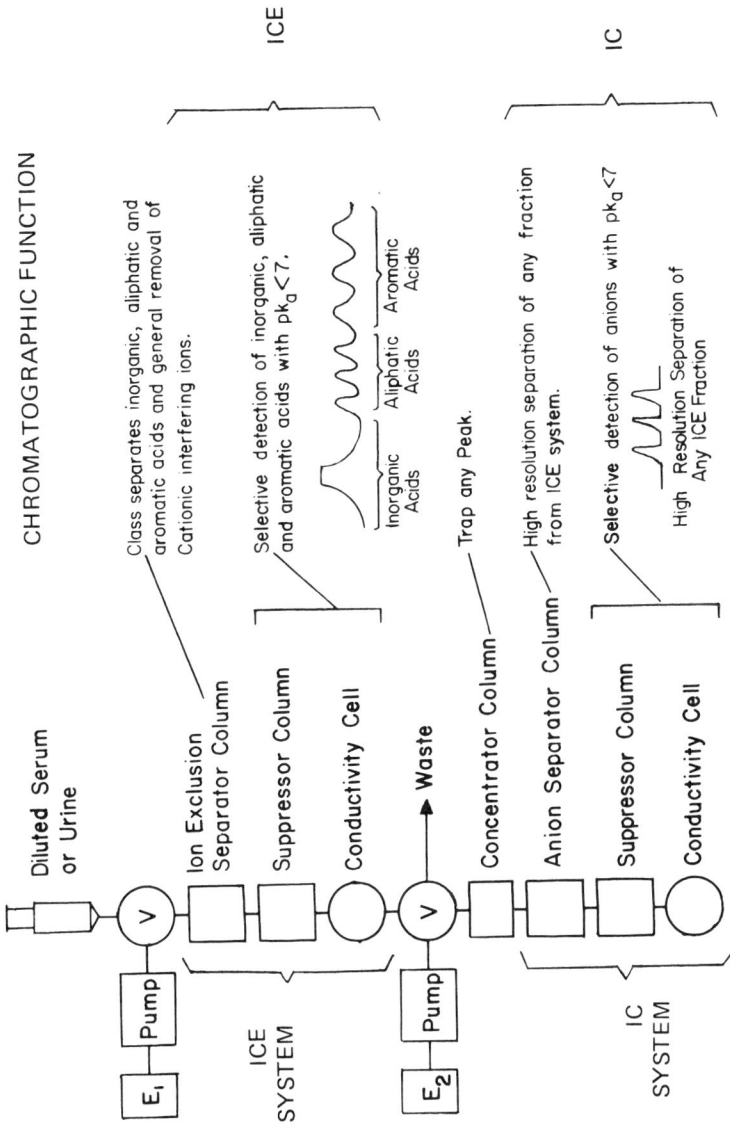

Figure 4. Schematic representation and chromatographic function for ion chromatography exclusion mode coupled to ion chromatography (ICE/IC). E_1 and E_2, eluant reservoirs; V, four-way valve. (From Ref. 39.)

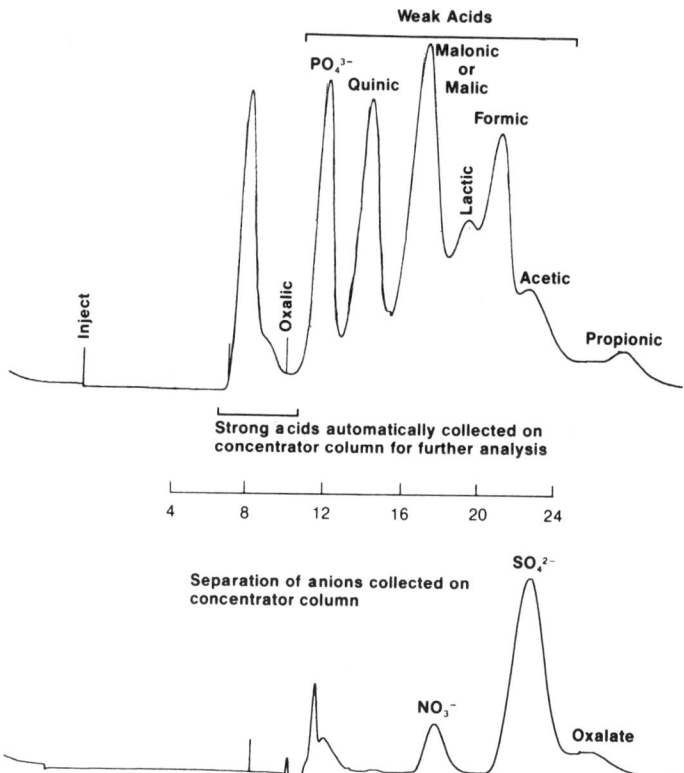

Figure 5. Determination of strong and weak acids in coffee extract. Instrumental conditions: ICE conditions: eluant, 0.01 M HCl; flow rate, 0.86 ml/min; separator column, 9 × 250 mm ICE separator; suppressor columns, 4 × 140 mm; injection volume, 100 µl; meter full-scale setting, 30 µS. IC conditions: eluant, 0.003 $NaHCO_3$/0.0024 Na_2CO_3; flow rate, 2.3 ml/min; separator column, 3 × 500 anion separator; suppressor column, 6 × 250 anion suppressor; injection, ICE void-volume peak concentrated on a 3 × 50 anion concentrator; meter full-scale setting, 30 µS. (From Ref. 38.)

which possess anions having lower equivalent conductance than Cl^-. Highly conducting H^+ in the eluant is exchanged for TBA^+, which has a much lower equivalent conductance. In this way, a low background conductivity is achieved, and the suppressor can be continuously regenerated.

V. EXPERIMENTAL CONDITIONS FOR ICE

Many factors affect the efficiency of ICE separations and can be varied to obtain optimal conditions for a particular analysis. Factors include the characteristics of the separator column, the nature and pH of the eluant, and the temperature of operation. The choice of detector is important in determining selectivity and sensitivity.

A. Columns

Recently, very stable microporous polystyrene-divinylbenzene materials of uniform size, improved rigidity, and lot-to-lot reproducibility have been developed [10,23,24]. Columns packed with these copolymers can be operated at pressures exceeding 200 atm (3000 psi) and so can be used at the eluant flow rates required in modern HPLC. Unlike silica-based packings, they are stable from pH 0 to 14, and have a very long life.

The columns most commonly used in ICE separations of organic acids, and other weakly acidic compounds, are packed with strongly acidic high-capacity sulfonated styrene-divinylbenzene in the hydrogen form. Commercially packed columns especially designed for separation of specific groups of compounds are listed in Table 3; these have been used in most of the applications that have been reported recently. Several suppliers market more than one organic acids column, the choice depending on the nature of the acids being analyzed and the selectivity that is required. Columns that allow resolution of acids having almost identical pK_a values (e.g., α-ketoglutaric and maleic acids), and a number of carbohydrate columns each optimized for analysis of a particular carbohydrate class, are also available. Many columns permit separation of components in mixtures of organic acids, alcohols, and carbohydrates, which until recently had not been possible because of coelution problems. For samples in which only a few of the species present are of interest, as is often the case in the food industry, shorter columns, which greatly reduce analysis time, have been introduced. New columns designed for specific separations will no doubt continue to be introduced, and therefore current information should be obtained before a column selection is made.

For separations of organic bases, columns packed with strongly basic copolymers with a quarternary ammonium functional group are commercially available. To date, there appears to be less interest in separating basic species on anion exchange columns than on separating acidic species on cation exchange columns.

Small particle size to facilitate diffusion is important in obtaining good resolution. In most recent applications, separations have been made on resins with particles in the 5- to 15-μm range. The degree of crosslinking (i.e., the percentage of divinylbenzene in the copolymer)

Table 3. Commercially Packed ICE Columns (Sulfonated Styrene-Divinylbenzene Copolymers)

Supplier	Column type	Description
Benson Co.	OA 850	Organic acid analysis
Bio-Rad Laboratories	Aminex HPX-87H Organic acid analysis	Organic acid analysis; also separates neutral species, carbohydrates, alcohols; high resolution
	Fast acid analysis	For specific carbohydrates, alcohols, and organic acids; high speed
	Fermentation monitoring	Better resolution between sugars than fast acid
	Fruit quality	For monitoring fruit quality indicators (glycerol, acetic acid, ethanol)
	Aminex HPX-87C, -87P, -87H (8% crosslinked) in Ca, Pb, H forms; Aminex HPX-42C, -42A (4% crosslinked), -65A (6% crosslinked) in Ca and Ag forms	Carbohydrate analysis; each tailored to a specific carbohydrate class
	Fast carbohydrate	For specific carbohydrates; high speed
Dionex Corporation	HPICE-AS1	For analysis of weakly ionized compounds with pK_a values 2-7; highest efficiency

Ion Chromatography Exclusion

	HPICE-AS2	Same as AS1; good general column
	HPICE-AS3	Low crosslinked resin for best separation of organic acids pK_a 2-4, and total CO_2
	HPICE-AS4	For separation of amino acids
Interaction Chemicals	ORH-801 organic acid	For separation of organic and inorganic anions
	ARH-601 fast organic acid	For separation of strong aliphatic acids and weak aromatic acids; high speed
	ION-300 organic acid (was FJA-801)	Separation of Krebs cycle acids and analysis of fruit juice; separates neutral species, carbohydrates, alcohols
	ION-310 (was FJA-101)	For monitoring fruit quality indicators (glycerol, acetic acid, ethanol)
	CHO-611 corn syrup	For analysis of corn syrup
	CHO-620 (Ca form)	For separation of oligosaccharides and sugar alcohols
	CHO-682 (Pb form)	For separation of di- and monosaccharides
Wescan Instruments	Anion exclusion	For organic acids and weak acid anions
	Anion exclusion/HS	For strongly retained weak acid anions

also affects the retention of weakly ionized species [2,3,18]. In general, for both strong and weak acids, the lower the degree of crosslinking, the longer will be the retention time of the acid. As crosslinking is decreased, ions more readily penetrate the resin, where they are held up. This effect was demonstrated by Wheaton and Bauman [2] for the separation of hydrochloric acid from acetic acid. The degree of crosslinking that will provide the best separation depends on the properties of the analytes. In their study of organic and inorganic acids, Harlow and Morman [3] found that Dowex 50W-X12 (12% crosslinked) provided the best resolution of very weakly ionized acids, whereas Dowex 50W-X2 (2% crosslinked) was better for the separation of stronger acids. At the present time, 8% crosslinked polymers are most commonly used, with lower crosslinking being preferred for separations of acids having pK_a values 2-4.

In addition to the copolymers, silica-based ion exchange materials and pellicular packings have been used for ICE, but they have the disadvantage of a restricted pH operating range [10,32]. Sephadex crosslinked dextran ion exchangers have also been employed [5,41], and because these gels contain fixed ionic functional groups similar to those present in styrene-divinylbenzene copolymers, ion exclusion and partition mechanisms can occur in an analogous manner. However, application of these materials to practical modern HPLC analysis has not been demonstrated.

B. Eluants

For separation of certain compounds, such as carbohydrates, water can be used as eluant and has the advantage of being nonconducting. However, for many determinations, dilute mineral acids give better results because they suppress the ionization of acidic analytes, allowing them to partition into the resin phase, which results in increased retention and improved separation [21]. At the same time, the acid continuously regenerates the cation exchange column with hydrogen ions. Dilute hydrochloric acid is used with the standard Dionex ICE system [42]. Dilute sulfuric acid is used with systems where UV, RI, and amperometric detection are employed [6,10,23,25,43].

As already mentioned, the Dionex Corporation has recently suggested using an octanesulfonic acid or tridecafluoroheptanoic acid eluant, together with an anion fiber suppressor, to avoid the need for their silver form suppressor. For the same reason, Itoh and Shinbori [44] have suggested using carbonic acid as eluant.

The concentration of the acid eluant used affects the separation. For many analyses, 0.01 N mineral acids have been recommended [6,10,23,25,37,42]. In some cases, depending on the composition of the sample, other concentrations have given better separations. For example, improved resolution of sulfate and oxalate was achieved when the acid concentration was increased to 0.03 N [37].

Addition of inorganic salts, or organic modifiers to the eluant improves separation for some samples [4,6,25,45]. Addition of salts such as ammonium sulfate increases retention of organic solutes by the resin. Organic modifiers such as acetonitrile, isopropanol, and ethanol have been used effectively, the choice being based on factors similar to those considered for reverse phase chromatography. Addition of acetonitrile in the case of high-molecular-weight, or aromatic, acids decreases the strong retention of these compounds on the resin, causing them to elute more rapidly [39,42]. Methanol causes greater resin volume changes than the other organic modifiers mentioned, but it can still be included if temperature and/or flow rate are adjusted [25]. Tanaka and colleagues [36,46,47] described the use of water containing varying amounts of acetone, or dioxane, as eluant for analysis of phosphates and the ammonium ion.

C. Detectors

Many different detection systems have been used with ICE separations, including UV/visible spectrophotometry, conductivity, electrochemistry, fluorometry, and refractive index measurement. Two detection systems (e.g., UV/amperometric and UV/refractive index) can be combined for added selectivity.

Ultraviolet light detection has been used successfully in many applications, including the separation of the citric acid cycle acids by Turkelson and Richards [21]. A wavelength of 210 nm is most frequently selected because this is the region where very weakly absorbing acetic acid has its maximum absorptivity. This detection method would appear very suitable for the ICE system because the eluants most commonly employed, water and dilute hydrochloric and sulfuric acids, do not absorb in the UV region, and so do not interfere with absorption of analyte. Unfortunately, many analytes that are particularly well separated by the ICE technique, such as aliphatic acids and alcohols, sugars, and certain inorganic ions, exhibit no UV absorbance themselves or, at best, weak absorbance at very low wavelengths. Also, UV-absorbing contaminants frequently cause noisy baselines at the low wavelengths used. However, despite these limitations, UV detection is used extensively with ICE.

Conductivity detection is also popular. If water is used as the eluant, ionizable analytes are readily detected. However, if hydrochloric acid is used instead of water, the highly conducting acid interferes with detection of the conductance of weakly ionized analytes. To overcome this problem, as already described (see ICE/IC), Rich et al. [34,37] introduced the suppressor column packed with cation exchange resin in the silver form, to remove hydrochloric acid from the eluant stream.

Electrochemical detectors, both coulometric and amperometric [48], have been used very successfully. They are well suited to ICE separations which are performed with isocratic elution and aqueous solutions. A requirement of this detection technique is that the compound of interest be electrochemically active, or capable of being coupled to an electrochemical reaction. The constant potential coulometric detector introduced by Takata and Muto [49] was used by them to analyze amino acids, carboxylic acids, phenols, and sugars. Tanaka and colleagues used this detection technique for ICE analyses of acids, the ammonium ion, and various phosphates [19,36,46,47,50]. Amperometric detection also provides excellent sensitivity and selectivity and has been shown to be very suitable for the detection of organic acids, alcohols, and phenols [6,25,51,52].

Refractive index monitors designed especially for HPLC offer significant advances in performance over older models and have proved particularly useful for detecting carbohydrates, alcohols, and other substances having little or no measurable UV absorption [6,43,53]. When a refractive index monitor is combined with a UV detector to monitor organic acids, it is possible to determine carbohydrates, alcohols, and organic acids with one sample injection.

Fluorometric detection offers excellent sensitivity and selectivity, and, when combined with a postcolumn reaction, has been used effectively with ICE separations of amino acids [54].

D. Temperature

Temperature is an important variable that is often overlooked. At elevated temperature, the efficiency and sensitivity of separations on styrene-divinylbenzene columns are improved [10,55]. Many ICE separations have been carried out at ambient temperature, but column temperatures up to 65°C have also been used for analysis of organic acids [25]. Carbohydrate columns are heated to 85°C to increase efficiency and decrease viscosity [6].

VI. DETECTION LIMITS

Organic acids can be detected in the parts-per-billion range [33,37, 43,56], and with preconcentration this limit can be further decreased. Pohl and Johnson [33] reported that the use of the Dionex system with a hydrochloric acid eluant, the silver-form suppressor, and conductivity detection allowed organic acids such as lactic, formic, and propionic to be detected at parts-per-billion levels. Jupille et al. [43]

found that for a 100-μl injection, detection limits for organic acids depended on the acid used as eluant, and were in the 0.1- to 10-ppm range. Phenols have been determined in the parts-per-billion range [52]. Similar limits have been attained for inorganic ions. Samples containing borates in the parts-per-billion range have been analyzed satisfactorily [57]. Detection of nanogram amounts of ammonia have been reported [36], and, with a 500-μl injection, amines have been determined at parts-per-billion levels [58].

VII. APPLICATIONS OF ICE

A. The Determination of Organic Acids

Detection and quantitation of organic acids is of great importance in biomedical research. Many biologically important products of carbohydrate, protein, and lipid metabolism are organic acids, and their study in biological fluids is vital to the understanding of many metabolic disorders [59]. In biochemical and biological research, there is also a need to monitor organic acids. Determination of organic acids is of increasing interest in the food and beverage industries [25,27,60,61], where the amount and type of organic acids present in products is critical in determining flavor and quality. In industrial and environmental waters, levels of organic acids are a matter of concern, and in the plating industry various organic acids need to be monitored. Examples of applications of ICE to these many analytical problems will be described.

Biomedical Applications

Buchanan and Thoene [62,63] profiled urinary organic acids by means of a two-column system consisting of a cation exchange column (Aminex HPX-87H) in tandem with an ODS reverse phase column. Addition of the reverse phase column greatly increased selectivity. They used both amperometric and UV detection and dilute sulfuric acid at pH 2.5 as eluant. Rich et al. [34,39] described the use of the Dionex ICE/ IC system for the determination of a large number of acids in serum, lactate and pyruvate in plasma, and vanillylmandelic acid in urine, and demonstrated the feasibility of using the system as a clinical method.

Woo and Benson [25,64], using an ORH-801 column, successfully analyzed many complex samples including urine (Figure 6) and cerebrospinal fluid. The eluant was 0.01 N sulfuric acid to which methanol was added, and UV or amperometric detection was employed.

Biological Applications

In biological studies, ICE is replacing time-consuming GC and less sensitive spectrophotometric methods. An example is the investigation

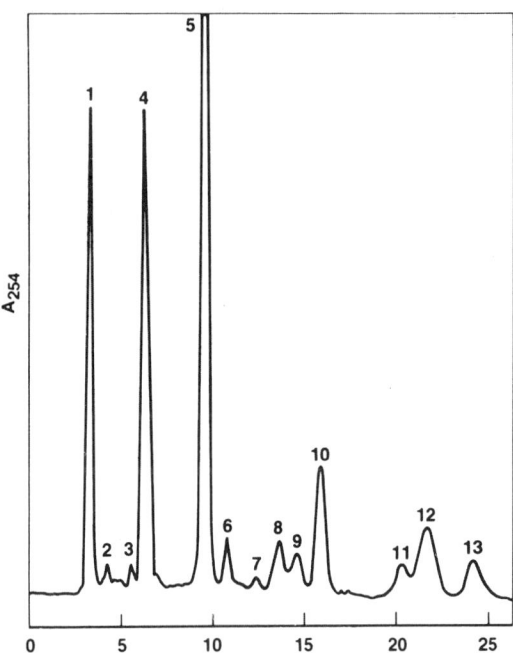

Figure 6. Analysis of human urine. Untreated urine sample filtered through 0.45-μm membrane prior to analysis can be injected onto the column for organic acid profiling in studies of metabolic pathways. Analysis conditions: column, Interaction ORH-801; temperature, 55°C; eluant, 0.01 N H_2SO_4/15% methanol; flow rate, 0.6 ml;min; pressure, 140 atm; pump, Waters 6000A; detector, Waters model 450 UV monitor (set at 254 nm). Sample peaks are (1) oxalic acid, (2) oxaloacetic acid, (3) α-ketoisovaleric acid, (4) ascorbic and α-keto-β-methyl-n-valeric acids, (5) β-phenylpyruvic acid, (6) uric acid, (7) α-ketobutyric acid, (8) homoprotocatechuic acid, (11) hydroxyphenylacetic acid, (12) p-hydroxyphenyl lactic acid, and (13) homovanillic acid. Peaks 9 and 10 are unknown. (From Ref. 64.)

of poly-β-hydroxybutyrate (PHB) in Rhizobium japonicum bacteroids, and its relationship to nitrogen fixation. PHB was converted to crotonic acid for determination by ICE, with the use of an Aminex-87 column, dilute sulfuric acid as eluant, and UV detection [65]. With a similar system, gluconic acid, its sodium salt, and other organic acids formed in biochemical or catalytic oxidation of glucose, have been separated [66].

Ion Chromatography Exclusion

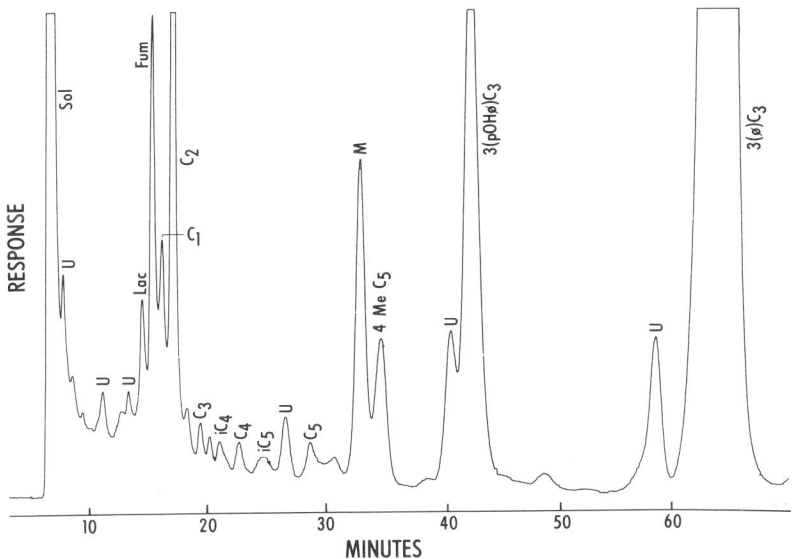

Figure 7. HPLC chromatogram of the free acids of *P. anaerobius*. U, unidentified components; Sol, solvent and unabsorbed components; Lac, lactic acid, Fum, fumaric acid; C_1, formic acid; C_2, acetic acid; C_3, propionic acid; iC_4, isobutyric acid; C_4, butyric acid; iC_5, isovaleric acid; C_5, valeric acid; M, PYG medium component; $4MeC_5$, 4-methylvaleric acid; $3(pOH\phi)C_3$, 3-(p-hydroxyphenyl)propionic acid; $3(\phi)C_3$, 3-phenylpropionic acid. The eluant was 0.007 M H_2SO_4-10.8% acetonitrile. (From Ref. 67.)

Guerrant et al. [67] found ICE to have definite advantages over gas-liquid chromatography (GLC) for determining the diagnostically useful short-chain acids produced in culture media by anaerobic bacteria. Figure 7 shows the separation of the acids present in a culture extract from *Peptostreptococcus anaerobius*. An Aminex HPX-87H column with a dilute sulfuric acid/acetonitrile eluant was used along with UV detection. Lebel and Yen [68] noted the superiority of ICE over other methods for determining the ionic species formed by sulfate-reducing bacteria. As shown in Figure 8, they monitored the production of sulfate, lactate, acetate, and carbonate in a mixed Desulfovibrio culture from sewage, via the Dionex chromatography system with an ICE column, a dilute hydrochloric eluant, and conductivity detection. Formate was added as internal standard.

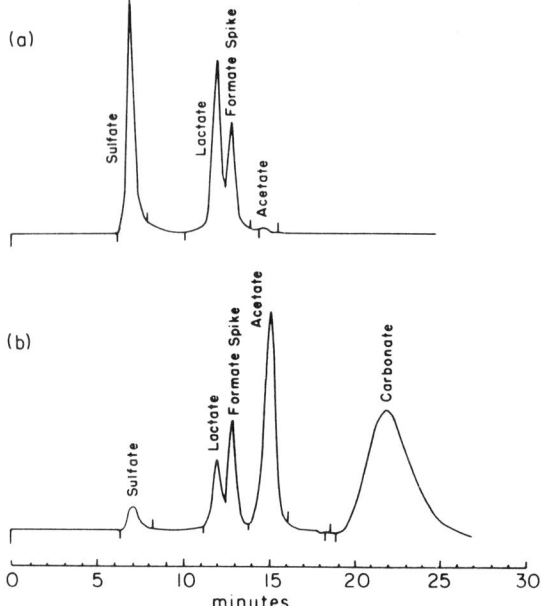

Figure 8. Ion chromatograph readings: (a) initial, (b) final (Dionex 2000i with ICE column connected to HP 3390A integrator). (From Ref. 68.)

Applications in the Food and Beverage Industries

The coupled ICE/IC system has been used for the simultaneous determination of organic acids and cations in wine (Figure 9) [69], and for the analysis of acids in coffee and other beverages [38,70], Jupille et al. [6,10] used an Aminex HPX-87H column with 0.01 N sulfuric acid as eluant and UV detection at 210 nm, for the separation of organic acids in whole milk and cultured sour cream. Using very similar conditions (Aminex HPX-87 column, 0.009 N sulfuric acid, UV detection at 220 and 275 nm), Marsili and colleagues [60, 61] reported the determination of organic acids in a variety of dairy products, including milk, butter milk, sour cream, yogurt, and cottage and cheddar cheeses. The nature and quantity of the acids present are important indicators of flavor, nutritional value, and bacterial activity. A typical chromatogram of sour cream is shown in Figure 10. Woo and Benson [23,25], using analysis conditions similar to those they employed for urine analysis (Figure 6), analyzed wine, beer, juices, and dairy products. Diet cola and grape

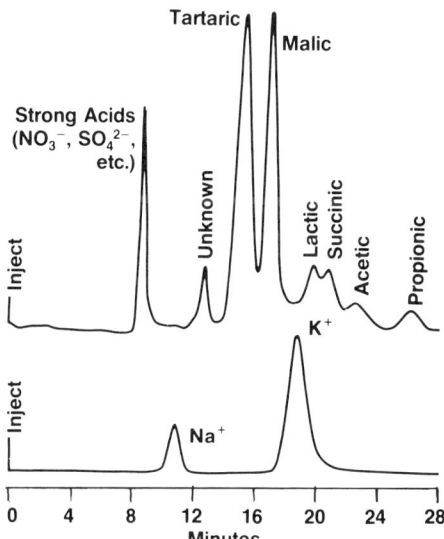

Figure 9. Red wine acids and cations. Instrumental conditions: same as Figure 5 except for IC condition: separator column, 6 × 250 mm cation; suppressor column, 9 × 250 mm cation; eluant, 0.005 M HNO_3. (From Ref. 69.)

drink have been analyzed by means of the Wescan Anion Exclusion column with 0.002 or 0.004 N sulfuric acid and conductivity detection [71]. Benzoic and sorbic acids are used as preservatives in many food products, and it is often necessary to monitor these acids to ensure that levels are maintained within legal limits. Rapid determinations of the acids have been achieved with the Bio-Rad fast acid analysis column [72].

Water Purity Applications

The ICE technique has been used to measure trace levels of organic acids in steam condensate samples [56]. Power plants routinely use IC for determining a large number of ionic species, but this technique does not give good separations of weak acids. With ICE, much better resolution has been achieved, and it has been possible to determine, simultaneously, dissolved carbon dioxide, hydrofluoric acid, formic and acetic acids, the weak acids typically found in steam condensates. For monitoring the weak acids, the Dionex ICE system, with both conductometric and UV detection, was employed. The pH of the eluant was adjusted to control separation.

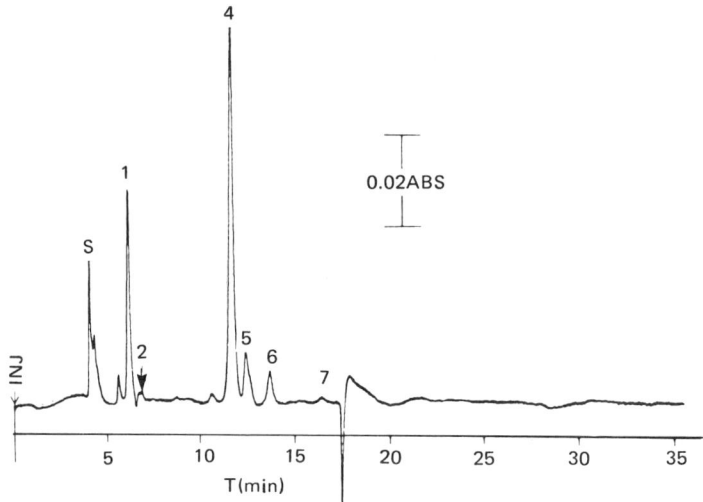

Figure 10. Typical chromatogram of sour cream. Chromatographic conditions: injection volume, 10 μl; mobile phase, 0.009 N H_2SO_4; temperature, 65°C; UV detection at 220 nm; flow rate, 0.7 ml/min. (1) Orotic, (2) citric, (3) pyruvic, (4) lactic, (5) uric, (6) acetic, (7) propionic. (From Ref. 61.)

Applications in the Plating Industry

A further industrial application of ICE is the determination of the organic acids used as mild chelating agents in electroless nickel and copper plating baths [73,74].

B. The Determination of Organic Compounds, Other than Carboxylic Acids

ICE was originally introduced for the HPLC analysis of organic acids. It has since been demonstrated that sulfonated polystyrene-divinylbenzene cation exchange resins also provide excellent separations of polar compounds such as alcohols and carbohydrates. Further, by use of specially optimized resins, carbohydrates, alcohols, and organic acids can be separated from each other.

Polar Compounds on Cation Exchange Columns

Good separations of sugars have been achieved by means of ICE columns. Scobell et al. [53] employed a cation exchange column in either the Ca^+ or Ag^+ form, degassed water as eluant, and refractive index detection, for the separation of sucrose, dextrose, and fructose, in medium invert sugar, and for studying the sugars formed in enzymatic hydrolysis of corn syrup. A weak complex between the sugar and the metal ion of the resin apparently increased retention and improved resolution.

Jupille et al. [6,10] described the separation on cation exchange resins of many polar organic compounds including aliphatic alcohols, amino acids, carbohydrates in sweeteners, and formaldehyde in environmental samples. These authors used the term *ion-moderated partition* (IMP) to refer to separations of polar nonionic compounds on ion exchange resins, the term including ion exclusion as one of several mechanisms other than ion exchange (e.g., normal phase partition, reverse phase partition, size exclusion, and ligand exchange) that occur on the resin. They separated carbohydrates on Aminex carbohydrate columns in a number of metal ion forms, using water as eluant, the exact choice of the column depending on the nature of the carbohydrates being analyzed. The superiority of the lead form (Aminex HPX-87P) over the calcium form (Aminex HPX-87C) for the separation of carbohydrates derived from cellulose hydrolysis is shown in Figure 11 [75]. The calcium form is ideal for most monosaccharides [24,75].

Jupille et al. [6,10] also separated aliphatic alcohols on an Aminex HPX-85H column, using dilute sulfuric acid as eluant, the elution order being methyl, ethyl, isopropyl, isobutyl, n-butyl, isoamyl, n-amyl, as shown in Figure 12. A more rapid separation can be achieved with the newer Bio-Rad fruit-quality column, which has been used for determining alcohols in wine [76]. The Dionex HPICE column has proved efficient for separations of alcohols and glycols in water, wine, whiskey, antifreeze, foods, and other matrices, when combined with pulsed amperometric detection [51].

Armentrout et al. [52] used strong cation-exchange resin columns, with dilute sulfuric acid/acetonitrile eluants and amperometric detection, to screen a variety of water samples for trace quantities of phenolic compounds. Takata and Muto [49], using a variety of eluants and coulometric detection, separated amino acids, phenols, and sugars on cation exchange columns. As shown in Figure 13, when combined with a postcolumn reaction and fluorescence detection, the Dionex ICE column has given excellent separations of amino acids in animal feeds [54].

Benson and Woo [24,77] used a column especially designed for fruit juice analysis to separate sugars, organic acids, and alcohols, as shown in Figure 14a. In certain applications in the fruit industry, speed is particularly important, and for these analyses short, fast columns have been introduced [24,76,77]. Figure 14b shows the very rapid (less than 3 min) determination of the fruit-quality

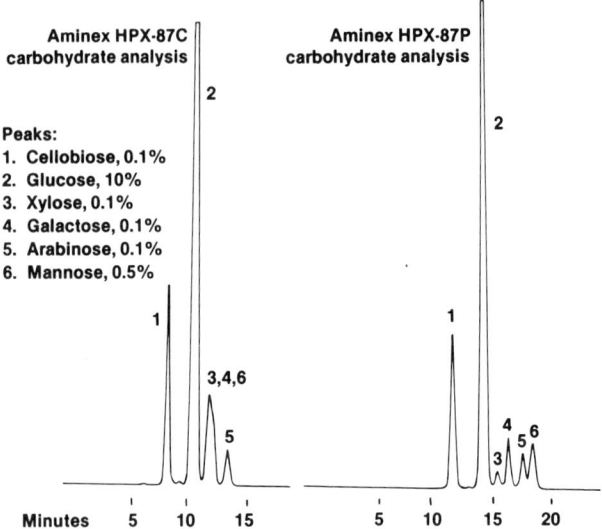

Figure 11. Effect of ionic form (lead versus calcium) for separation of carbohydrates. (From Ref. 75.)

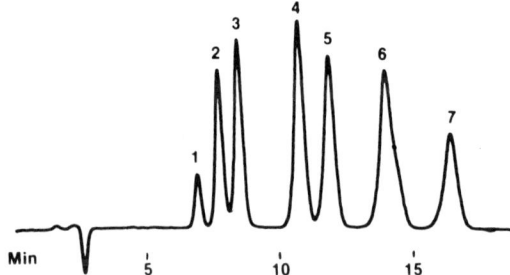

Figure 12. Separation of aliphatic alcohols on Aminex HPX-85H. The separation is based on reverse phase partition. (From Ref. 6.)

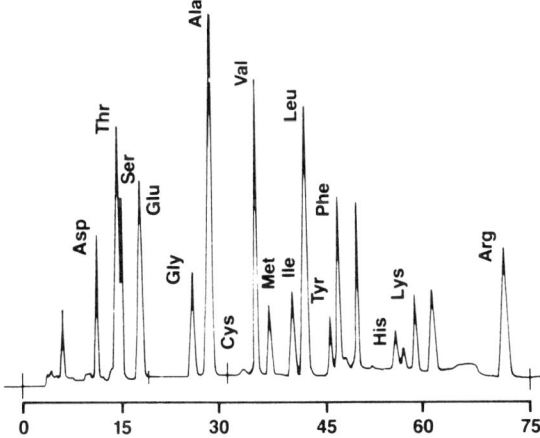

Figure 13. Free amino acids in animal feeds. Separation: HPICE/AS4, detection: opti-ion/fluor (with PCR). (From Ref. 54.)

indicators glycerol, acetic acid, and ethanol, in the presence of large quantities of sugars. Rapid analyses of this kind are needed during the grape harvest.

Basic Organic Compounds on Anion Exchange Columns

There have been few reports of applications of ICE to the analysis of weak organic bases, although strongly basic anion-exchange columns, by a combination of ion exclusion, ion exchange, and partition chromatography, can provide good separations of amines and other bases. For example, trimethylamine, which is an indicator of spoilage in fish, can be determined by ICE [78]. Tanaka et al. [58] separated a mixture of aliphatic amines on a hydroxide-form anion exchange resin, using water or water/acetone as eluant and coulometric detection, but these authors did not describe practical applications.

C. The Determination of Weak Inorganic Acids and Bases

ICE has been used to analyze tap water and industrial waste waters, for fluoride and bicarbonate [71,79-81]. Pohlandt [79] described the determination of bicarbonate and fluoride in tap water via a cation exchange column coupled to a suppressor column, very dilute hydrochloric acid as eluant, and conductometric detection (Figure 15). The same system was used to separate hypophosphite, citrate, lactate, and cyanide. Carbonate has also been analyzed via conductivity

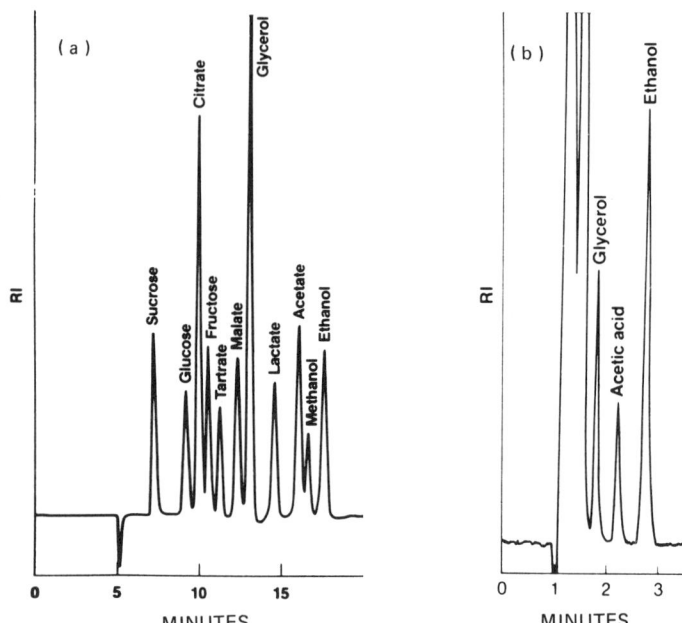

Figure 14. (a) Fruit juice analysis. Analysis conditions: column, FJA-801, 0.65 × 30 cm (Interaction); eluant, 0.01 N sulfuric acid; flow rate, 0.4 ml/min; temperature, 37°C; detection, refractive index. (b) High-speed fruit juice analysis. Analysis conditions: same as for (a) except for column, FJA-101, 0.46 × 12 cm (Interaction); flow rate, 0.5 ml/min; temperature, 50°C. (From Ref. 24.)

detection, with water rather than hydrochloric acid as eluant; in this case, a conductivity suppressor column is not required [80]. Among early applications of the dual ICE/IC system were determinations of chloride, chlorate, sulfate, and carbonate in caustic, and sulfate, formate, acetate, and carbonate in brine [1,37]; for these separations, water was used as eluant.

The weakly dissociated species azide, cyanide, and sulfide have been successfully analyzed with ICE; the chromatographers took advantage of the fact that these species are readily oxidized and thus can be detected amperometrically [82].

Borate concentrations in a number of ores have been determined with the Dionex ICE/IC system, mannitol being added to the hydrochloric acid eluant to form the borate-mannitol complex, which has a higher conductance than boric acid alone [57]. Borate could be

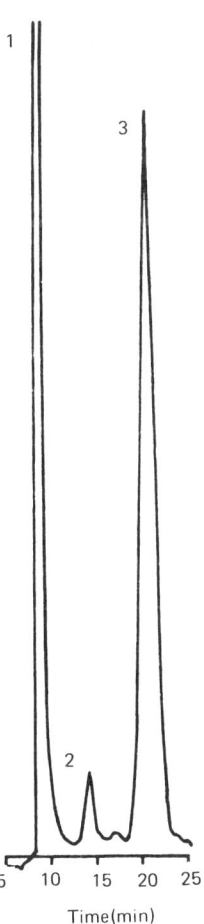

Figure 15. Determination of fluoride and bicarbonate in tap water. Column: AG50W-X4 300 × 9 mm; eluant; 0.001 M HCl; detection, suppressor column conductometric. (1) Sulfate, (2) fluoride, (3) carbonate-bicarbonate. (From Ref. 79.)

determined without interference from chloride, or other strongly ionized species, and analysis time was shorter than was possible with conventional IC. A knowledge of the quantity of borate buffer present in certain nickel and nickel alloy plating baths is important, and ICE/IC has replaced time-consuming titration procedures for this determination [73].

Soluble silica in a variety of matrices, including automobile radiator coolant, has been analyzed via ICE [83]. Because the analyte was too weak an acid for adequate conductivity detection, a postcolumn reaction with molybdic acid was introduced. The silicomolybdic acid that formed was reduced to molybdenum blue before being detected spectrophotometrically. A Dionex ICE column with 0.2 mM hydrochloric acid as eluant was used.

Tanaka and Ishizuka [47] separated phosphate from strong acid anions in industrial waste water and domestic sewage samples. They employed a strongly acidic cation-exchange resin, elution with acetone-water, and coulometric detection. Using the same elution and detection procedure, these authors described the determination of the ammonium ion in sewage and river water via an anion exchange column [36]. They also reported the simultaneous determination of nitrate, nitrite, and ammonium ions in biological nitrification-denitrification process water [50]. A cation exchange column, with UV detection, was used for analysis of nitrate and nitrite ions and was coupled to an anion exchange column, with coulometric detection, for ammonium ion determination. The eluant for both columns was 10% methanol-water.

The ammonia added to feed water in the electrical power industry interferes with IC determination of trace quantities of sodium and potassium. Using ICE, Zarifa [84] was able to solve this problem. The metal cations were excluded and so eluted rapidly, whereas weakly ionized ammonia was retained on the column.

VIII. CONCLUSION

The development of polymeric materials suitable for modern HPLC has focused attention on their suitability for chromatographic separations involving not only ion exchange but also ion exclusion and partition mechanisms. These very stable materials are particularly valuable for determining analytes in the complex matrices encountered in biological samples and in the food and beverage industries.

ICE has become the accepted technique for organic acids analysis and is being used increasingly for determinations of other organic species such as carbohydrates and alcohols, and for weakly ionized inorganic compounds. The simplicity of the technique and the relatively inexpensive equipment and materials required make ICE an attractive method for chromatographers in many fields.

REFERENCES

1. W. Rich, F. Smith, L. McNeill, and T. Sidebottom, Ion exclusion coupled to ion chromatography: Instrumentation and applications

in *Ion Chromatographic Analysis of Environmental Pollutants*, Vol. 2 (J. D. Mulik and E. Sawicki, eds.), Ann Arbor Science Publishers Inc., Ann Arbor, MI, 1979, pp. 17-29.
2. R. M. Wheaton and W. C. Bauman, *Ind. Eng. Chem. 45*, 228 (1953).
3. G. A. Harlow and D. H. Morman, *Anal. Chem. 36*, 2438 (1964).
4. J. S. Fritz, D. T. Gjerde, and C. Pohlandt, in *Ion Chromatography: Chromatographic Methods* (W. Bertsch, W. G. Jennings, and R. E. Kaiser, eds.), Huthig, Heidelberg, New York, 1982, pp. 185-199.
5. H. Waki and Y. Tokunaga, *J. Chromatogr. 201*, 259 (1980).
6. T. Jupille, M. Gray, B. Black, and M. Gould, *Am. Lab. 13*, 80 (1981).
7. D. Reichenberg and W. F. Wall, *J. Chem. Soc.* 3364 (1956).
8. D. W. Simpson and R. M. Wheaton, *Chem. Eng. Prog. 50*, 45 (1954).
9. W. C. Bauman and J. J. Eichorn, *J. Am. Chem. Soc. 69*, 2830 (1947).
10. R. Wood, L. Cummings, and T. Jupille, *J. Chromatogr. Sci. 18*, 551 (1980).
11. R. M. Barrer, *J. Soc. Chem. Ind. 64*, 130T, 133T (1945).
12. C. S. Cleaver and H. G. Cassidy, *J. Am. Chem. Soc. 72*, 1147 (1950).
13. F. F. Cantwell and S. Puon, *Anal. Chem. 51*, 623 (1979).
14. J. S. Fritz, D. L. Du Val, and R. E. Barron, *Anal. Chem. 56*, 1177 (1984).
15. R. M. Wheaton and W. C. Bauman, *Ann. N.Y. Acad. Sci. 57*, 159 (1953).
16. J. F. Thompson and C. J. Morris, *Arch. Biochem. Biophys. 82*, 380 (1959).
17. D. L. Buchanan and R. T. Markiw, *Anal. Chem. 32*, 1400 (1960).
18. P. Jandera and J. Churacek, *J. Chromatogr. 86*, 351 (1973).
19. K. Tanaka, T. Ishizuka, and H. Sunahara, *J. Chromatogr. 174*, 153 (1979).
20. H. Hyakutake and T. Hanai, *J. Chromatogr. 108*, 385 (1975).
21. V. T. Turkelson and M. Richards, *Anal. Chem. 50*, 1420 (1978).
22. M. Richards, *J. Chromatogr. 115*, 259 (1975).
23. D. J. Woo and J. R. Benson, *LC 1*, 238 (1983).
24. J. R. Benson and D. J. Woo, *J. Chromatogr. Sci. 22*, 386 (1984).
25. D. J. Woo and J. R. Benson, *Am. Lab.* Jan. (1984).
26. S. I. Goodman and S. P. Markey, *Diagnosis of Organic Acidemias by GC-MS*, Alan R. Liss, New York, 1981.
27. R. Schwarzenbach, *J. Chromatogr. 251*, 339 (1982).
28. L. R. Snyder and J. J. Kirkland, *Introduction to Modern Liquid Chromatography*, 2nd ed., John Wiley and Sons, New York, 1979, pp. 294-295.

29. L. R. Snyder and J. J. Kirkland, *Introduction to Modern Liquid Chromatography*, 2nd ed., John Wiley and Sons, New York, 1979, p. 779.
30. L. R. Snyder and J. J. Kirkland, *Introduction to Modern Liquid Chromatography*, 2nd ed., John Wiley and Sons, New York, 1979, Chapter 11.
31. T. D. Rotsch and D. J. Pietrzyk, *J. Chromatogr. Sci. 19*, 88 (1981).
32. L. R. Snyder and J. J. Kirkland, *Introduction to Modern Liquid Chromatography*, 2nd ed., John Wiley and Sons, New York, 1979, Chapter 5.
33. C. A. Pohl and E. L. Johnson, *J. Chromatogr. Sci. 18*, 442 (1980).
34. W. Rich, E. Johnson, L. Lois, B. Stafford, P. Kabra, and L. Marton, Organic acids by ion chromatography, in *Liquid Chromatography in Clinical Analysis* (L. Marton and P. Kabra, eds.), Humana Press, Clifton, NJ, 1981, Chapter 17.
35. S. A. Bouyoucos, *Anal. Chem. 49*, 401 (1977).
36. K. Tanaka, T. Ishizuka, and H. Sunahara, *J. Chromatogr. 177*, 21 (1979).
37. Dionex Corporation, Sunnyvale, CA, Ion Chromatography Exclusion Mode (ICE) and Ion Chromatography (IC): Complimentary (sic) Ion Exchange Modes.
38. Dionex Corporation, Sunnyvale, CA, Application Note No. 19, April, 1979.
39. W. Rich, E. Johnson, L. Lois, P. Kabra, B. Stafford, and L. Marton, *Clin. Chem. 26*, 1492 (1980).
40. Dionex Corporation, Sunnyvale, CA, Technical Note 17, Oct., 1984.
41. H. Waki and Y. Tokunaga, *J. Liq. Chromatogr. 5*, 105 (1982).
42. Dionex Corporation, Sunnyvale, CA, Ion Chromatography Systems, LPN 32084, August, 1981.
43. T. H. Jupille, D. W. Togami, and D. E. Burge, *Ind. Res. Dev., 25*, 151 (1983).
44. H. Itoh and Y. Shinbori, *Chem. Lett. 12*, 2001 (1982).
45. W. Rieman and H. F. Walton, *Ion Exchange in Analytical Chemistry*, Pergamon Press, Oxford, 1970, Chapter 9.
46. K. Tanaka and T. Ishizuka, *J. Chromatogr. 190*, 77 (1980).
47. K. Tanaka and T. Ishizuka, *Water Res. 16*, 719 (1982).
48. P. T. Kissinger, Electrochemical detection in liquid chromatography and flow injection analysis, in *Laboratory Techniques in Electroanalytical Chemistry* (P. T. Kissinger and W. R. Heineman, eds.), Marcel Dekker, New York, 1984, Chapter 22.
49. Y. Takata and G. Muto, *Anal. Chem. 45*, 18641 (1973).
50. K. Tanaka, *Bunseki Kagaku, 31*, T106 (1982).

51. Dionex Corporation, Sunnyvale, CA, Product Information, LPN 32276, March, 1983.
52. D. N. Armentrout, J. D. McLean, and M. W. Long, *Anal. Chem.* 51, 1039 (1979).
53. H. D. Scobell, K. M. Brobst, and E. M. Steel, *Cereal Chemistry* 54, 905 (1977).
54. Dionex Corporation, Sunnyvale, CA, *The Capabilities Edge in Ion Analysis*, LPN 32213, March, 1982.
55. L. R. Snyder and J. J. Kirkland, *Introduction to Modern Liquid Chromatography*, 2nd ed., John Wiley and Sons, New York, 1979, pp. 426, 450.
56. W. A. Byers, S. L. Anderson, and W. M. Hickam, *Proc. Int. Water Conf., Eng. Soc. West. Pa.* 44, 436 (1983).
57. J. B. Wilshire and W. A. Brown, *Anal. Chem.* 54, 1647 (1982).
58. K. Tanaka, T. Ishizuka, and H. Sunahara, *J. Chromatogr.* 172, 484 (1979).
59. R. A. Chalmers and A. M. Lawson, *Organic Acids in Man*, Chapman and Hall, London, New York, 1982.
60. R. T. Marsili, *J. Chromatogr. Sci.* 19, 451 (1981).
61. R. T. Marsili, H. Ostapenko, R. E. Simmons and D. E. Green, *J. Food Sci.* 46, 52 (1981).
62. D. N. Buchanan and J. G. Thoene, *J. Liq. Chromatogr.* 4, 1587 (1981).
63. D. N. Buchanan and J. G. Thoene, *Anal. Biochem.* 124, 108 (1982).
64. D. J. Woo and J. R. Benson, *Am. Clin. Prods. Rev.*, Jan. (1984).
65. D. B. Karr, J. K. Waters, and D. W. Emerich, *Appl. Environ. Microbiol.* 46, 1339 (1983).
66. E. Rajakyla, *J. Chromatogr.* 218, 695 (1981).
67. G. O. Guerrant, M. A. Lambert, and C. W. Moss, *J. Clin. Microbiol.* 16, 355 (1982).
68. A. Lebel and T. F. Yen, *Anal. Chem.* 56, 807 (1984).
69. Dionex Corporation, Sunnyvale, CA, Applications Note 21, June, 1979.
70. Dionex Corporation, Sunnyvale, CA, Application Note 25, September, 1980.
71. Wescan Instruments Inc., Santa Clara, CA, The Wescan Ion Analyzer. No. 4, Fall 1982.
72. Bio-Rad Laboratories, Richmond, CA, Bulletin 1159, 1984.
73. Dionex Corporation, Sunnyvale, CA, Application Note 49, June, 1983.
74. Dionex Corporation, Sunnyvale, CA, Application Note 50, June, 1983.
75. Bio-Rad Laboratories, Richmond, CA, HPLC Columns for Carbohydrate Analysis, 1984.

76. Bio-Rad Laboratories, Richmond, CA, Bulletin 1173, 1984.
77. Interaction Chemicals, Inc., Mountain View, CA, Ion Chromatography, March, 1985.
78. Bio-Rad Laboratories, Richmond, CA, HPLC Organic Acid and Base Analysis, January, 1985.
79. C. Pohlandt, Natl. Inst. Metallurgy, Randburg, S. Africa, Report No. 2107, 1981.
80. Wescan Instruments, Inc., Santa Clara, CA, The Wescan Ion Analyzer, No. 7, Winter 1984.
81. K. Tanaka, *Bunseki Kagaku 32*, 439 (1983).
82. Wescan Instruments, Inc., Santa Clara, CA, The Wescan Ion Analyzer, No. 5, Winter 1983.
83. J. Stillian, Trace analysis via post column chemistry and IC: Silica and ppb Ca and Mg in brines, presented at the Pittsburgh Conference, 1984.
84. M. R. Zarifa, Doctoral dissertation, American University, Washington, DC, 1984.

6
Approaches to Ionic Chromatography

Purnendu K. Dasgupta
Department of Chemistry and Biochemistry
Texas Tech University
Lubbock, Texas

I. INTRODUCTION

The 1975 report by Small, Stevens, and Bauman [1] of a facile, sensitive, conductimetric approach to chromatographic analysis of ions through the combination of a novel, ion exchanger stationary phase and postcolumn conversion of the eluant counterion was a major milestone in ionic analysis and analytical chemistry itself, rivaling in importance the development of atomic absorption spectroscopy [2]. In more than one sense, the paper of Small et al. is a landmark, not only because of the novelty it had introduced, but particularly because of the resurgence of interest it created in the general area of chromatographic analysis of ions. It is to this latter end that the present chapter is devoted, to acquaint the reader with the variety of approaches available to the enterprising chromatographer for sensitive analysis of ions, other than the current "mainstream" activities in ion chromatography (IC) delineated in the foregoing chapters. Perhaps, the sole exception is the first section, on membrane-based suppression systems, which, while being widely applied in mainstream IC, is vital to the practice of suppressed ion interaction chromatography with conductimetric detection. One preamble is necessary, which is that the practice of postcolumn chemical conversion of the eluant counterion to reduce the background conductivity of the eluant, widely referred to as *eluant suppression*, is protected by patents in most countries [3]. To some researchers, that is a productive state of affairs, because the incentive to develop equally or perhaps even more powerful techniques for chromatographic analysis of ions remains undiminished. New ideas with old tools, old ideas with powerful new tools, and above all, new ideas with new tools are rapidly changing the evolving facets of the modern practice of IC; it is an exciting time to be an ion chromatographer.

Before the contents of this chapter are discussed, its title deserves comment. Small has commented on the origin of the term *ion chromatography*. The title of the original paper [1] left little room for ambiguities. (For the reader who is hampered by the intervening decade, the title read: "Novel Ion Exchange Chromatographic Method Using Conductimetric Detection.") Small and Solc [4] stated that although a variety of names, including *conductimetric chromatography* and *eluant suppression chromatography*, were experimented with, the Dow workers eventually settled on the term *ion chromatography*. Considering that we operate under a trade name registration system which grants a certain manufacturer of filter media the sole rights to the use of the numerals "41," the use of the name "Ion Chromatograph" as a registered trademark for the first commercial instrument that embodied the invention in Ref. 1 can hardly be begrudged. That the work of Small et al. was an important impetus to explore alternatives to accomplish the same ends, namely chromatographic analysis of ions, is clearly revealed by the continued use of the term *ion chromatography* as the single-column methods were developed and refined. When such chromatography is conducted, as is most commonly the case, on ion exchange stationary phases, there is no difference whatsoever, as the authors of the 1984 biennial review on column liquid chromatography in *Analytical Chemistry* [5] have observed, with the practice of ion exchange chromatography. Consequently, the net result is confusion. To make matters worse as regards nomenclature, the use of ion interaction chromatography with nonpolar reverse phase supports is steadily gaining acceptance for ionic analysis by chromatography. This is certainly *not* ion exchange chromatography; whether it is also *ion chromatography*, as designated by some authors, depends on one's consideration of a term that has heretofore remained undefined in a strict sense, and also on the recognition that "ion chromatography" may not necessarily be carried out on a "Ion Chromatograph." Molnar et al. pursued and compared all three principal avenues of ionic analysis by chromatography--namely ion exchange chromatography with and without eluant suppression, and also ion interaction chromatography. All the techniques were aptly covered by the generic name *high performance liquid chromatography of ions*. The same course has been taken by Haddad and Heckenberg in a recent review [7], although other equally enlightened authors continue to use the term *ion chromatography* to cover all the approaches [8]. Because an "Ionic Chromatograph" is not immediately apparent on the horizon, the present author feels more secure in referring to the chromatographic analysis of ions as *ionic chromatography*, regardless of the mode of separation or detection. The ionic species that are considered here are mostly inorganic, along with some organic ions of relatively small size. Although many large organic or biomolecules, including proteins, are ionic or ionizable, their chromatographic separation and analysis are subject to very different conditions and considerations and are therefore not included.

Below, the organization of this chapter is briefly outlined. Basic theory and operation of suppressed and nonsuppressed IC have already been discussed in foregoing chapters. In Section II, the theory and operation of membrane-based suppressors and postsuppressors are discussed, beginning with considerations on ion transport through an ion exchange membrane, continuing with considerations on mass transport to and from the membrane, and overall exchange efficiency. Designs of actual membrane-based suppressors next follow, continuing with a discussion of preparation and properties of ion exchange membrane tubes and the choice of suitable regenerants. Electrodialytic suppression is briefly discussed, followed by an account of membrane-based postsuppressors used to remove CO_2.

Section III deals with indirect detection methods; conductimetric and other electroanalytical methods as well as optical detection methods are considered. Section IV is devoted to the use of ion interaction reagents (IIRs) in IC. The dynamic ion exchange model, the ion-pair and the ion interaction models are considered next, in that order. Although chromatography with immobilized IIR systems is more akin to chromatography on conventional ion exchangers, a brief account of this is included in this section. A more detailed account of dynamic ion interaction chromatography follows, after which is presented a discussion of the use of chelating ligands forming metal complexes that are easily electrochemically or optically detected.

Section V is concerned with the chromatographic analysis of metal species. Although organometallic compounds are frequently nonionic, several examples of the chromatographic analysis of organometallics are included here because their analysis is most often considered along with other ionic metallic species. Applications of conductimetric, electrochemical, atomic spectrometric, UV/VIS, and other detection methods are described. Considerations are outlined on the operation and care of instrumentation when chelating ligands are used in the eluant. Separations on ion exchangers, normal phases, reverse phase, and other bonded phases and size exclusion-type stationary phases are described. Less common stationary phases such as chelating supports including those containing crown ethers, other complexing moieties, and unusual ion exchangers for chromatography and preconcentration are described.

Section VI describes the three principal routes to better detectability: trace enrichment, postcolumn reaction, and gradient elution. In the first subsection, extractive and dialytic preconcentration are briefly considered, with principal emphasis on enrichment on stationary phases. For the latter, plumbing configurations and applications are described. The discussion of postcolumn reaction (PCR) systems begins with those PCRs that do not involve reagent addition, and continues with the more conventional types that do. Reaction requirements and residence times, which are important factors in PCR and mixer designs, are next considered, followed by an account of various types of membrane-based PCRs, much of it based on the author's own experience. The subsection concludes with applications of PCR schemes, and the section ends with an account of the limited number of studies conducted to date on gradient elution in conductimetric IC.

Section VII, the concluding section in this chapter, is merely an effort of the author to gaze into his cloudy crystal ball, to prepare, so to speak, the banners for the "coming attractions."

II. MEMBRANE-BASED ELUANT COUNTERION SUPPRESSION SYSTEMS

The use of a packed suppressor column in IC, as originally reported by Small et al. [1], displays several drawbacks. Besides the need for frequent regeneration, large band broadening occurs with packed column suppressors that permit operation with eluants of usual ionic strength for an appreciable period of time, thus limiting attainable resolution of early-eluting ions. Variable retention of easily protolyzed ions is also observed, through retention of the protolyzed molecular form (e.g., retention of acetate as acetic acid; see Chapter 5 on ion chromatography exclusion), resulting in difficulties in both identification and quantitation. In some cases, such as with nitrite, other difficulties, due to undesirable side reactions with the suppressor packing, have also been reported [9]. Trace analysis is also hampered because of the so-called water and carbonate dips, which appear in the chromatogram and migrate as a function of suppressor column exhaustion. The accurate quantitation of analyte ions that elute in this region of the chromatogram therefore becomes difficult at low levels.

Ion-exchange hollow fibers for continuous ion exchange were invented by Rembaum et al. [10] in work sponsored by the National Aeronautics and Space Administration. These fibers were prepared by introducing a polymerizable liquid monomer into the walls of a porous fiber and then polymerizing the monomer within the walls of the fiber to form solid, insoluble, crosslinked, ion exchange resin polymers. Such fibers were shown to have high ion-exchange capacities, practical wall permeabilities, and good mechanical strength even with very thin wall dimensions. The utilization of hollow cation-exchange fibers as suppressors in anion chromatography was introduced in 1981 by Stevens, Davis, and Small [11]. Figure 1 is a scheme of the first reported membrane-based suppressor. Figure 2 illustrates the actual device. The principle of operation for such a suppressor with a sodium carbonate eluant and a dilute sulfuric acid regenerant is shown in Figure 3. The cation exchange fiber bears negatively charged sulfonate groups and permits the passage of oppositely charged ions (cations) while retarding the passage of similarly charged ions (anions). When, on the outer side of the fiber, an ample supply of a proton-containing solution is maintained, that flows continuously and carries the exchanged permeated sodium ion out to waste, all the influent sodium ion is exchanged for hydrogen ion, if the length of such a membrane tube is sufficient. According to the Gormley-Kennedy equation [12], which theoretically embodies the rate of transport to the walls of a cylindrical tube, from a fluid flowing through the tube, under fully developed laminar flow conditions:

$$f = 1 - 0.8191 \exp\left(\frac{-3.657\pi DL}{Q}\right) - 0.0975 \exp\left(\frac{-22.3\pi DL}{Q}\right)$$

$$- 0.0325 \exp\left(\frac{-57\pi DL}{Q}\right) - \cdots \qquad (1a)$$

or, for very small values of $\pi DL/Q$:

$$f = 4.07\left(\frac{\pi DL}{Q}\right)^{2/3} + \frac{2.4\pi DL}{Q} + 0.446\left(\frac{\pi DL}{Q}\right)^{4/3} + \cdots \qquad (1b)$$

where f is the fraction of the influent concentration transported to the wall in a tube of length L with a volumetric fluid flow rate of Q, and D is the diffusion coefficient of the species of interest (e.g., Na^+ in the above example). For most cases, Eqn. (1a) with only the first two terms shown on the right-hand side of the equation is an adequate approximation of the infinite series. The important aspect of the Gormley-Kennedy equation is that the transport rate to the wall is independent of the diameter of the tube, whereas band dispersion for a Gaussian band of effluent analyte is acutely dependent on tube diameter [13]. Because one of the desirable characteristics of any postcolumn system is minimum band dispersion, the tube should ideally be of very narrow diameter. Thus, the simplest and likely ideal design of such a suppressor is a single length of very narrow-bore ion-exchange membrane tube, with provisions for flowing an appropriate regenerant solution outside the membrane tube. Availability of membrane tubes, and structural strength (burst pressure limit) considerations, as well as physical design of the device often require compromises, as shown in the multiple-strand design originally employed by Stevens et al.

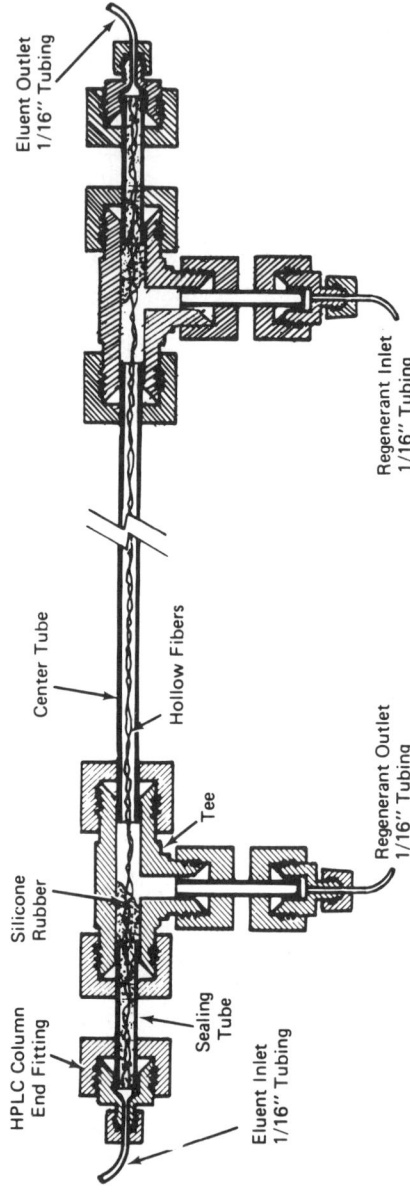

Figure 1. Schematic drawing of the first reported hollow-fiber suppressor. (From Ref. 11.)

Approaches to Ionic Chromatography

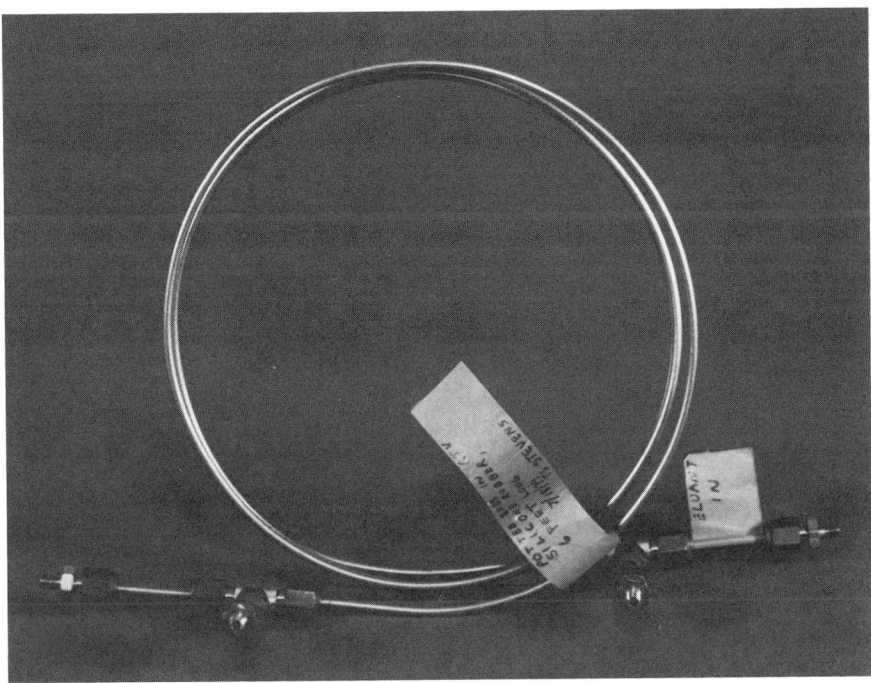

Figure 2. Actual prototype of the multiple-fiber hollow-fiber suppressor shown in Figure 1. (Courtesy of T. S. Stevens.)

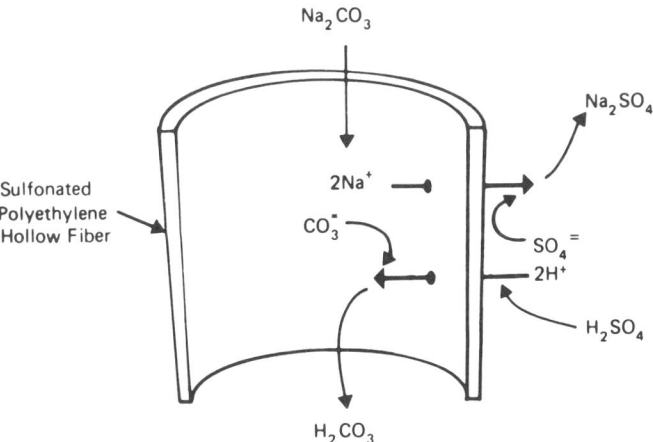

Figure 3. Principle of operation of the hollow-fiber suppressor. (From Ref. 11.)

Largely because of the unavailability of very narrow-bore membrane tubing of sufficient structural strength, alternative designs have been developed for membrane-based suppressors, although at least two commercially available membrane suppressors are of the hollow-fiber type [14]. When one is choosing appropriate membrane material for constructing such a suppressor, the permeability characteristics of the membrane must be considered. Aside from the rate of mass transport to the wall, which is determined by the nature of the hydrodynamics of the velocity field of the fluid flowing through the tube, the overall transport rate may well be limited by the rate of transport through the wall, which is determined by the intrinsic permeability of the membrane and its thickness. For the typical case of a solution with an ionic strength of 1-100 mM, the ion exchange rate over the initial part of the tube is limited by the wall transport rate, i.e., the ionic flux to the membrane is greater than the transport capacity of the membrane. This situation continues until the concentration of the ion to be exchanged falls to a point where the flux brought to the wall is equal to the ability of the wall to transport the material. Past this point, the exchange rate becomes limited by the mass transport rate to the wall. To improve wall transport rates, the membrane should be thin walled, the limits being dictated by structural strength considerations. Narrower-bore tubes are desirable from this respect also, because the burst pressure rating increases with decreasing tube internal diameter for otherwise identical membrane thickness. However, too thin a membrane may result in excessive penetration of the undesired regenerant counterion into the inner flowstream, as will be considered in Section II.G (Choice of a Suitable Regenerant). The use of a membrane suppressor is not limited to the analysis of inorganic ions; applications for the analysis of organic ions have been described [15].

A. Theoretical Considerations on Mass Transport Through the Membrane

If mass transport through the wall is the rate-limiting factor in the observed ion-exchange rate, the system characteristics are simple to describe theoretically. The affinity of an ion for an ion exchanger is described by its selectivity coefficient S, which is defined most commonly with respect to the affinity of H^+ or OH^- for the ion exchanger (for cation and anion exchangers, respectively). For example, the selectivity coefficient of sodium ion for a cation exchanger, denoted as S_{Na^+}, is defined as:

$$S_{Na^+} = \frac{X_{Na^+} [H^+]}{X_{H^+} [Na^+]} \qquad (2)$$

where X_{Na^+} and X_{H^+} represent the fraction of ion exchange sites occupied in the ion exchanger by Na^+ and H^+, respectively, and $[Na^+]$ and $[H^+]$ represent, respectively, the molarity of Na^+ and H^+ in the solution with which the ion exchanger is in equilibrium. S is not, in a rigorous sense, a constant (S_i increases at very low values of X_i) but will for the sake of simplicity be regarded as a constant in this treatment. Recognizing that the concentration of ion exchange sites is fixed for a given membrane material, Eqn. (2) may be rewritten as:

$$S_{Na^+} = \frac{M_{Na^+}[H^+]}{[(\rho/EW) - M_{Na^+}][Na^+]} \tag{3}$$

where M_{Na^+} is the concentration of Na^+ in the exchanger phase in moles per centimeter cubed, EW is the equivalent weight of the exchanger, and ρ its density; and ρ/EW represents the total exchange site concentration in moles per centimeter. Rearrangement of Eqn. (3) yields a more convenient form:

$$M_{Na^+} = \frac{S_{Na^+}(\rho/EW)[Na^+]}{[H^+] + S_{Na^+}[Na^+]} \tag{4}$$

Consider now an infinitesimally small length of the membrane tube of dL cm, such that by passage through this section, dC molar Na^+ is exchanged at the wall for H^+. We assume C >> dC, so that the entire length dL of the segment may be assumed to be in equilibrium with C molar Na^+. For simplicity, it is further assumed that the concentration of Na^+ outside the membrane (and therefore at the outer surface of the membrane) is essentially zero. If the thickness of the membrane, t, is small, the transport flux of Na^+, Q, through the membrane is given by Fick's first law [16]:

$$Q = M_{Na^+} \frac{D_m}{t} \tag{5}$$

where Q is expressed in moles per square-centimeter per second, t is expressed in centimeters, and D_m is the diffusion coefficient of Na^+ in the membrane in square-centimeters per second. Values of D_m are available for only a few membrane matrices and a few ions. Yeager [17] reports a value of 9.44×10^{-7} cm^2/sec for Na^+ for the perfluorosulfonate cation exchanger membrane, Nafion [18]. For relatively permeable membranes such as Nafion, values of D_m for various ions are likely to range from 10^{-8} to 10^{-6} cm^2/sec, depending on the size of the ion. Apparently the membrane matrix, the density of ion exchange sites (related to EW), and the resulting hydrophilicity all affect the diffusivity of the permitted ion and cannot generally be predicted a priori.

The flux through the membrane for the segment under consideration is given by:

$$Q' = 2\pi r M_{Na^+} \frac{D_m dL}{t} \quad (6)$$

where $2\pi rdL$ is the available membrane surface area of the membrane through which the flux occurs; and r may be taken as the average of the inner and the outer diameter of the membrane (logarithmic mean of o.d. and i.d. is preferred). At steady state, with a solution flow rate of F cm^3/sec, the decrease in concentration of Na$^+$ by ion exchange at the wall, $-dC$, is:

$$-dC = 1000 \frac{Q'}{F} \quad (7)$$

and the effluent concentration C' is given by:

$$C' = C - dC \quad (8)$$

Using a new value of C equal to C', and iterating calculations through Eqns. (4) through (8), we can evaluate the unexchanged concentration after any number of such segments. However, the concomitant change in [H$^+$] needs to be taken into account. If the influent solution is NaX, where HX is an acid with dissociation constant, K, then from charge-balance requirements, at any point in the system we have:

$$[Na^+] + [H^+] = [X^-] + [OH^-] \quad (9)$$

Upon substitution of [Na$^+$] = C, where C is the current unexchanged concentration of Na$^+$, and with the recognition that:

$$[X^-] = C_0 \frac{K}{K + [H^+]} \quad (10)$$

where C_0 is the original concentration of Na$^+$, Eqn. (9) yields:

$$[H^+] + (C + K)[H^+]^2 + (KC - K_w - KC_0)[H^+] = KK_w = 0 \quad (11)$$

Similar equations may be derived for salts of multiprotic acids, e.g., Na$_2$CO$_3$ or any mixtures thereof (e.g., Na$_2$CO$_3$ + NaHCO$_3$). Equation

Approaches to Ionic Chromatography

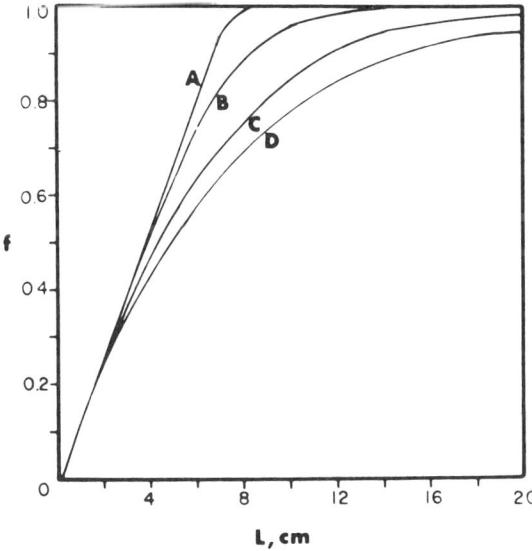

Figure 4. Effect of the dissociation constant of HX on the wall transport-limited ion exchange of NaX. (A) pK = 8, (B) pK = 5, (C) pK = 2, (D) strong acid. Influent concentration, 0.01 M; flow rate, 2 ml/min. (From Ref. 31.)

(11) may be solved by any method of solving for the real root(s) of a polynomial, including the Raphson-Newton procedure [19]. By solving Eqn. (11) at each iteration step and feeding back this value in Eqn. (4) for the subsequent cycles, the numerical solution can be obtained. As an example, let us calculate for the case of a perfluorosulfonate cation exchanger membrane tubing, Nafion 811x. The dimensions are as follows: radius, 350 μm (.035 cm); thickness, 75 μm (.0075 cm). The density ρ is 1.98 g/cm^3, and the equivalent weight of this membrane material is 1100. The selectivity coefficient S_{Na^+} for Nafion membrane has been measured [20] to be 1.22. The numerical solution for the case above shows a stable value at iteration steps of dL = 0.01 cm, and thus no further reduction in dL is necessary. The results are shown in Figures 4-6 for various conditions. Figure 4 shows the effect of the dissociation constant of HX on the wall transport-limited exchange of NaX (the fraction exchanged is designated on the ordinate as f) with 10 mM influent NaX, flow rate 2 ml/min. Note that although only 8 cm of tubing is necessary for complete exchange of Na$^+$ for H$^+$ when pK of HX is 8, the solution is far from completely exchanged, even with 20 cm of tubing, when HX is a strong acid. (In reality, much longer lengths of tubing will be

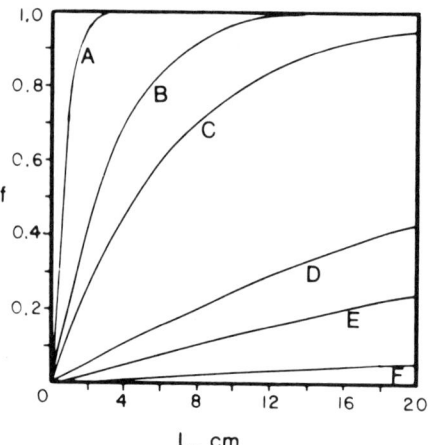

Figure 5. Effect of the influent concentration on the wall transport-limited ion exchange of NaX (HX is strong acid): Curves A-F: 0.001, 0.005, 0.01, 0.05, 0.1, and 0.5 M influent NaX. Flow rate, 2 ml/min. (From Ref. 31.)

required to achieve complete exchange in both cases of simple hollow tubing because mass transfer to, and not through, the wall will become rate limiting.) When HX is a strong acid, the solution pH drops much more abruptly with the ion exchange process, and the resulting $[H^+]$ competes with the Na^+ for the ion exchange sites (Eqn. 4); also M_{Na^+} decreases, and sodium transport rate is inhibited as compared to the weak acid case. This dependence on pK of the acid is particularly fortunate, because in anion chromatography, alkaline eluants, typically salts of weak acids, are invariably used. The dependence of the exchange rate on pK decreases as the selectivity coefficient increases (e.g., for Cs^+ as compared to Na^+), but typically ions that have large affinities for the ion exchanger (large selectivity coefficients) also display smaller diffusion coefficients, more than offsetting the benefit of a larger selectivity coefficient. Note that the variation of pH with the fraction exchanged, f (the ordinate in Figure 4), is exactly the same as that during the titration of NaX with a strong acid. Consider also that if HX is a strong acid, Eqn. (11) simplifies to:

$$[H^+] = C_0 - C \tag{12}$$

and an analytical solution is possible. Equations (4) through (7) combine to give:

Approaches to Ionic Chromatography

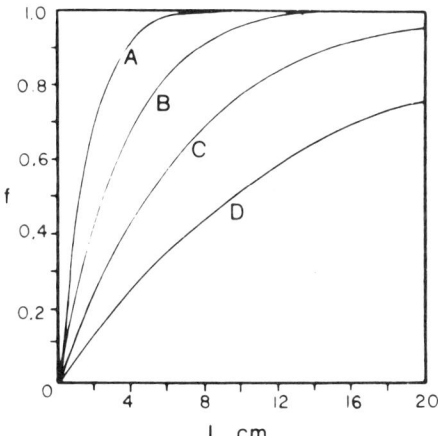

Figure 6. Effect of flow rate on the wall transport-limited ion exchange of NaX (HX is strong acid): Curves A-D: 0.5, 1.0, 2.0, and 4.0 ml/min. Influent concentration was 0.01 M. (From Ref. 31.)

$$-dC = \frac{1}{[H^+] + SC} C\theta dL \qquad (13)$$

where

$$\theta = 2000 \frac{\pi r D_m S \rho}{FtEW} \qquad (14)$$

Substitution of Eqn. (12) into Eqn. (13) and integration yields:

$$C = C_0 \exp -\left(\frac{S-1}{f} + \frac{\theta L}{C}\right) \qquad (15)$$

Figure 5 shows the effect of the influent concentration on the wall transport-limited rate of ion exchange (HX is assumed to be a strong acid) for 5-500 mM NaX at a flow rate of 2 ml/min. Clearly, at high concentrations, the wall transport rate is likely to be the limiting factor.

Figure 6 shows the effect of varying the flow rate of the influent solution (HX is assumed to be a strong acid) upon the wall transport-limited ion exchange rate of 10 mM influent NaX. Note that variation of D_m, t, r, or EW leads to similar results, provided of course that these parameters are independent of each other.

Considerations on an anion exchange membrane tube, which functions as a suppressor for cation chromatography, are essentially identical. Other models for permitted ion transport through an ion exchange membrane are available in the literature [705], but generally are more complex.

B. Mass Transfer to the Wall

For dilute solutions, which are used as eluants in most IC applications, mass transfer to the wall is likely to be a very important factor. The Gormley-Kennedy equation (Eqn. 1) describes idealized mass transport to the walls of a hollow linear tube. For small-diameter hollow linear membrane tubing with typical solvent delivery systems, the mass transport to the wall is generally observed to be somewhat more efficient than predicted by the Gormley-Kennedy equation. The requirement of a finite length to attain fully developed laminar flow, turbulence at inlet/outlet, pump pulsations, and turbulence induced by the surface roughness of the membrane--all contribute to the enhanced transport rates. This rate, however, is not especially efficient, and there are a number of possible ways to improve the mass transfer characteristics to the wall, as compared to a hollow linear tube. One relatively simple solution for small-bore tubing is to physically coil the membrane tubing into a small-diameter helix. The helical structure leads to the development of a secondary flow perpendicular to the axial flow [21-26]. This secondary flow flattens the parabolic velocity profile observed in a linear tube and leads to reduced axial dispersion and increased radial mass transport to the walls of the tube. Following the pioneering work of Dean [21,22], the most important parameter in helical flow, $Re(d_t/d_c)^{1/2}$, is designated the *Dean number* (Re is the *Reynold's number*, d_t and d_c are the diameters of the tube and the coil, respectively). The differences between linear and helical flow increase with increasing Dean number. The rigorous analytical solution for the hydrodynamics of helical flow is quite complex. Although numerical methods have recently become available [27,28], certain approximations are necessary, and such methods apply to a limited, albeit useful, range of Dean numbers. In general, helical flow will lead to increased mass transport to the wall, and the effect will be more pronounced at high fluid velocities (small-bore tube) and with decreasing radius and pitch (spacing between turns) of the coil.

Mass transport to the wall may also be enhanced by interrupting laminar flow; any consequent increase in band dispersion must be carefully minimized, however. One configuration that reportedly accomplishes this is knitted or woven tubing [29]. Although details are not available in the literature at this time, "braided" polytetrafluoroethylene (PTFE) tubing has been described [706].

C. Mass Transport from the Wall

Adequate flow hydrodynamics and concentration of regenerant solution is necessary to prevent the mass transport from the outer wall of the membrane from becoming a limiting factor. The maximum permissible degree of regenerant penetration, which must be decided for a specific application, sets an upper limit on the regenerant concentration for any given regenerant. Other than the convenience of gravity flow and practicality of maintaining the flow rate, there are no restrictions on increasing the flow rate. The regenerant solution may also be cycled through a large-bed exchanger. The flow geometry in the jacket carrying the regenerant is obviously important in determining how efficiently the regenerant flow is utilized. When the influent ionic concentration and the membrane permeability are high, mass transport from the outer wall may become a significant factor. Results for the overall exchange efficiency observed with 3 mM $NaHCO_3$ + 2.4 mM Na_2CO_3 as eluant as a function of regenerant concentration is presented in Table 1. These data represent an interplay between two opposing factors on the effluent conductance: (a) regenerant counterion penetration and (b) inefficiency in mass transport at the outer wall. Thus, the data in the second column, where

Table 1. Effect or Regenerant Concentration on Ion Exchange Efficiency[a]

Regenerant (H_2SO_4) concentration (mM)	Conductance (μS) at the indicated flow rate (ml/min)			
	0.5	1.0	2.0	4.0
12.5	0.62	1.77	7.72	8.25
25	0.70	0.88	3.05	6.45
50	0.99	0.73	2.30	4.85
100	2.80	1.07	2.39	4.83

[a]Influent solution, 2.4 mM Na_2CO_3 + 3 mM $NaHCO_3$; Nafion 811x tube length, 50 cm; filament diameter, 0.66 mm; coil diameter, 4 mm; regenerant flow rate, 5 ml/min; direct conductance of this solution, 20.8 μS; completely exchanged (packed column) conductance, 0.690 μS.
Source: Ref. 31.

the mass transport efficiency is not limiting, show increased conductance with increasing regenerant concentration; this is due to regenerant penetration. In contrast, the data in the last column show the opposite trend; here mass transfer efficiency is the limiting factor. Note also that for carbonate-containing eluants, quantitative cation exchange through membrane-based suppressors often leads to an effluent conductance significantly lower than that obtained with a packed-bed exchanger. This is due to facile loss of H_2CO_3 through the membrane. The rate of loss increases with increasing distance from the inlet, as the amount of free H_2CO_3 increases, and no simple method can be used to calculate the fraction exchanged from effluent conductance data.

For obvious reasons, regenerant flow is usually maintained countercurrent to the flow direction of the solution of interest. Westbrook [30] has outlined the necessary principles involved in membrane-based countercurrent mass transfer.

D. Overall Exchange Efficiency

The rate of mass transport from the wall need not be a limiting factor in the overall exchange efficiency through proper choice of regenerant, its concentration, and rate of flow; effects due to this problem may be obviated in a membrane suppressor that is properly designed and operated. In general, overall exchange efficiency will be limited by mass transport through the wall for the initial part of the device, and then, as the flux brought to the wall falls below the transport capacity of the wall, will be limited by the rate of mass transport to the wall. Although the hydrodynamics, which governs transport to the wall, is too complex for all but the hollow linear configuration to be accurately described mathematically at the present time, it is reasonable to assume that the same fraction, k, of the influent flux, is capable of reaching the wall during passage through a segment of length dL. The calculation for fraction ion exchanged, f, as performed in Section II.A then needs to be modified for the complete device only in that:

$$-dC = kCdL \qquad (16)$$

At each iteration step, either Eqn. (7) or Eqn. (16) is used, depending on which provides a lower (limiting) value for -dC. The net effect of decreasing k for a given set of wall transport parameters is then manifested in divergence from wall transport-limited exchange curves (as in Figures 4-6) at lower and lower values of L. Note that k depends on the assumed value of dL, the highest permissible (but not attainable, since this would indicate that all the flux entering the segment sees the wall) value being 1/dL.

E. Actual Membrane-Based Suppressors

Filament-Filled Helical Membrane Suppressors

In putting the helical design of a membrane tube to practice, several problems are encountered. By and large, thin-walled ion-exchange membrane tubes are not rigid enough to hold a helical structure. If it is coiled tightly on a support rod and the support rod is left in place, the surface of the membrane in contact with the support is inhibited with respect to transport of the permeated ion away from the outer surface of the membrane, and this surface thus becomes essentially unavailable.

The insertion of a filament inside the membrane tube is an effective solution to this problem [31,32]. Any suitable inert filament of appropriate diameter, that will hold the helical shape once coiled, is utilizable. The author's laboratory has made extensive use of nichrome wires and monofilament nylon lines for this purpose. Besides supporting the helical shape, the inserted filament fills up a significant portion of the void volume of the tube and thus directly reduces residence time and band dispersion. A secondary reduction in band dispersion is observed because the annular space through which the liquid must now flow, represents a much smaller effective hydraulic diameter; this then results in a greater fluid flow velocity at the same volumetric flow rate, and thus enhances the centrifugal transport when the membrane is coiled. The flow velocity of a fluid flowing at 1 ml/min through an annular gap consisting of a 0.66-mm-diameter filament separated from the membrane by 0.04 mm is of the order of 20 cm/sec, nearly an order of magnitude higher than what would be observed for the same membrane tube in the absence of the filament.

A question commonly encountered by the author regarding filament-filled membrane helices is, what happens to the surface of the membrane in contact with the filament? The question itself is based on an erroneous assumption. The flowing fluid essentially inflates the membrane around the filament to create its own passageway; there is no gap without any fluid flow. Consequently, the fluid pressure renders the filament self-centering, i.e., it does not touch the wall!

The deterioration of chromatographic performance caused by a given amount of extracolumn band dispersion is determined primarily by the analytical column efficiency. The columns initially used in IC [1] were relatively inefficient. However, the development of more efficient columns by Stevens and Langhorst [33] have improved the current practice and, in the process, have made the problem of band dispersion in the suppressor relatively more important. The original hollow-fiber suppressor, because of the relatively large band dispersion it introduced, could not take appropriate advantage of these more efficient columns. If hollow membrane tubes are used as suppressors, the internal diameter should not exceed 400 µm. Commercially available membrane tubes such as Nafion have minimum diameters nearly twice this value (Nafion 811x, the smallest-bore Nafion tube currently available, is approximately 700 µm in internal diameter), and some means must be provided by which the band dispersion in tubes of such diameter is reduced. For the filament-filled helical configuration, the annular gap between the filament and the membrane can be reduced to a point where the band dispersion is very small and high-efficiency columns can be used without deleterious effects on the early-eluting peaks. With Nafion 811x cation exchanger membrane tubes and monofilament nylon fishing lines, filaments that actually exceed the dry i.d. of the membrane tube have been successfully inserted inside the membrane tube. Special techniques are obviously necessary to achieve this; the approach succeeds because the burst pressure of this tubing is sufficiently high ($\geqslant 400$ lb/in.^{2}) to tolerate flow at appreciable rates through the narrow annular gap. Nafion 811x, in the dry state, exhibits an internal diameter of 625 µm with a wall thickness of 75 µm. Like any other ion exchange membrane tubing, it is hydrophilic and expands somewhat upon wetting.

In the following, methods to make leakproof connections to membrane tubing, capable of withstanding several hundred pounds per square inch, are first discussed, followed by filament insertion procedures. The swell-seal method [34,35] takes advantage of the hydrophilic expansion of ionomeric membrane tubes to achieve a good seal. A small hole, just large enough or even slightly smaller than the dry outer diameter (o.d.) of the membrane tube is drilled through a plastic fitting such as a male nut provided with $1/4 \times 28$ threads (plugs used as caps for $1/4 \times 28$-threaded column inlets are suitable). With the help of a fine-wire loop, the membrane tube is pulled through the hole in the fitting. Upon wetting, the tubing expands outward and provides a good seal. A second approach, which involves making a flange and sealing with a washer and male nut of appropriate size, is common practice with polytetrafluoroethylene (PTFE) tubing. The same technique may be applied with some ionomeric membrane tubes also, provided that washers and connecting fittings with sufficiently small holes are used. Typical commercial fittings and washers supplied with them have bores that are unacceptably large. To produce a flange, an open end of the membrane tube is brought close to a small flame, whereupon it softens, rolls back on itself, and produces a ringlike flange. A third and most generally applicable technique involves the insertion of a small segment (~ 1 cm) of microbore PTFE tubing into the membrane tubing. The membrane tubing is then inserted into a connecting PTFE tubing (typically 1.5 mm o.d.; the tubing inner bore is enlarged at the end with a suitable tool to

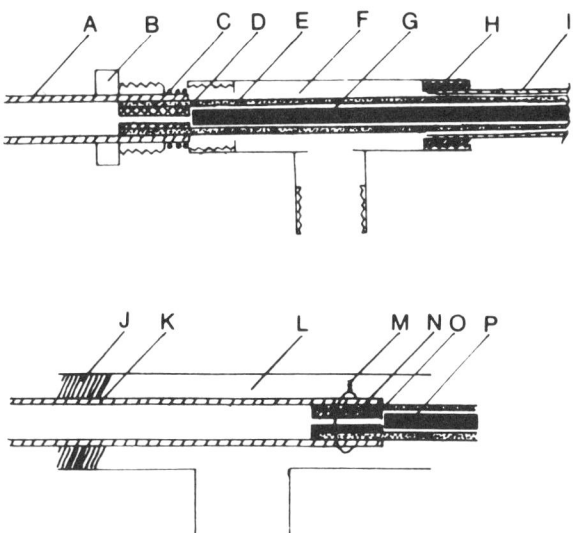

Figures 7, 8. Inlet/outlet designs for the annular helical suppressor: (A/K) inlet/outlet PTFE tube; (B) male nut; (C) O rings; (D, O) microbore PTFE insert; (E, N) Nafion membrane tubing; (F) $1/4 \times 28$ -threaded tee; (L) polypropylene tee; (G, P) nylon monofilament; (M) nichrome wire crimp; (J, H) adhesive filler; (I) jacket tubing. (From Ref. 31.)

accommodate the membrane tubing) to no further depth than the microbore insert. Sealing is then accomplished by compression with O rings (as shown in Figure 7) or by crimping with 28-gauge nichrome wire (as shown in Figure 8).

There are several available techniques for inserting filaments in to membrane tubing. With Nafion 811x membrane tubes as an example, for relatively small lengths ($\leqslant 50$ cm) of membrane tube, a smaller filament (e.g., 30-lb-strength fishing line; STREN, Du Pont; diameter, 0.56 mm) may first be inserted relatively easily while the tubing is thoroughly wet. The larger-diameter filament (0.66-mm diameter, 40-lb-strength fishing line) is next spliced on to the end of the narrower filament with cyanoacrylate adhesive, and the joint is polished with fine-grade abrasive paper to permit facile passage. The larger filament is then pulled through. For longer lengths of membrane tubing, this method is not suitable. After insertion of the microbore PTFE tubing and establishment of a leak-free connection at one end of the membrane tubing, this end is connected to a pump, and water is pumped through the membrane tube. The filament is pushed from the open end of the membrane tube inward. A flow rate of 1 ml/min produces a pressure drop of about 200 lb/in.^{2}, which adequately expands the tubing for smooth passage of the filament. The flowing fluid also provides lubrication, which aids the insertion. When the filament is all the way in and butting the microbore insert, the membrane tubing and filament is gripped from the insertion end, and the pressure is relieved by discontinuing flow. Excess filament is then cut off flush with the end of the membrane tube. Gentle stretching of the membrane tube then produces enough void space at the insertion end for inserting a small section of microbore PTFE tubing for connections, as at the other end. With some types of membrane tubes such as Nafion, a substantially simpler alternative to the above approaches is feasible, which is based on the capability of membranes such as Nafion to absorb very large quantities of polar organic solvents such as the lower alcohols. When immersed for a few minutes in methanol (or better, boiled for 15 min in 95% ethanol), Nafion tubes swell and expand substantially, significantly more than that achieved by wetting with water. With the membrane thoroughly wet with organic solvent, a filament may be inserted. When the solvent is allowed to evaporate off, the membrane shrinks back, producing a very narrow annular gap. Annular gaps as small as 40 μm can be obtained with Nafion 811x. An annular helical membrane suppressor is illustrated in Figure 9.

The Packed-Bead Membrane Suppressor

A different approach to reduce band broadening in tubular flow was pioneered by Poppe et al. [36-39] for other applications and predates the filament-filled helix approach. This approach involves filling the tube with inert beads having diameters somewhat larger than the inner radius of the tube so that two beads cannot occupy parallel lateral positions. Consequently, a zigzag pattern is created, as shown in Figure 10. This has been referred to as the *pearl string reactor* or the *single bead-string reactor* (SBSR). Such a configuration reduces the holdup volume in the tube and creates significant turbulence in the flow path, leading to improved mass transfer to the walls of the tube. This concept was first utilized in IC by Stevens et al. [34] in 1982, toward the fabrication of a bead-packed, low-dispersion, membrane-based IC suppressor. Poly(styrene-divinylbenzene) copolymer beads of 500 μm diameter are suspended in water (containing 1% Brij-35, a nonionic surfactant). One end of a 150-cm-long Nafion 811x tubing is pinched partially (in the actual device, the tube passes through a slot cut in a Plexiglass support rod, and the slot provides the necessary pinching) such that it acts as a bed support to prevent the escape of beads. The other end is then immersed in the bead suspension, which is sucked into the tube by applying

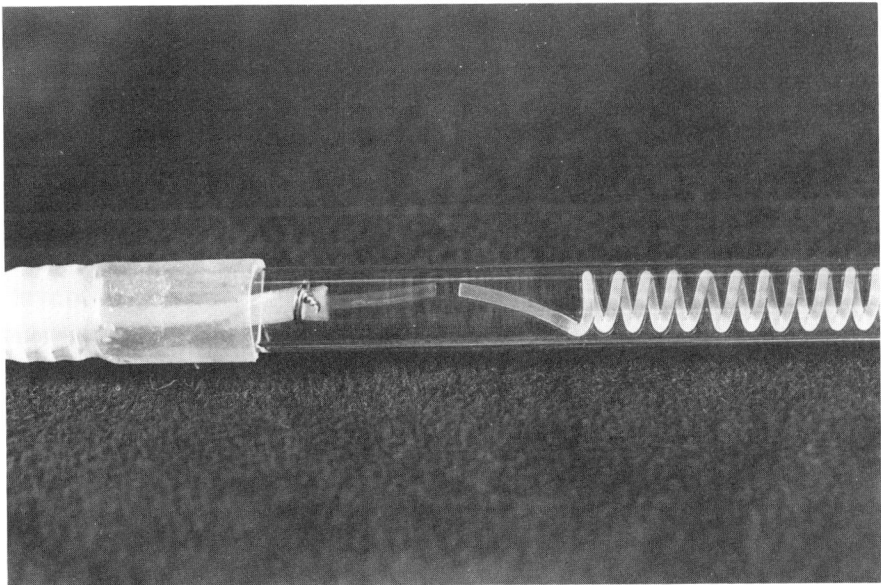

Figure 9. Filament-filled annular helical suppressor. A gap between the microbore inlet tubing and filament has been intentionally left for visual clarity.

vacuum. Any jams formed partway down the tube may be cleared by pinching the tube. The bead-filled tube is then wrapped around a support cylinder. The support cylinder is approximately 2.5 cm in diameter, too large for any significant contribution of centrifugal flow to further aid in mass transport, assuming that centrifugal flow can occur in a packed-bead geometry. The design of the complete device is shown in Figure 11. This particular device was constructed of Plexiglass parts. The first commercial membrane suppressor, marketed by Dionex Corporation, is essentially identical in device design,

Figure 10. A single bead-string reactor.

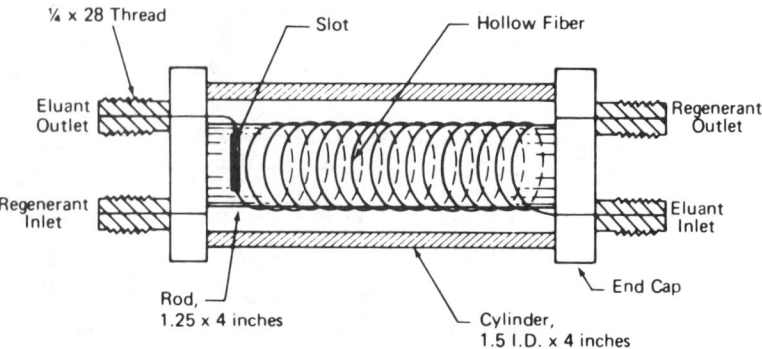

Figure 11. Drawing of a spiral-wound fiber suppressor. (From Ref. 34.)

materials, and dimensions, except that (a) poly(vinyl chloride) is used as the construction material; (b) glass, rather than synthetic polymers, is used for packing; and (c) the membrane tubes are pinched at both ends by metal crimps to act as bed support. Standard anion chromatograms on higher-efficiency columns introduced by Stevens and Langhorst [33], using an unpacked hollow-fiber suppressor, a packed suppressor column, and a packed-bead fiber suppressor, are shown in Figures 13 to 15, respectively. The superior performance of the packed-bead suppressor (depicted in Figure 12) is clearly evident [34].

Stevens et al. have pointed out that if the bead diameter is reduced by a factor of four, the band dispersion will be reduced by a factor of 2 whereas the pressure drop will rise by a factor of 16. If small enough packing is used, radial transport would be reduced to the point where a significant fraction of fluid initially flowing down the center of the packed bed would never contact the wall of the tube [39]. The authors thus reached the conclusion that the use of a relatively small packing creates more problems (i.e., reduced radial transport and increased pressure drop) than it solves (reduced band dispersion). Further, these authors suggest that the optimum diameter of beads for this purpose is 60% of the internal diameter of the tube, for tubing diameters at least up to 2 mm.

There are a few disadvantages to the packed-bead suppressor. With continued use, pressure expands the elastomeric Nafion membrane, allowing the mobile beads to pack down densely and less uniformly, leaving occasional large voids in the tube. Thus, dead volume and band dispersion characteristics tend to deteriorate with use. Also the mobility of the beads and the elasticity of the membrane

Approaches to Ionic Chromatography

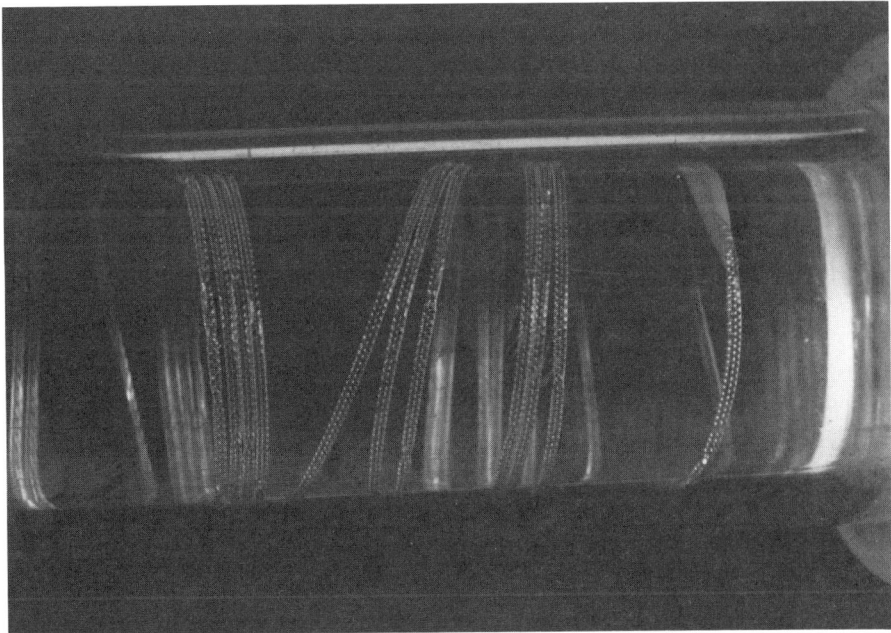

Figure 12. Close-up view of the packed-bead suppressor with the outer shell removed. (Courtesy of T. S. Stevens.)

leads eventually to parallel juxtapositions, which contribute to the pressure-induced rupture of the membrane at pressures lower than those tolerated by the filament-filled helix configuration. Note, however, that the pressure drop for identical flow rates are much larger for the filament-filled helical devices. Helical flow produces an increased pressure drop even for hollow tubes, compared to their linear counterparts [23,24].

Recent Advances in Membrane Suppressor Technology

Recently, Stevens, Jewett, Bredeweg, Westover, and Small [40] conducted a thorough study to determine the optimum design of a membrane-based suppressor. It was found that the problems encountered with the first-generation packed-bead suppressor, which mostly arise from the mobility of the packing beads and the elastomeric nature of the membrane, can be essentially eliminated if the outside of the membrane tube is packed with beads as well. Such packing not only prevents movement of the beads inside the membrane; it also enhances

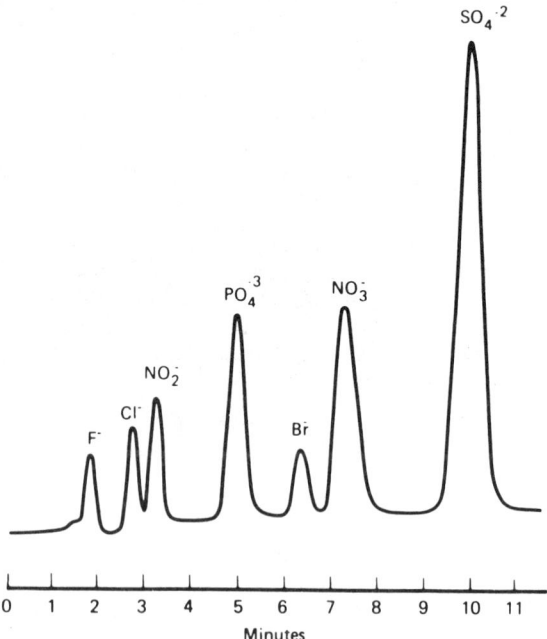

Figure 13. Chromatogram obtained using an unpacked, hollow-fiber suppressor. (From Ref. 34.)

mass transfer of the regenerant ion to the outer surface of the membrane and mass transfer of the permeated ion away from the outer surface of the membrane, resulting in a device with very substantially superior performance characteristics. In the preferred configuration for a cation exchange membrane suppressor, Nafion 811x is inserted into a PTFE jacket tubing (3 mm o.d., 2.4 mm i.d.) which terminates into suitable connecting tees such as those in Figure 1. The space inside and outside the membrane tubing are packed with 500-μm- and 250-μm-diameter poly(styrene-divinylbenzene) beads, respectively. The performance characteristics of this and of various other devices tested by Stevens et al., as well as those of the filament-filled helical device [32], are listed in Table 2. The study of Stevens et al. also establishes that a greater amount of material can be exchanged if the packing beads are smooth, spherical ion-exchange resins of the appropriate type (i.e., cation exchange resin beads for use with Nafion 811x), compared to packing with nonfunctional beads of the same diameter, in otherwise identical devices.

Approaches to Ionic Chromatography

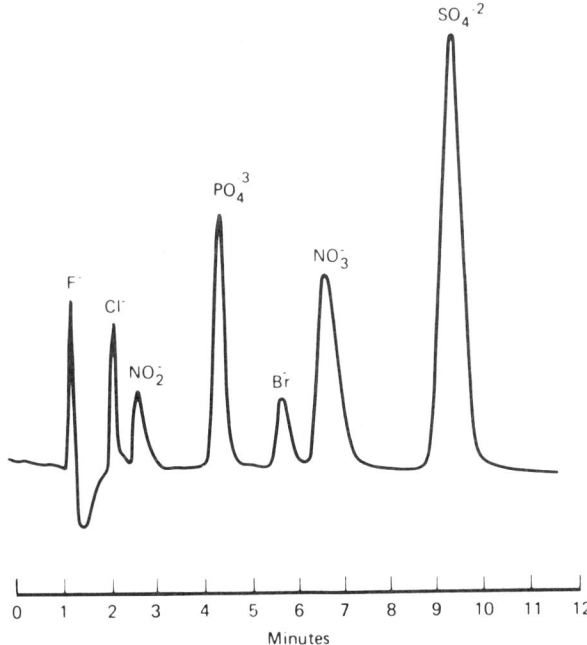

Figure 14. Chromatogram obtained using a conventional suppressor column. (From Ref. 34.)

There appears to be no difference in this improvement when a conventional high-capacity (5.2 meq/g) resin is substituted for a surface-sulfonated resin of very low capacity (0.01 meq/g). With surface-sulfonated resins of very rough topography, exchange rate increases even more, but band dispersion also increases, such that this type of resin is not considered attractive. It is not eminently clear as to why device performance improves if ion exchange beads are substituted for their inert counterparts.

Another interesting factor studied by Stevens et al. is the effect of an ultrasonic field on the mass transfer efficiency. Whereas the mass transport to the wall is improved significantly for devices that are not of a particularly efficient configuration, no improvement is noticeable for devices that are already operating at high efficiency, and furthermore, band dispersion increases with an ultrasonic field for all the devices.

It is likely that the performance of membrane-based IC suppressors will continue to improve in the foreseeable future with respect

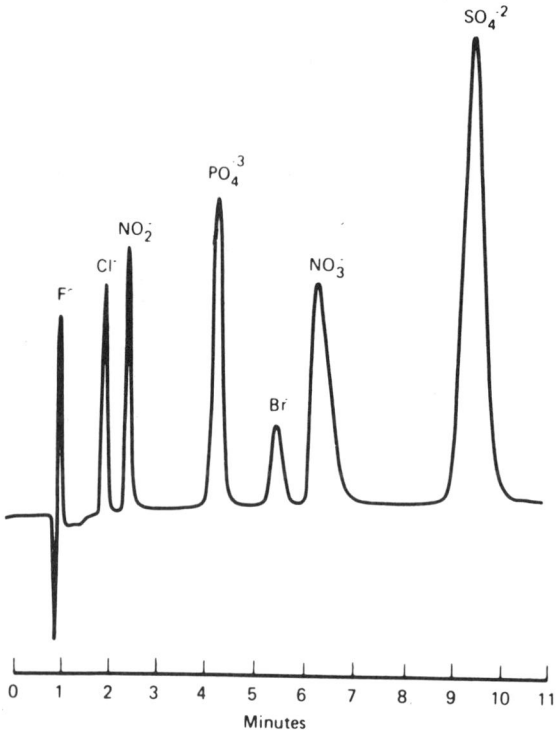

Figure 15. Chromatogram obtained using a packed bead-fiber suppressor. (From Ref. 34.)

to decreasing dispersion for any given application. The concept of reversing the geometry of the flow paths is particularly interesting [41]. In this configuration, the membrane tube is inserted in the dry state inside a reasonably closely fitting PTFE or stainless-steel 316 jacket tube; during operation, the regenerant solution flows inside the membrane tube while the solution to be ion exchanged flows in the annular channel. The annular gap decreases to a small value as the tubing is wetted, because of hydrophilic expansion; this gap can be intentionally decreased further by restricting the outlet end of the membrane tube and increasing the pressure drop through the inner flow channel; the gap may also be reduced with a flexible jacket tube, via application of radial compression on the latter from its exterior, as carried out in radially compressed chromatographic columns [42]. Work in progress at the author's laboratory involves, on the other hand, dual-membrane devices, in which one membrane tube is inserted inside another membrane tube, and a narrow gap is created

Table 2. Performance Characteristics of Nafion Membrane Suppressors of Various Designs

Device description[a]	Band dispersion (μl)[b]	Necessary length (m)[c]
1. Hollow tube, no modifications	800	6.1
2. Wire insert, 575-μm-diameter nichrome	380	4.3
3. Wire insert in waved configuration (bends every 3 cm, 575-μm nichrome)	430	3.7
4. Hollow tube, pinched and waved by construction at every 2.5 cm	930	6.4
5. Hollow tube in ultrasonic field of medium strength	810	4.3
6. Hollow tube in ultrasonic field of high strength	990	3.4
7. Packed inside with 500-μm beads of poly(styrene-divinylbenzene)	110	1.3
8. Packed inside with 500-μm PSDVB beads, in moderate ultrasonic field	60	0.52
9. Packed inside with 500-μm and outside with 250-μm PSDVB beads	50	0.58
10. Device 9, in moderate ultrasonic field	70	0.58
11. Nylon monofilament (660-μm diameter), filled helix; coil diameter, $\sim$2 mm	40	0.75

[a] All based on Nafion 811x.
[b] Band dispersion are often measured in different ways. Refs. 34 and 40 compute it as the difference between the band volumes (triangulated base width for a gaussian band) with and without the suppressor device, whereas Ref. 32 computes it in the more traditional manner used in chromatography, as the square root of the difference between the squares of the band volumes obtained with and without the suppressor device for an injected plug [13]. For comparison purposes, the dispersion value for device 11 is reported following the method of Ref. 13.
[c] This is the length necessary to completely exchange a solution of 3 mM $NaHCO_3$ + 2.4 mM Na_2CO_3 (the most commonly used eluant in traditional anion chromatography) flowing at 2.3 ml/min (the most commonly employed flow rate in traditional anion chromatography).
Source: Data for devices 1-10 taken from Ref. 40; device 11 data, from Ref. 32.

by radial expansion of the inner tube or by radial compression of the outer tube, or by yet other methods. The device can be further rendered into a small-diameter helix by support provided by a filament (coiled subsequently) inserted within the inner membrane tube [43]. Such a device potentially offers the maximum membrane area available for ion transport per unit band dispersion and may be of particular utility in ion exchanging solutions where mass transfer through the wall is a greatly limiting factor (for example, when large lipophilic ions are used as ion interaction reagents in the eluant [44, 45]). Typically, both membrane tubes are of the same type (e.g., both cation exchangers), and the solution to be ion exchanged flows in the central annular channel whereas the regenerant solution flows countercurrent to this flow through the inner membrane tube and through the jacket surrounding the entire device (Figure 16). One interesting possibility with this configuration, although not of particular interest to IC, is the complete deionization of a solution by use of two types of membrane tubes with corresponding regenerants flowing through the appropriate flow channels (e.g., anion exchanger inserted inside cation exchanger, acidic regenerant flowing outside outer membrane tube, alkaline regenerant flowing through inner membrane tube, and the solution to be deionized flowing in the annular space). In any type of dual tubular membrane ion exchanger, ion exchange resins may be used advantageously as packing on either side of the annular channel [710].

Another dual-membrane design, based on very thin sheet membranes, has been recently introduced by Dionex Corporation [46]. Two very thin membrane sheets (thickness, 50 µm) are placed next to ion exchange support screens and sandwiched tightly against each other. The liquid to be ion-exchanged flows between the two membrane sheets, whereas the regenerant flows outside the membrane. The typical pressure drop is very small, about 5 psi. This is fortunate and necessary, because the device is not likely to withstand excessive back pressures. For a dead volume as small as 40 µl, the device can exchange an impressive amount of influent flux, e.g., 0.1 M NaOH at 2 ml/min, because of the large membrane area available and facile membrane transport. The dead volume on the regenerant side is also quite small. This small dead volume and the presence of the ion exchange screen, which aids in mass transport, contribute to efficient utilization of the regenerant. However, to minimize regenerant penetration, special considerations are necessary with regard to the precise choice of the regenerant, because of the very small membrane thickness.

It is clear, however, that in spite of considerable ingenuity on the part of a number of researchers in the concept and design of novel and attractive membrane-based suppressors, the simplest will ultimately prevail. Nafion fibers have been thermally stretched to

Approaches to Ionic Chromatography

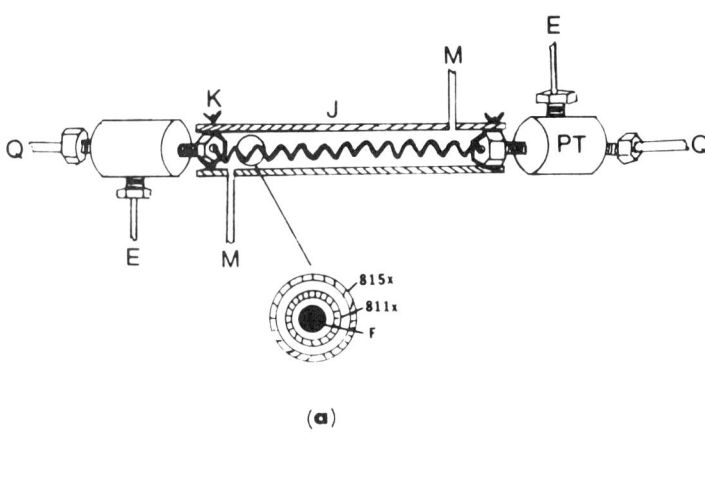

(a)

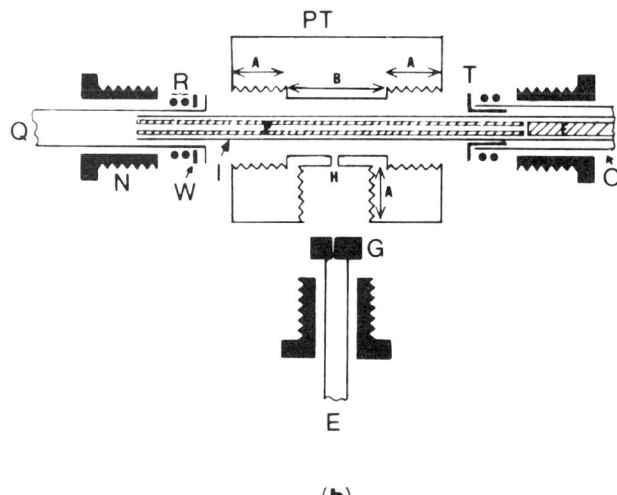

(b)

Figure 16. (a) A complete dual-membrane annular helical suppressor; Q, PTFE inlet/outlet tube for inner channel regenerant; E, PTFE tube for column effluent outlet/inlet; M, PVC inlet/outlet tube for outer channel regenerant; K, wire crimp; J, Tygon jacket tube; PT, polypropylene tee. Inset shows cross section of filament-filled dual-membrane assembly. (b) Details of the polypropylene tee and connections: W, washer; G, gripper fitting; other legends as above. Dimension A is 1 cm; dimension B is exaggerated, and should be as small as possible, hole H is of 0.5-mm bore. Drawings are not to scale; gaps are exaggerated for clarity. (From Ref. 43.)

400-μm i.d. for long lengths [47] and to 200-μm i.d. for very short lengths [48]. Thermal stretching is difficult; work in this laboratory shows that by stretching appropriately solvent-swelled (see above) Nafion 811x tubes, very narrow bore tubes can be obtained. Connections to such narrow-bore tubing can still be made by the general approach described earlier, namely via insertion of a small segment of a rigid support tube (e.g., appropriate-gauge syringe needle tubing) and subsequent insertion into connecting PTFE tubing, followed by sealing with O rings or crimps. The technology for converting almost any type of polymer tube into ion exchange membrane tubes exists [49], and porous polymer hollow fibers of 100-μm i.d. and 25-μm wall thickness with a burst pressure rating exceeding 475 lb/in.^{2} have been available for some time [50].

F. Availability, Preparation, and Properties of Ion Exchange Membrane Tubes

It was mentioned earlier that Rembaum et al. [10] prepared ion exchange membrane tubes from porous polymer tubes. In their work, Stevens et al. [11] chose to treat custom extruded polyethylene tubing with 10% chlorosulfonic acid in dichloromethane. The reported ion exchange capacity (EW, 1000) is unusually good for an unreactive matrix of this type. In subsequent work, Stevens et al. [34] chose commercially available Nafion tubing, which remains the only ion exchange tubing commercially available [51]. No small-bore anion exchange membrane tubing is commercially available at this time. However, almost any type of polymer tubing may be converted into ion exchange membrane tubing by proprietary radiation grafting procedures [52]. Segments of tubes are immersed in a solution of vinylbenzyl chloride in dichloromethane and irradiated with γ-radiation from a Cobalt-60 source (typical dosage, 1.0-1.2 Mrads). Under these conditions, the polymer forms macromolecular radicals. In the presence of styrene functional groups, a (poly)styrene-type sidechain graft results. To produce cation exchangers, the grafted tube is treated with chlorosulfonic acid, followed by hydrolysis in boiling water to convert the sulfonyl chloride formed into free sulfonic acid. To produce anion exchangers, the grafted tube is treated with a trialkylamine (typically triethylamine or trimethylamine) in dichloromethane, whereupon the benzyl chloride moiety is converted to quaternary ammonium functional groups. The reaction scheme is depicted in Figure 17. The ion exchange capacity obtained is generally substantially greater for the cation exchangers than for the anion exchangers, reflecting primarily the efficiency with which the desired postgrafting reactions take place. Typical x-ray diffraction data (Ni-filtered Cu K_α radiation) for radiation-grafted PTFE membranes show that the crystallinity of the original polymer decreases substantially upon grafting, presumably due to the disruption of the crystalline sites

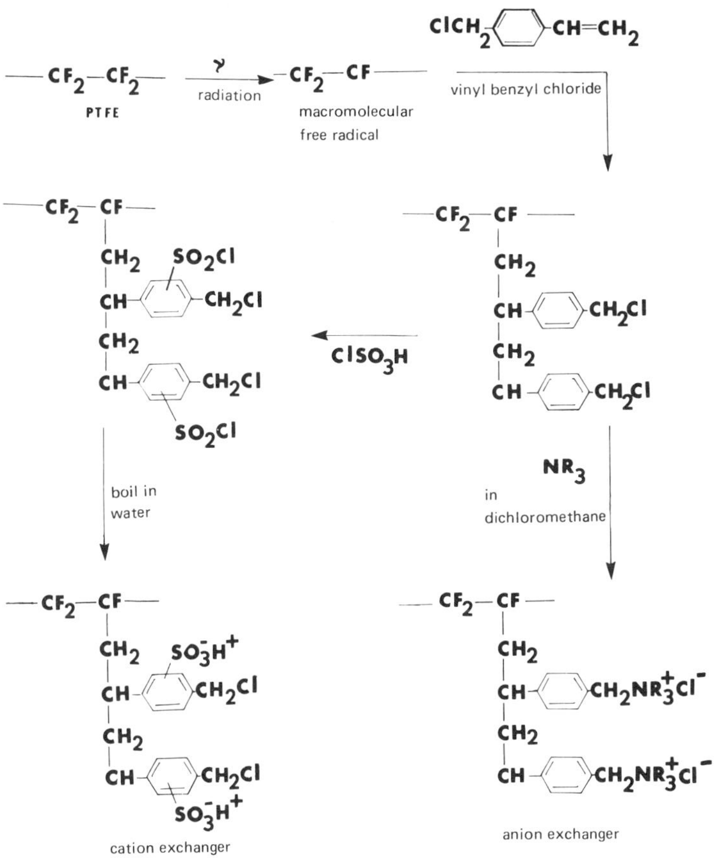

Approaches to Ionic Chromatography 221

Figure 17. Scheme showing the preparation of radiation-grafted ion exchangers. (Courtesy of J. Lee and V. D'Agostino.)

by the grafted side chains [53]. Upon functionalization of the side chains into ion exchange groups, the crystallinity increases again (but does not quite reattain that of the original polymer), presumably due to cluster formations by the ionic side chains. Ionic clusters in ionomeric membranes are well documented [54].

Characteristics of different types of ion exchange membrane tubes investigated by the author are shown in Tables 3 to 4. It is worthwhile to point out first that the number of significant figures in the results reported in these tables pertains to the reproducibility of a given experiment [55]. Because the membrane tubes vary considerably in wall thickness and bore along their length, the actual results from one particular sample to another of the same type frequently

Table 3. Membrane Tube Parameters

Type	Material	Internal diameter (cm)	Wall thickness (cm)	Mean surface area/ unit length (cm)	EW[a]	Percent water adsorption
		A. Cation exchangers				
1	PTFE[b]	0.102	0.005	0.336	520	32.6
2	Nafion 815x	0.092	0.017	0.342	1100	8.8
3	Nafion 811x	0.062	0.013	0.236	1100	8.8
4	Nafion 811x(s)[c]	0.046	0.009	0.173	1100	8.8
5	PEVA[d]	0.051	0.021	0.226	570	43.0
6	PE[e]	0.030	0.004	0.107	1000	NA[f]
		B. Anion exchangers				
7	PTFE	0.101	0.005	0.333	2800	3.1
8	PTFE	0.050	0.005	0.173	3300	4.8
9	PEVA[g]	0.030	0.024	0.170	1700	21.1
10	PEVA[h]	0.030	0.030	0.188	570	190.0

[a] EW, equivalent weight.
[b] Poly(tetrafluoroethylene) tube, obtained from Zeus Industrial products, Raritan, NJ.
[c] Nafion 811x stretched by solvent swelling.
[d] Poly(ethyl vinyl acetate) tube, "microline tubing," Thermoplastic Scientific, Warren, NJ.
[e] Poly(ethylene) tube.
[f] Data not available, described in Stevens et al. [11].
[g] Obtained from Dionex Corporation, as cation fiber suppressor, CFS-1.
[h] This tubing was extensively grafted; the resulting hydrophilicity was so great that the extensive swelling led to a very fragile membrane.
Source: Ref. 55.

Table 4. Permitted Ion Transport Rates

A. Transport of Cu^{2+} through cation exchangers[a]					
Membrane tube type (Table 3)	1	2	3	4	5
Relative permeability (neq/cm·min)	0.52	2.14	2.18	1.88	3.78

B. Transport of NO_3^- through anion exchangers[b]			
Membrane tube type (Table 3)	7	8	9
Relative permeability (neq/cm·min)	0.65	0.89	0.66

[a] 100-cm-long membrane tube; external solution, 10 mM $CuSO_4$; internal solution, 50 mM K_2SO_4; rate, 0.5 ml/min. Effluent copper concentration determined by a spectrophotometric detector.
[b] 50-cm-long membrane tube; external solution, 10 mM KNO_3; internal solution, 5 mM K_2SO_4, rate, 0.5 ml/min. Effluent nitrate concentration was determined by a spectrophotometric detector.
Source: Ref. 55.

differ by as much as 30%. In comparing the behavior of different membrane types, no firm conclusions are possible unless the differences are substantially larger than the above variation.

The data in Table 4 represent the relative permeability of a permitted ion (Cu^{2+} for cation exchangers and NO_3^- for anion exchangers). In determining these values, the simplest of diffusive transport models have been assumed, based on Fick's first law [16]. At a given concentration gradient across the membrane, the rate of ion transport is considered directly proportional to available surface area (the figures reported in Table 3 are the averages of the respective inner and outer surface areas) and inversely proportional to the wall thickness. The relative permeability is thus calculated to be CFt/Ls, where C is the concentration (N) of the transported ion in the effluent, F is the flow rate through the tube, t is the wall thickness, L is the length of the tube, and s is the surface area of the tube per unit length. The adequacy of the model is evident from the close comparability of tubes 2-4 (all made from Nafion EW 1100). Considering its favorable equivalent weight, the low permeability of the PTFE based cation exchanger is unexpected. The relative permeabilities of the poly(ethyl vinyl acetate (PEVA)- and PTFE-based anion exchangers are virtually indistinguishable--an unexpected result in view of the distinctly more favorable EW of the PEVA tube.

In choosing a tube material for application as a membrane suppressor, a compromise between decreasing wall thickness and the consequent undesirable decrease in the burst pressure limit should be sought. The PTFE tubes represented by the dimensions in Table 3 are very thin walled; their structural strength is also very limited. The manufacturing tolerance for the production of very thin-walled tubes is also substantially larger than the relative tolerances involved with greater wall thickness [56] and results in susceptibility to rupture for the thin-walled tubes at the weaker portions of the tube.

Finally, not only structural considerations are important for the longevity of a membrane suppressor. Some ions are so strongly retained by ion exchange membranes that they effectively "poison" the exchange sites. For example, Nafion is particularly susceptible to fouling by calcium and other alkaline earth metals.

G. Choice of a Suitable Regenerant

Ions similarly charged to the membrane matrix (e.g., anions for cation exchanger membranes bearing negatively charged sulfonate groups) are retarded with respect to transport through the membrane (in a simplistic sense, by the electrostatic field), and such ions are referred to as *Donnan forbidden* [57]. The barrier to the forbidden ion is not sufficient to eliminate completely the penetration when the concentration differential across the membrane is high. The situation is aggravated with very thin membranes because the penetration rate increases. When one is ion-exchanging a sodium carbonate eluant for H^+, using a cation exchange membrane suppressor with sulfuric acid as regenerant, significant penetration of the anion occurs with all but very dilute solutions of sulfuric acid. The penetration rate of a forbidden ion depends on the ionic size, which governs its diffusive transport across the membrane matrix. Chloride is smaller than perchlorate; therefore, use of HCl rather than $HClO_4$ as regenerant leads to greater counterion penetration. If one uses a very large forbidden ion, the penetration rate can be drastically reduced, as demonstrated by Japanese workers [47,48,58]. For any given forbidden ion, the penetration rate is inversely related to the membrane thickness and the density of available exchange sites, and is directly related to the available surface area.

Theoretical Considerations on the Penetration Rate of a Forbidden Ion

Helfferich [59] determined that the concentration of the forbidden ion in the ion exchange phase is related to the equilibrium concentration of the forbidden ion in the surrounding solution as:

Approaches to Ionic Chromatography

$$\frac{\bar{m}}{m} = \left(\frac{\bar{m}_R}{2zm}\right)^2 + \left[\left(\frac{\gamma_\pm}{\bar{\gamma}_\pm}\right)^2 \left(\frac{\bar{a}_w}{a_w}\right)^{\bar{v}/v_w}\right] - \frac{\bar{m}_R}{2zm} \tag{17}$$

where

- $\bar{m}$ = molality of the ion in the ion exchanger
- m = molality of the ion in the surrounding solution
- $\bar{m}_R$ = molality of ion exchange sites in the ion exchanger
- $\gamma_\pm$ = mean activity coefficient of the ion in solution
- $\bar{\gamma}_\pm$ = mean activity coefficient of the ion in the ion exchanger
- a_w = activity of water in the solution
- $\bar{a}_w$ = activity of water in the ion exchanger
- $\bar{v}$ = partial molar volume of the electrolyte constituted by the forbidden ion
- v_w = partial molar volume of water
- z = charge magnitude of the ion

For a limited range of concentration of a given forbidden ion and with a given ion exchange membrane, $\bar{m}_R$, $\gamma_\pm$, $\bar{\gamma}_\pm$, a_w, $\bar{a}_w$, $\bar{v}$, v_w, and z may all be regarded as essentially constant, thus leading to:

$$\frac{\bar{m}}{m} = \left(\frac{Q_1}{m^2} + Q_2\right)^{1/2} - \frac{Q_3}{m} \tag{18}$$

where Q_1, Q_2, and Q_3 are positive constants. Transposition, then squaring both sides, and rearrangement yields:

$$m^2 = \left(\frac{1}{Q_2}\right)\bar{m}^2 + 2\left(\frac{Q_3}{Q_2}\right)\bar{m} + \frac{Q_3^2 - Q_1}{Q_2} \tag{19}$$

Consider a membrane tube immersed in a solution containing the forbidden ion. The permitted ion present with the forbidden ion will be H^+ or OH^- for typical anion or cation chromatography applications, respectively. Water is pumped through the membrane tube, and the effluent conductivity is measured. If the water flow rate through the tube is sufficient, such that the concentration of the permeated ion in the water stream is negligible as compared to that in the external solution, the concentration of the ion at the inner membrane surface may be considered negligible as compared to that at the outer surface of the membrane. Thus, the concentration differential across the membrane may be taken to be equal to the concentration of the ion at the outer surface of the membrane.

The premise of negligible ion concentration at the inner surface of the membrane is verifiable. Under this condition, the penetration rate will be independent of the water flow rate through the tube, and thence the effluent concentration in the water stream will be inversely related to the flow rate. In dilute solutions, conductivity is directly related to ionic concentration. With the experimental flow rates used in the study [55], the effluent conductance was always found to be in exact inverse relationship with the flow rate through the tube.

The diffusive transport of the ion through the membrane is directly proportional to the concentration differential across the two surfaces of the membrane and inversely proportional to the membrane thickness. For a given membrane, the penetration rate at a given inner flow rate is then directly proportional to $\bar{m}$. Because at a given flow rate, the effluent conductance is a direct measure of the penetration rate, Eqn. (19) takes the experimentally verifiable form:

$$m^2 = \alpha A^2 + \beta A + \gamma \qquad (20)$$

where A is the measured effluent conductance, α and β are positive constants, γ is a constant of either sign depending on the relative magnitudes of Q_3 and Q_1, and m is the concentration of the forbidden ion that surrounds the membrane.

Experimental Results: Penetration of Forbidden Ions

The anion penetration through a cation exchanger Nafion tube is shown in Table 5 in terms of observed conductance of the effluent. The different anions vary in their equivalent conductance, and penetration rates in molar units are not strictly parallel to the conductance results. Table 5 also shows the results for cation penetration through a PTFE-based anion exchanger tube as a function of the concentration of NaOH in which the tube was immersed. The fit of the data to Eqn. (19) is shown in Figure 18. The coefficients α, β, and γ are listed in Table 5 along with the regression coefficient. Clearly, the data are well represented by Eqn. (20).

A critical examination of the data in Table 5 pertaining to forbidden anion penetration through a cation exchanger reveals some interesting features. Even though sulfate is not particularly smaller than perchlorate, the penetration rate of sulfate is lower than that of perchlorate because the doubly charged sulfate ion is subject to a higher electrostatic barrier. The even smaller penetration rate obtained with dodecylbenzenesulfonate [60] may be attributed to its very large size. Naphthalenedisulfonate [61] is both large and doubly charged, and its effluent conductance is less than that of dodecylbenzenesulfonate. Naphthalenetrisulfonate [62] was tested to push this approach even further, and within the limits of attainable purity of this compound, successfully demonstrates how penetration can be minimized. To virtually eliminate the penetration of

Table 5. Concentration Dependence of Forbidden Ion Penetration

A. Cation exchange membrane[a]

$HClO_4$ (N)	0.020	0.025	0.050	0.100	
Conductance (µS)	0.300	0.450	1.70	6.66	
$\alpha = 1.92 \times 10^{-6}$	$\beta = 1.50 \times 10^{-3}$		$\gamma = -4.88 \times 10^{-5}$		$R^2 = .999$
H_2SO_4 (N)	0.020	0.025	0.050	0.100	
Conductance (µS)	0.199	0.249	0.665	2.32	
$\alpha = 1.19 \times 10^{-5}$	$\beta = 4.50 \times 10^{-3}$		$\gamma = -4.95 \times 10^{-4}$		$R^2 = .999$
DBSA[b] (N)	0.100	0.200	0.300	0.400	0.500
Conductance (µS)	0.359	1.22	2.64	4.60	7.20
$\alpha = 6.00 \times 10^{-5}$	$\beta = 3.56 \times 10^{-2}$		$\gamma = -3.10 \times 10^{-3}$		$R^2 = .999$
NPS2[c] (N)	0.100	0.200	0.300	0.488	
Conductance (µS)	0.102	0.234	0.418	0.889	
$\alpha = 7.96 \times 10^{-2}$	$\beta = 2.11 \times 10^{-2}$		$\gamma = -1.28 \times 10^{-2}$		$R^2 = .999$
NPS3[d] (N)	0.100	0.200	0.400	0.600	0.800	1.00
Conductance (µS)	0.059	0.089	0.209	0.408	0.687	1.04
$\alpha = 1.33 \times 10^{-2}$	$\beta = 9.94 \times 10^{-1}$		$\gamma = -4.85 \times 10^{-3}$		$R^2 = .999$
PSSA[e] (N)	0.050	0.200			
Conductance (µS)	nd[f]	0.066			

B. Anion exchange membrane[g]

NaOH (N)	0.010	0.020	0.050	0.100	0.200
Conductance (µS)	0.027	0.066	0.310	1.16	3.81
$\alpha = 6.80 \times 10^{-4}$	$\beta = 7.93 \times 10^{-3}$		$\gamma = -9.54 \times 10^{-5}$		$R^2 = .999$

[a] 85-cm-long Nafion 811x(s) (type 4; Table 1), influent water, 0.5 ml/min.
[b] Dodecylbenzenesulfonic acid. This material contains a small amount of H_2SO_4. Addition of $Ba(OH)_2$ (1% of total acidity) reduces the penetration conductance of the 0.500 M acid to 1.30.
[c] Naphthalene-2,6-disulfonic acid.
[d] Naphthalene-1,3,6-trisulfonic acid.
[e] Poly(styrenesulfonic acid), purified by dialysis; average MW, 70,000
[f] Not detectable.
[g] PTFE base anion exchanger membrane (50 cm; type 8; Table 3); influent water, 0.5 ml/min.
Source: Ref. 55.

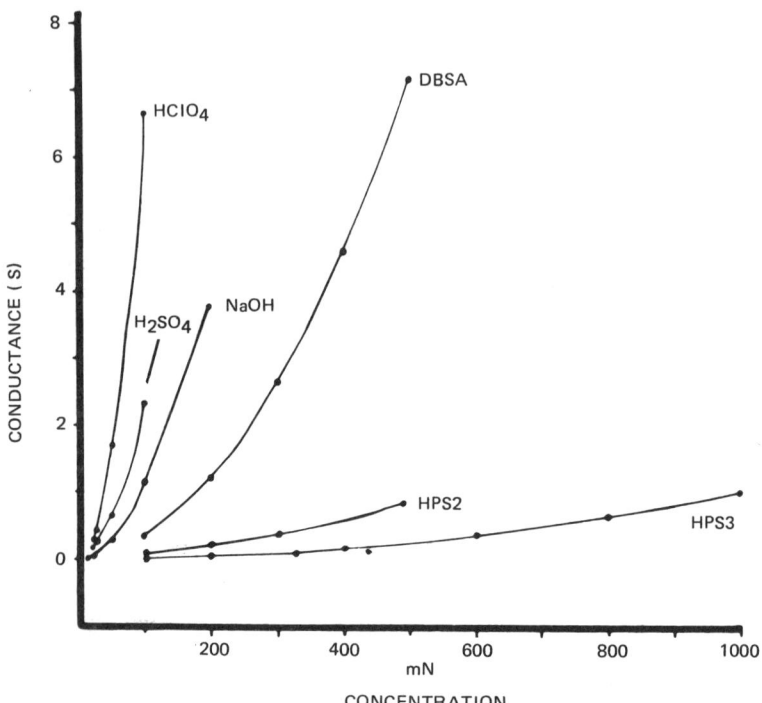

Figure 18. The fit of Eqn. (20) to experimental forbidden-ion penetration data as a function of forbidden ion concentration. The solid lines depict the equation represented by the coefficients in Table 6 whereas the points represent the experimental values. DBSA, dodecylbenzenesulfonic acid; NPS2, naphthalenedisulfonic acid; NPS3, naphthalenetrisulfonic acid. (From Ref. 55.)

the undesired counterion, such an ion should be very large and be multiply charged. Both of these criteria are satisfied by a polymeric ion, and indeed penetration rates with purified poly(styrenesulfonate) [63] are not measurable at concentrations of 0.05 N and below. This result pertains to a typical batch of purified poly(styrenesulfonic) acid; lower penetration rates have been obtained with further purification, which suggests that penetration can be essentially eliminated with appropriately pure reagent. Dodecylbenzenesulfonate results, as such, are enigmatic, because this material is a typical anionic surfactant, and, at the concentrations employed, the bulk of the material should be present as micelles. The monomeric ion concentration should be approximately equal to the critical micelle concentration (CMC). The

CMC for surfactants containing a 12-carbon alkyl chain are typically on the order of 1 mM [64]. Apparently, the primary material that contributes to the observed conductance for dodecylbenzenesulfonate is the small amount of sulfuric acid present in commercial dodecylbenzenesulfonic acid as impurity. Addition of $Ba(OH)_2$ to the commercial acid (to the extent of 1% neutralization of acid equivalents), and using it as such without removing the suspended $BaSO_4$, reduces the observed conductance due to penetration by more than a factor of 5 for the 0.5 M acid. Addition of greater amounts of $Ba(OH)_2$ results in increased effluent conductance due to the transport of free Ba^{2+} across the membrane.

The use of poly(styrenesulfonic) acid or dodecylbenzenesulfonic acid as H^+-rejuvenating agents provides another advantage that is not immediately obvious. Solutions of these compounds behave as liquid ion exchangers, thus reducing the concentrations of the free cations that permeate out to the regenerant side. Consequently, they permit slow and efficient use of regenerant flow. Even systems where the membrane tube is simply immersed in a large volume of the regenerant solution work quite well.

Table 6 shows the behavior of different membranes toward a given forbidden ion: sulfate for cation exchangers and sodium for anion exchangers. For both cation and anion exchangers, the PEVA matrix appears to be the most susceptible to forbidden ion penetration. This is contrary to expectations based on the equivalent weights (EWs) reported in Table 3. The behavior of cation exchangers versus anion exchangers may be compared after accounting for the different concentration gradient employed in the two cases (0.025 N H_2SO_4 versus 0.100 N NaOH). Based on the data in Table 5 for a type 4 membrane tube, the relative permeability with 0.100 N H_2SO_4 will be approximately 10 times greater as compared to the data in Table 6. Taking into account the much lower EW of the cation exchangers and that sulfate is doubly charged, no drastic differences between the two types of exchangers are apparent. In Table 7, the permeabilities of different cations through an anion exchange membrane is shown. The trend is predictable. Tetrapropylammonium hydroxide is available in pure form and makes an excellent OH^--regenerant. In contrast, significant efforts are necessary to prepare poly(diallyldimethylammonium) hydroxide. Few cationic surfactants are available as the hydroxides, and even these contain large amounts of carbonates or halides. Whether the particular application justifies the use of polymeric or micellar hydroxides is obviously up to the user.

Finally, a comparison of the data in Table 6 with those in Table 4 reveals an important membrane parameter--relative permeability to a permitted ion versus that to a forbidden ion. There is little difference among the cation exchanger types, but among the anion exchangers, it is clear that the PEVA exchanger does not perform as

Table 6. Effect of Membrane Type on Forbidden Ion Penetration

A. Cation exchange membrane[a]

Membrane tube type (Table 3)	1	2	3	4	5
Conductance (μS)	0.338	0.283	0.310	0.283	0.420
Relative permeability, (peq/cm·min)	3.00	8.44	10.2	8.80	23.3

B. Anion exchange membrane[b]

Membrane tube type (Table 3)	7	8	9
Conductance (μS)	4.86	2.32	2.19
Relative permeability, (peq/cm·min)	70.9	65.3	300

[a] Each membrane tube, 100 cm long; external liquid, 0.025 N H_2SO_4; internal, water, 0.5 ml/min.
[b] Each membrane tube, 100 cm long; external liquid, 0.100 N NaOH; internal, water, 0.5 ml/min.
Source: Ref. 55.

well as its PTFE counterpart. However, it should be noted that this particular PEVA exchanger was specially formulated to have a high selectivity for OH^-, so that it could function well as a OH^--form suppressor, at the cost of some counterion penetration [65].

Normally, regenerant penetration contributing to no more than 50% of the background conductance will not significantly deteriorate attainable limits of detection because the penetration rate is quite constant. However, the use of regenerant concentrations that lead to penetration conductances greater than the exchanged effluent conductance does begin to affect detectability.

H. Electrodialytic Postcolumn Conversion

It is possible to operate a membrane-based suppression system without using chemical regenerants. If an inert metallic wire is inserted inside the membrane tube and the membrane tube is surrounded by an inert metallic tube, and a unidirectional electrical potential is established across the two metallic components, an electrodialysis assembly is formed. In the case of a cation exchanger membrane, the

Table 7. Cation Penetration Through Anion Exchangers: Cation Dependence[a]

Cation	Li^+	Na^+	NMe_4^+	NEt_4^+	NPr_4^+	$PDDMA^{n+}$ [b]
Relative permeability, (peq/cm·min)	180	58.1	11.9	5.22	1.85	nd[c]

[a] 50-cm-long PTFE anion exchanger (type 8, Table 3); external solutions, 0.085 N hydroxides; internal solution, water, 0.5 ml/min.
[b] Poly(diallyl dimethyl ammonium); MW, ~200,000.
[c] Not detectable.
Source: Ref. 55.

outer cylinder should be placed at a negative potential relative to the inner electrode, establishing a motive force for cation permeation to the outer dialyzate stream. Barring undesirable electrochemical conversion of an analyte species, the approach may be attractive. The concept, by itself and as a supplement to the use of chemical regenerants, has been patented in Japan [66,67] and in Europe [68,69], as well as in the United States [68,69].

I. Membrane-Based Postsuppressors

As has been mentioned earlier, ion exchange of carbonate-containing eluants through a properly designed and operated cation exchanger membrane suppressor leads to an effluent conductance significantly less than that obtained with a conventional packed-column suppressor [31]. This difference stems from the loss of H_2CO_3 through the membrane wall. In general, this further decrease in background conductivity is beneficial. It is the author's experience, however, that for solvent delivery systems based on reciprocating pumps that are not adequately damped, significant pressure pulsations appear at the membrane suppressor inlet, and CO_2 loss (which is dependent on the pressure experienced by the membrane) pulsates in phase with the pressure surges. The exact dimensions of the elastomeric membrane-based device also changes with changes in inlet pressure and may affect exchange efficiency. Both of these factors increase baseline noise, offsetting the advantage of a somewhat lower mean background conductance. For otherwise comparable exchange efficiency, filament-filled helical membrane suppressors operate at a much higher pressure than their packed-bead counterparts and consequently show greater CO_2 loss. If, however, CO_2 can be nearly quantitatively removed, the background conductivity approaches that of pure water and becomes relatively immune to pressure pulsations experienced by the system. Ion exchange membranes are not optimum for rapid transport of CO_2; in their studies, Siemer et al. [70,71] have shown that the use of CO_2-permeable tubing such as silicone rubber or that of a porous polymer, following the suppressor (conventional or membrane-based), may lead to significant advantages.

The original work [70] was carried out with porous PTFE tubes (Gore-Tex tubing, W. L. Gore and Associates, Elkton, MD). The smallest-bore tube of this type that is commercially available is 1 mm in internal diameter, 400 μm in wall thickness, with a mean pore size of 2 μm (type TA; surface porosity, 50%) or 3.5 μm (type TB; surface porosity, 70%); such tubes are used for a variety of applications including vascular implants. The material is hydrophobic, and water or aqueous solutions do not readily pass through the small pores because of the necessity of overcoming surface tension. The barrier, however, is relatively limited; water entry pressure for unsoiled clean tubes ranges from 12.7 (for type TA) to 7.1 (for type TB) $lb/in.^2$. The entry pressures are much lower for solvents of lower surface tension, and as such, mixed aqueous solvents cannot be used. Relatively long lengths, required (~ 10 m) to achieve a near quantitative removal of CO_2 ($\geq 90\%$), result in undesirably large band dispersion with tubes of this diameter. Incorporation of a single bead-string reactor design is difficult to implement, both because of problems of availability of inert beads of appropriate size and, more importantly, because of the excessive pressure drop from such a configuration resulting in liquid leakage through the pores. Alternatives to increase mass transport to the walls with concomitant decrease in band dispersion were sought; as a result, a number of ingenious ways to pack the membrane tube were developed. Introduction of a 0.7-mm-diameter fishing line produces some improvement, which is discernibly greater when the filament is notched. The best results were obtained with a small-diameter fishing line, with knots (half-hitches) as close as possible, resulting in a pseudo-SBSR. The making of such an insert, bearing several thousand handmade knots, is extremely time consuming, however. One other successful alternative is a pair of twisted stainless-steel wires. The performance of various CO_2 removal devices is shown in Table 8, and some of the inserts used are shown in Figure 19.

As in the case of the membrane-based suppressors, the ideal configuration for a membrane-based postsuppressor is a very-narrow-bore hollow tube with highly permeable walls. Silicone rubber is highly permeable to CO_2, and narrow-bore tubing of this material (310 μm i.d. and 600 μm o.d.) has been successfully used by Siemer and Johnson [71] with negligible band dispersion. Narrower-bore polypropylene

Table 8. Background Suppression with Different Postsuppressors[a]

Tubing	Filling	Suppression
1 m TB 001[b]	empty	30
1 m TB 001	empty, coiled	45
3 m TB 001	empty	74
10 m TB 001	empty, coiled	89
1 m TA 001[c]	empty	30
1 m TA 001	0.7-mm nylon fishing line	53
1 m TA 001	notched 0.7-mm nylon fishing line	68
1 m TA 001	twisted stainless steel wire, 2 × 0.4 mm	70
1 m TA 001	knotted 0.28-mm nylon fishing line	77
2 m TA 001	Gore-Tex Teflon thread Y 10175	89
2 m TA 001	twisted stainless steel wire, 2 × 0.4 mm	90

[a]Eluant, 3 mM $NaHCO_3$, 2.4 mM Na_2CO_3; flow rate, 2.5 ml/min.
[b]Gore-Tex porous PTFE tube, 1-mm i.d., 400-μm wall, 3.5-μm pores, 70% porosity.
[c]Gore-Tex porous PTFE tube, 1-mm i.d., 400-μm wall, 2-μm pores, 50% porosity.
Source: Ref. 70.

tubes (100- to 400-μm i.d., 25-μm wall) with very small pores (0.03- to 0.05-μm average pore diameter, surface porosity 20-40%) are commercially available [50]. Because the pores are much smaller, the water entry pressure in these tubes is as high as 150 psi. Such tubes should make excellent postsuppressors for the removal of CO_2. In dealing with *porous* tubing (Nafion and silicone rubber tubing are *not* porous; the diffusive loss of CO_2 occurs by permeation through the wall, and the wall resistance to transport is much larger than their porous counterparts), it should be noted that the water entry pressure merely denotes the necessary pressure required to establish a liquid connection between the inside and the outside of the membrane. Once such a connection is established, liquid leakage may continue at a pressure much lower than the water entry pressure. Therefore, if liquid leakage begins, the membrane tube must be removed and washed with a volatile solvent that wets the membrane

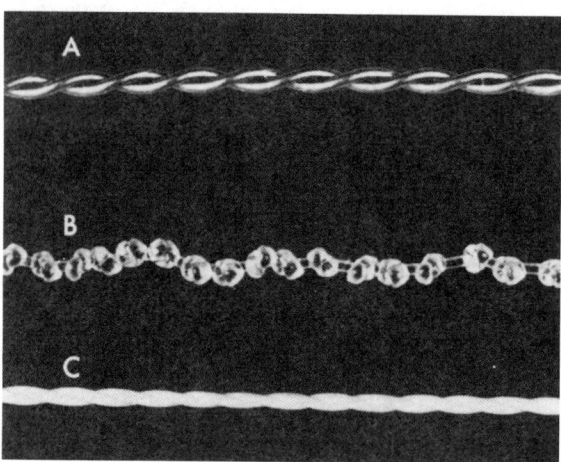

Figure 19. Mass transport devices for the porous-membrane postsuppressor: (A) Two twisted stainless-steel wires, each 400 μm in diameter; (B) knotted 0.28-mm nylon fishing line; (C) Teflon thread (Gore-Tex Y10175, ca. 0.5 mm in diameter). (From Ref. 70.)

easily (e.g., methanol), and then this solvent must be evaporated from the pores before it is put back into operation.

Aside from considerations on transport to and through the wall, transport of the permeated CO_2 away from the wall (or providing a fast sink for it) is important. In their work with porous PTFE tubing, Siemer et al. [70] simply allowed air, from which CO_2 has been removed by Ascarite, to flow around the membrane tubing and carry out the permeated CO_2. In later studies, Siemer and Johnson [71] found it advantageous to immerse the silicone rubber membrane tube in a KOH solution. With a 3.6-m-long hollow tube immersed in 0.1 N KOH maintained at 78°C, quantitative removal of CO_2 was achieved. The removal is more efficient at elevated temperatures; only 73% removal was observed when the KOH solution was maintained at room temperature (23°C). Although the authors suggest that the increase in removal efficiency is due to an acceleration of the process of dehydration of H_2CO_3 and a decrease in the solubility of CO_2(aq) at elevated temperatures, it is more likely that the increase in diffusion coefficient with temperature alone is primarily responsible. In contrast to the hydration of CO_2(aq), dehydration of H_2CO_3 is a relatively fast process, with a half-life of 58 msec [72]. In view of the essentially zero concentration of free CO_2 in a KOH solution, the decrease in solubility of CO_2 with temperature is unlikely to be a major factor. (The chemistry and kinetics of hydration and dehydration of CO_2 is a complex process; the interested reader should refer to the excellent discussion contained in Ref. 16.) Improvement in device performance at elevated temperatures is common also to membrane-based suppression systems; Japanese investigators [47,48,58] have found it advantageous, for example, to operate their Nafion membrane-based suppressor at 40°C.

Siemer et al. have pointed out other advantages of a postsuppressor CO_2 removal system. The so-called water and carbonate dips [709] are virtually completely eliminated, greatly facilitating quantitation of analyte peaks (e.g., F^-, Cl^-, IO_3^-, etc.) that tend to overlap these dips. Also, the analyte signal increases significantly (45% enhancement for Cl^- and F^-, for example), as compared to a system without postsuppression. This improves detectability above and beyond the improvement resulting from a lower baseline and decreased baseline noise. This enhancement of analyte signal is caused by the fact that in a conventional system, when a strong acid anion elutes from the column, the resulting drop in pH suppresses the dissociation of carbonic acid concurrently present, and the analyte signal therefore represents the net difference. There is no such change in background conductance accompanying elution of an analyte when the background effluent is essentially pure water. Consider for example, a conventional system without postsuppression, that uses 2.4 mM Na_2CO_3 + 3 mM $NaHCO_3$ as eluant. The result is that 5.4 mM H_2CO_3 is the background effluent. With 6.3 as the pK_1 for H_2CO_3 [16], the effluent contains 52 μM each of H^+ and HCO_3^-, which are the primary contributors to the observed conductance. If one assumes the limiting equivalent conductance of H^+ and HCO_3^- to be 349.8 and 44.5, respectively [73], multiplies by the respective concentrations, and finally adds, the overall conductance should be 2.04×10^{-2} in arbitrary units. When a 1.5-ppm Cl^- (42 μM) sample elutes from the column, if one assumes that at the peak the effluent chloride concentration is the same as that originally injected, the effluent contains 42 μM HCl and virtually the same amount of H_2CO_3 as before, i.e., 5.38 mM. The ionic composition of such a solution is easily calculated from charge and mass balance considerations to be: $[H^+]$ = 77 μM, $[HCO_3^-]$ = 35 μM, and $[Cl^-]$ = 42 μM. The overall conductivity in the same units for this solution (the limiting equivalent conductance of Cl^- being 76.35) is 3.17×10^{-2}. The net change is thus 1.13×10^{-2} units, and represents the analyte signal. In contrast, when the background is pure water, the net change due to 42 μM HCl effluent should be 1.80×10^{-2} units, by the same method of computation—an enhancement of ~59%. Because some dispersion is unavoidable in the chromatographic process itself (i.e., 1.5 ppm injected Cl^- is unlikely to elute with quite the same peak concentration), the agreement between theory and the observed experimental result of 45% enhancement for this concentration of chloride injected [71] is reasonable.

One note of caution is in order concerning the use of membrane-based postsuppressors. Significant loss of signals from analyte ions that are derived from volatile weak acid gases (such as H_2S, HCN, SO_2, HNO_2) may occur through the same mechanism that removes CO_2.

The postsuppressor for CO_2 removal is particularly useful for carrying out gradient elution in IC, which is considered in Section VI.C.

III. INDIRECT DETECTION METHODS

Whenever possible, sensitive direct detection is unquestionably the preferable approach. By *direct detection* is meant that there is some easily measurable property that is associated with the analyte ion, very significantly different from the eluant background, which is detected as the analyte ion elutes from the system. Optical properties (absorbance, fluorescence, refractive index), and electroanalytical properties (conductance, permittivity, electrochemical reducibility or oxidizability, electrode potential) represent some parameters routinely measured for detection of eluted analyte ions. Indeed, for highly optically absorbing ions, if a suitably transparent eluant system is available, few approaches will equal the sensitivity and simplicity of direct optical absorbance detection. However, many ions of interest, sulfate for example, do not display any optical absorption of significant utility. It was the triumph of suppressed conductimetric IC, as introduced by Small et al. [1], that showed how analyte anions derived from low pK acids can be detected easily and sensitively when the salt of a high pK acid is used as the eluant and all cations are converted to H^+ prior to conductimetric detection. Anions derived from very weak acids, e.g., carbonate, borate, or cyanide, cannot be detected with good sensitivity via such an approach because the corresponding acids do not ionize well. Consider now that the detection of an eluted ion merely requires that there is some difference in the property being measured between that of the analyte ion and the eluant ion. Whereas direct detection generally refers to the situation where the transduced value of the monitored property is high for the analyte ion and low for the eluant ion, indirect detection refers to the reverse situation.

It is possible to use indirect detection in many cases where the inherent sensitivity of the system is so high that in spite of limitations posed by the baseline noise, achievable limits of detection (LODs) are excellent, and no comparable direct detection methods exist. It must be noted that increased baseline noise with high-background methods is due to nonideal conditions, e.g., thermal fluctuations, flow fluctuations, etc., which can be reduced but cannot be completely eliminated. Note that in direct detection, measuring a small absorbance signal against a low background essentially means measuring a small decrease in a large photomultiplier standing current. One of the most sensitive detection methods used in gas chromatography is electron capture detection, which also involves the measurement of a small decrease in a large standing current.

It is worthwhile to consider why direct detection is preferable if the absolute difference in the measured property between the analyte and the eluant ion is the same, and only the roles are reversed. An identical difference in the measured property means that the slope of a calibration plot or the sensitivity for both cases is the same. *Sensitivity* may be defined as the change in the measured property in some arbitrary units per unit change in injected analyte concentration. Methods with identical sensitivities may, however, represent very different achievable LODs. LOD is governed, aside from sensitivity, by baseline noise. All other things being equal (i.e., identical detection technology), baseline noise is most often related to the absolute magnitude of the baseline or background signal (at the primary transducer). Let us compare for example two situations:

1. The analyte ion property is 100 units per equivalent, whereas the eluant ion property is 10 units per equivalent. Direct detection is being employed, and the eluant background concentration is 1 meq/L. The background signal is thus 0.01 unit.
2. The analyte ion property is 10 units per equivalent, whereas the eluant ion property is 100 units per equivalent. Indirect detection is being used, and the eluant background concentration is 1 meq/L. The background signal is thus 0.1 unit.

Sensitivities (change from the background value due to the elution of a unit concentration of the analyte ion) are the same in both cases. In the effluent band containing the analyte, the analyte ion merely replaces the eluant ion, whereas the total ionic concentration (in equivalents per liter) in the effluent remains unaltered at all times; this is considered in more detail later. The sensitivity in both cases is 90 units/eq/L of eluted analyte ion; its absolute sign in the two cases is opposite. Now assume that the solvent delivery system is pulseless or nearly so, and that the primary source of the background noise is temperature fluctuations. Assume further that the dependence of the measured property on temperature is 1%/°C and the whole system is thermostatted to ±1°C. The noise in case 1 is then 10^{-4} units, whereas the absolute value of the noise in case 2 is 10^{-3} units. The guidelines recommended by the American Chemical Society define *limit of detection* (LOD) as a sample signal corresponding to three standard deviations above blank, or roughly $S/N = 3$, whereas *limit of quantitation* (LOQ) is defined as ten standard deviations above blank [74]. The LOD in case 1 is then 3×10^{-4} units or 3.3 µeq/L of the analyte ion, whereas in case 2 it is 3×10^{-3} units or 33 µeq/L of the analyte ion.

The above discussion does not imply that indirect detection methods always yield less desirable LODs as compared to direct detection. This would be true *if* and *only if* the eluant and the detection and the chromatographic system were identical for the two cases.

Indirect detection methods generally rely on the principle that the effluent ionic concentration (in equivalents per liter, eq/L) from a ion exchange column remains constant at all times. For example, if 5 mM NaE (E^- is the eluant ion) is being used as the eluant, then when a sample ion S^- elutes from the column, the concentration of the eluant ion E^- drops concomitantly by an equivalent amount. If at the peak of the elution band, the concentration of S^- is 1 mM, the effluent at that point has a composition of 4 mM NaE and 1 mM NaS; both the counterion concentration and the total effluent ionic concentration (eq/L) remains unaltered. Thus, if a strongly optically absorbing eluant ion is being used, the emergence of a nonabsorbing ion is indicated by a dip in the baseline. Whether or not the transduced signal is electronically inverted to produce a peak instead of a dip is immaterial. More important is the fact that to achieve low level detection, the detector electronics must permit a substantial baseline offset (to the full extent of the baseline signal present), which is not necessarily available in all commercial detectors. Many commercial detectors permit a maximum offsetting capacity of 200% of the full-scale measurement range; this may well be inadequate for indirect detection at trace levels. The necessary electronics may however be constructed very simply and relatively inexpensively with operational amplifiers; a schematic is shown in Figure 20.

Although indirect detection methods are applicable also to systems employing ion interaction reagents (IIR) and nonpolar stationary phases, the premise of constant effluent ionic strength does not hold in such systems, and it is essentially impossible to predict a priori the nature of response from a given analyte ion. This is discussed in greater detail in Section V. In many instances, the differences in indirect and direct detection methods are less clear-cut than simplistic considerations allow for. Single-column ion chromatography (SCIC), as discussed in Chapter II, is a case in point. In cation SCIC, when an acid is being used as the eluant, the sample signals arise from the replacement of the highly conducting H^+ by less conducting metal ions in the effluent band; it is clear that this is best classified as indirect detection [75]. An exactly analogous situation exists for anionic SCIC

Approaches to Ionic Chromatography

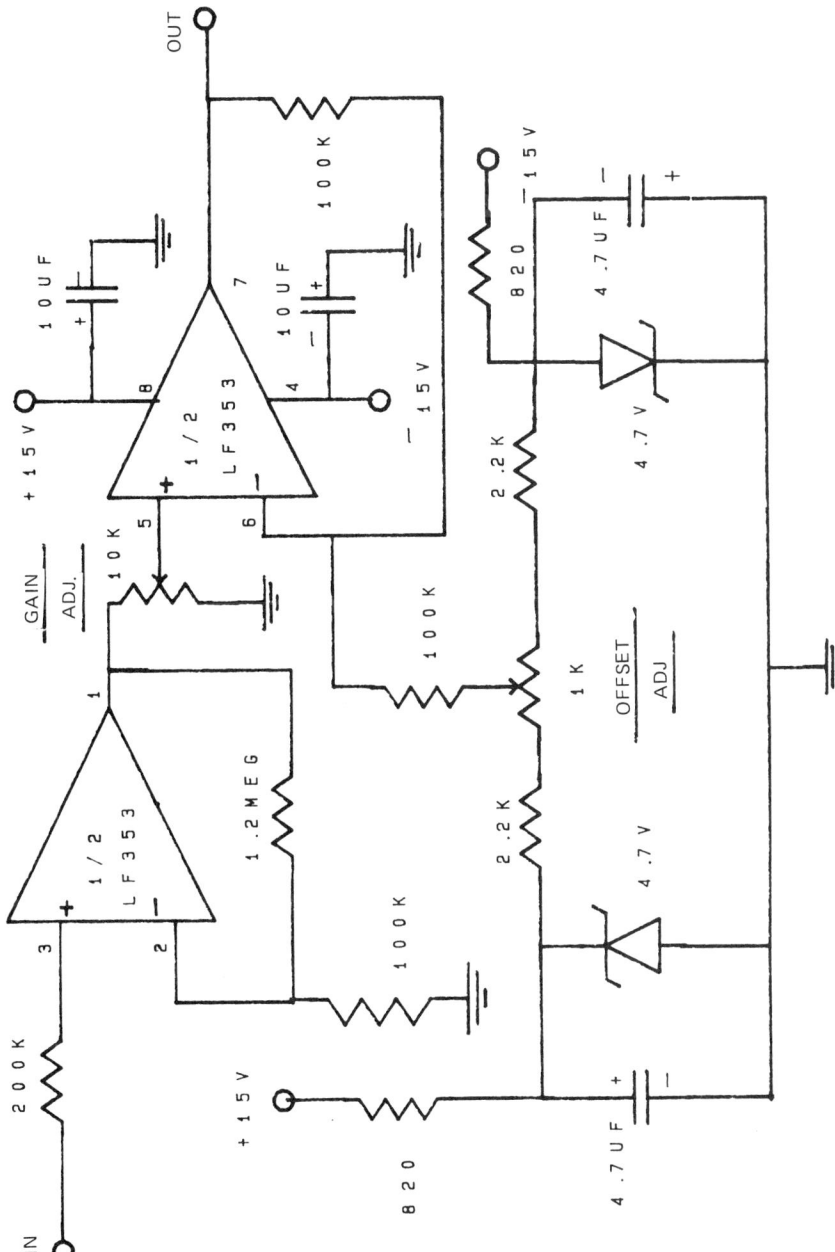

Figure 20. Schematic of a baseline offset amplifier for indirect detection methods. (Courtesy of E. L. Loree.)

when the highly conducting OH⁻ ion is used as the eluant [76-78] and the detection is essentially indirect. When a phthalate [79], borate [80], tartrate/malate [81], or ethylenediaminetetraacetate (EDTA) and related ligands [82] are used as eluant for anion SCIC, however, most analyte ions of interest have equivalent conductances larger than these eluant ions, and as such, these cases should be considered as direct detection.

A. Applications: Indirect Conductimetric Detection with Ion Exchanger Stationary Phases

One of the early reports in indirect conductimetric detection is by Pinschmidt [83]. To detect anions derived from high-pK acids (e.g., cyanide, borate, carbonate), the salt of a low-pK acid (NaCl) was used as the eluant; this was followed by postcolumn cation conversion to H^+. Such a system can hardly be considered "suppressed"; the second column converts the eluant background from NaCl to HCl, which results in an increase, rather than a decrease, in the background conductance. Replacement of the highly conducting HCl by a weak acid like HCN constitutes the basis for detection. In the actual system, NaOH was added to the NaCl eluant to maintain an alkaline pH, such that the weak acid anions of interest were adequately ionized. Pinschmidt also showed that by using a salt of an acid of moderate pK (e.g., Na_3PO_4) as eluant, emergence of weak acid annions is registered as dips on the background conductance whereas elution of strong acid anions such as sulfate or nitrate is indicated by a positive excursions from the baseline. Pinschmidt called these approaches *resistivity detection*; the difference from *indirect conductimetric detection*, as referred to here, is essentially a matter of semantics.

It has been argued in a recent review [84] that the inclusion of the suppressor converts a conductimetric detection system from a bulk-property detector to a solute-specific detector. This reasoning is rather stretched. A single-column system employing indirect conductimetric detection responds differently to ions of different equivalent conductance (bearing in mind the principles outlined in the foregoing section for indirect detection with an ion exchange column), in much the same fashion as a "suppressed" system responds differently to anions derived from acids with different pK values as well as to different equivalent conductances of the anions. It is certainly true, however, that inclusion of the suppressor in a conductimetric detection scheme makes the response more selective.

For indirect conductimetric detection without postcolumn conversion of the counterion (SCIC), a H^+ eluant for cation chromatography and an OH⁻ eluant for anion chromatography are clearly the eluants of choice from a standpoint of sensitivity. H^+ and OH⁻ ions display by far the largest equivalent conductances among cations and anions, respectively. Although this assures good sensitivity and unidirectional analyte signals, the eluting power of H^+ or OH^- ions is relatively limited. To keep the background conductance at a reasonable value, the eluant concentrations must also be kept reasonably low. This necessitates a low-capacity column (such that the analyte ions are eluted in an acceptably short period of time) and makes the analysis of strongly retained ions difficult. Decreasing column capacities also has some lower limit, because too low a column capacity may lead to frequent column overloading. Optimized chromatographic analysis of alkali metal cations with a 1.25-1.5 mM nitric acid eluant has been reported by Fritz et al. [75]. This paper is also an excellent source for a more detailed explanation of conductimetric response behavior of analyte ions in such systems. In the same study, chromatographic analysis of alkaline earth metals with the use of 1 mM ethylenediammonium dinitrate as eluant (the larger more strongly retained eluant cation is necessary to elute the divalent ions) was also demonstrated. The theory of separations with basic eluants and applications thereof were first described in the monograph on ion chromatography by Fritz, Gjerde, and Pohlandt [76]. The utility of a KOH eluant for anion chromatography has been established admirably by Okada and Kuwamoto [77]. LODs in the low parts-per-billion (ppb) range were attained for F^-, Cl^-, Br^-, NO_2^- and NO_3^-, with the use of 2 mM KOH as eluant. Higher KOH concentrations were necessary to elute more strongly held ions such as sulfate. Attainable sensitivities were significantly better than those obtained with the more commonly used phthalate eluant. Retention data for a large number of organic anions were also reported. In more recent work [78], this approach has been extended to IC determinations of weak acids. This approach would appear to be more versatile for detecting a greater variety of weak acids as compared to potentiometric detection, which has been strongly advocated recently for detecting ions such as sulfide or cyanide [85], or as compared to indirect detection methods that translate cyanide into iodide [707]. In the present author's experience, great care should be exercised in preparing and storing pure hydroxide eluants free from CO_2 contamination. Absorption of CO_2 leads to formation of CO_3^{2-} in the eluant which greatly alters both retention times (carbonate is far stronger than hydroxide in eluting power) and response behavior.

Recently, complex metalloanions, e.g., ethylenediaminetetraacetatocobaltate(III) $[Co(EDTA)]^-$, have also been successfully used as eluants for indirect conductimetric detection [86].

The stability of the baseline conductance signal that ultimately governs the attainable LOD is dependent on the detector/cell design as well as on effective thermostating. The conductivity detector, the mainstay of IC, is undergoing rapid and significant changes.

The bipolar pulse detection mode [87] has already been incorporated into commercial detectors [88]. A dual differential detector has been studied [89], and an ingenious cell/detector design has become commercially available [90]. The evolution in general design philosophies of conductivity/permittivity detectors and their combinations continues [91-96]. The recent combined conductivity/permittivity detector reported by Alder et al. [96] was able to attain a LOD of 50 ppb Cl⁻ in a 100-µl injected sample on an SCIC phthalate eluant system, even without major thermostating efforts; detection of this level on some commercial detectors may require 20 times this sample volume under comparable conditions [97]. Although electronic corrections can compensate for temperature fluctuations to a degree, the prime factor contributing to conductance detector instabilities is most often thermal drifts and has been so recognized [98,99]. At least one manufacturer markets active thermostating systems [100] and strongly advocates their use. At the very least, the system should be thermally insulated. A new detector combining fluorescence (excitation wavelength, λ_{ex}, 254 nm), conductivity, and UV (254 nm) detection in a single cell has appeared [101]. Although the design necessarily involves compromises and does not claim the utmost of performance in any given mode, the relative inexpensiveness of the combined functionalities is attractive.

Indirect conductimetric detection is hardly limited to simple samples; azide in human serum has been successfully determined by conductimetric SCIC [102].

B. Other Indirect Electroanalytical Detection Methods

Direct potentiometric detection employing electrodes that are selective for the analyte ion of interest is easily practiced [103-105]. Nonselective ion-exchange, membrane-based electrodes are also relatively easily fabricated [106]. In similar work carried out at the author's laboratory on Nafion membrane-based electrodes, we concluded that the sensitivities are inadequate for trace analysis; the results of the study described in Ref. 106 also indicate this. Liquid membrane-based electrodes may often show significantly nonselective response [107].

An unusual approach to indirect potentiometric detection of anions such as halides or thiocyanate by indirect potentiometric detection has been reported by Russian scientists [108]. The separation is carried out on a cadmium-form cation exchange column using cadmium acetate or nitrate as eluant. Because these anions form cadmium complexes of various strengths, they elute from the column at different times. The bottom of the column bed contains a layer of AgCl. The eluting ions alter the silver ion concentration effluent from this bed, and this change is sensed by an Ag/AgCl electrode.

Approaches to Ionic Chromatography

The general indirect route to potentiometric detection is to use an electrode that is sensitive to the eluant ion and thus monitors the emergence of the sample ion by registering a decrease in the eluant ion concentration in the effluent bed. The earliest example of this technique is likely the work of Deguchi et al. [109], who used a chloride ion-selective electrode and a NaCl eluant to monitor the emergence of various ions. More recently, potentiometric detection has been extensively exploited by Schmuckler et al. [110]. The eluant ion-selective electrode is constructed by forming an intimate coating of the silver salt of the eluant ion upon a bare silver wire. This electrode is used in conjunction with a conventional Ag/AgCl reference electrode in a flow cell with a high-input impedance voltmeter as the detector. The eluant ion is deposited on the silver wire by electrodeposition: The silver wire is immersed in the eluant and is connected as the anode; electrolysis (6-volt applied potential for 1 hr) is then conducted against a platinum cathode. In this work [110], 1-5 mM salicylic acid was used as eluant, and detection of Cl^-, Br^-, I^-, and SCN^- was demonstrated. In general, the electrode exhibited Nernstian response behavior over a large concentration range. It is not eminently clear whether the electrode responds to a change in the salicylate concentration or directly responds to halide and pseudohalide ions.

An electrode of the first kind consists of a metal in contact with its own cation [111]. An electrode such as this responds to changes in its free cation concentration. Consequently, one can use an approach wherein a solution of the free cation is continuously mixed with the column effluent. As species that complex the cation elute from the column, concentration of the free cation is reduced, and this change is then sensed by the electrode. This approach was used by Loscombe et al. [112] for the detection of diamines and amino acids; these workers employed a copper electrode and introduced copper sulfate in a postcolumn tee. Actually even when copper ion is not intentionally added, a small amount of copper ions is automatically formed at the electrode surface by autooxidation. Haddad and Alexander [113] have utilized this property to detect alkaline earth metals. The method is based on the dependence of the potential of a bare copper wire, functioning as an electrode, upon the concentration of chelating ligand ions such as tartrate, citrate, or glutamate which are present. Elution of alkaline earth metals decreases the concentration of the free ligand present and alters the potential of the sensing electrode. Separation and detection of Mg^{2+}, Ca^{2+}, Sr^{2+} and Ba^{2+} with 1 mM ethylenediammonium tartrate as the eluant was demonstrated on a silica-based cation exchange column. Subsequently these authors reported the successful detection of several organic acid anions, using phthalate, phosphate, or citrate eluants [114]. The most recent paper in this series addresses the detection of inorganic anions by a copper electrode

[115]. Two basic detection principles are involved: (a) reduction of free copper concentration at the electrode surface when a complexing anion, e.g., cyanide, elutes, or (b) increase in free copper concentration at the electrode surface when a strong oxidant anion, e.g., chlorate, bromate, etc., elutes. This paper is also an excellent source of the principles and operations involved in using such an electrode as a detector.

Coulometric/amperometric detection of electroactive species has been widely reported in the literature. Most of these detection methods are based on secondary rather than primary processes and hence are better classified as indirect methods. For example, the detection of cyanide by a silver electrode-based coulometric/amperometric cell does not involve the oxidation of cyanide but the oxidation of the electrode material to form the argentocyanide complex. Such methods involving controlled potential coulometric detection of ions such as CN^-, SCN^-, $Fe(CN)_6^{3-}$, $Fe(CN)_6^{4-}$, $S_2O_3^{2-}$, S^{2-}, etc. at a silver working electrode, maintained at +0.24 V, versus a saturated calomel electrode (SCE) was reported by Yoshinori et al. almost two decades ago [116]. Interestingly, $C_2O_4^{2-}$, PO_4^{3-}, and BrO_3^- were not detected by this system. A recent paper describing amperometric measurements of CN^-, Br^-, I^-, and S^{2-} at a silver electrode along similar lines has been published by Rocklin and Johnson [117]. A mercury-based electrode system for determining S^{2-} and CN^- has been described by Bond et al. [118]. A mercury-based polarographic detector was used by Russian workers to detect $EDTA^{2-}$, Cl^-, Br^-, I^-, SCN^-, SO_3^{2-}, $Fe(CN)_6^{3-}$, and $Cr_2O_7^{2-}$, with the use of a perchloric acid eluant [119].

Indirect coulometric detection of anions is also possible via pH changes accompanying analyte elution. Tanaka et al. [120] have introduced p-quinone in a postcolumn reactor to mix with the column effluent. The coulometric detector responds to changes in pH in such a system because of the involvement of the hydrogen ion in the reduction of quinone to hydroquinone. A more detailed exploitation of this appears in the work of Girard [121]. This system takes advantage of the fact that in a conventional carbonate-eluant, suppressed anion chromatography system, not only does elution of an analyte ion derived from a low-pK acid change the effluent conductance, but also the elution of the strong acid causes a drop in the effluent pH. Many electrochemical equilibria are pH dependent. If an electrochemically active compound, the redox properties of which are pH dependent, is introduced via a postcolumn tee to mix with the column effluent and the flow stream is then monitored with a coulometric detector, an indirect coulometric detection scheme is established. In his work, Girard [121] used a solution of 0.01 M quinone, 0.001 M hydroquinone, 0.1 M KCl (the KCl serving as the supporting electrolyte) as the postcolumn reagent, taking advantage of the pH-dependent reduction behavior of quinone (Q) to hydroquinone (H_2Q):

$$Q + 2H^+ + 2e^- \rightarrow H_2Q \tag{21}$$

In fact, the background current with almost any electrochemical detector and with an aqueous eluant is pH dependent, without the benefit of additional electroactive reagents (hydrogen ion itself is of course, the simplest electroactive species). In commercial suppressed IC systems, the most common location of the electrochemical detector is immediately after the separator column. If the detector is instead located after the suppressor column, species that are not electroactive are monitored indirectly via pH changes, as reported by Tarter [122,123]. Relative merits of these detection schemes (or for that matter, direct monitoring of pH after the suppressor column with or without introduction of a response linearizing buffer, see Ref. 124) are unclear.

C. Indirect Optical Detection Methods

The stability and sensitivity of optical detectors, especially absorbance detectors, make them the most widely used class of detectors used in liquid chromatography. Direct optical detection, whenever applicable, will be rarely equaled by other methods. For example, methods for determining NO_2^-/NO_3^- both of which display strong optical absorption, abound [125-128]. Indirect detection methods have been introduced for dealing with ions that have no useful optical absorption. Small and Miller [129] have shown, in exemplary detail, the principle and practice of such detection schemes. The salient considerations on the pertinent parameters are as follows: Because the noise is related to the absolute value of the baseline signal, high sensitivity and low LODs are achieved only with low concentrations of highly absorbing ions as eluants. An absorbance background of 0.2-0.8 absorbance units (AU) is likely to yield the best results. If the eluant concentration is fixed, the background absorbance may be adjusted by choosing an appropriate wavelength for monitoring; for this reason a variable-wavelength detector is especially useful. Of course, maximum sensitivity is attained when the absorbance is monitored at the absorption maximum of the eluant ion. The lower limit of the eluant concentration is dictated by imposing some convenient arbitrary upper limit on the retention time(s) of the analyte ion(s) of interest, because retention times will increase as the eluant concentration decreases. The relative affinities of the eluant ion versus the analyte ions for the ion exchange sites is an important parameter. Eluant ions that are particularly useful in indirect optical detection include nitrate, iodide, phthalate, sulfobenzoate, and trimesate; the eluting power of these ions increases in the cited order. Whereas iodide and nitrate are useful eluants for relatively weakly retained anions, more powerful eluant ions are required for convenient separation of strongly retained analyte anions. The magnitudes of the charges on the eluant ion and the sample ion govern the retention behavior in a predictable fashion. From basic ion-exchange theory, it is derivable that:

$$t_R [E]^x = \text{constant} \tag{22}$$

where t_R is the corrected retention time (observed retention time minus the time necessary for an unretained material to elute, calculatable from column void volume) attained with an eluant concentration of [E], and x is the ratio of the charge magnitude of the sample ion to that of the eluant ion. For example, if I^- is the eluant ion and SO_4^{2-} is the sample ion, the value of x is 2; if phthalate is the eluant ion and SCN^- is the sample ion, the value of x is 0.5. Equation (22) appears to hold well for a large number of sample/eluant ion combinations; note, however, that the constant in Eqn. (22) is a constant only for a given sample/eluant ion/column combination. A simple way to consider Eqn. (22) is to realize that the eluting power of any given eluant is proportional to its concentration raised to the power of the charge magnitude of the eluant ion. One practical implication is obvious: If the retention time is known for one particular concentration of an eluant, the retention time for the same sample ion at another concentration of the same eluant can be easily evaluated. Typical eluant concentrations in these systems are 1-10 mM. Attainable separations and detection limits are represented by Figures 21 and 22. Cation separation and detection can be accomplished with a cation exchange column and an absorbing cation, e.g., Cu^{2+}, as the eluant. By connecting two columns of different ion exchange capacities in series and switching at the appropriate time, separation of Na^+, K^+, Mg^{2+} and Ca^{2+} with a single eluant was demonstrated (Figure 23). One particularly interesting aspect of indirect optical detection is the potential of simultaneous cation and anion analysis. Small and Miller, using a cation exchange and an anion exchange column serially connected, showed how a mixture of NaF, RbCl, and $MnCl_2$ can be resolved into the five component ions, using a 5 mM copper nitrate eluant (Figure 24). Obviously, considerations on the eluting power of both the eluant cation and anion and their respective absorption behavior are critically important. With real samples, the probability of coelution of a cation with an anion clearly exists, and may cause difficulties in interpreting the results. One potential way to circumvent this problem is to devise an eluant system wherein there is no overlap between the characteristic absorption of the eluant cation and the eluant anion at the respective wavelengths used to monitor them simultaneously and separately (with two different detectors or with a dual-wavelength detector). The merit of this approach is clear in the recent work of Iskandarani and Miller [713].

Approaches to Ionic Chromatography

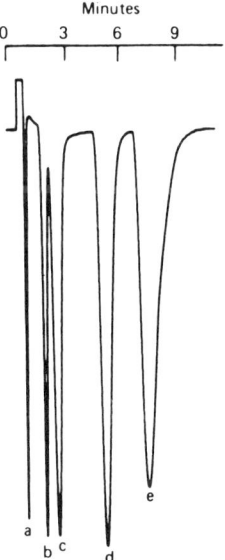

Figure 21. Indirect optical detection of several anions: (a) carbonate, 1.8 µg; (b) chloride, 1.4 µg; (c) phosphate, 3.8 µg; (d) azide, 5.0 µg; (e) nitrate, 10 µg. Column: 4 × 250 mm surface agglomerated 40-µm cation exchanger coated with 0.6-µm colloidal anion exchanger. Eluant: 1 mM sodium phthalate, 1 mM boric acid, pH 10. Flow rate: 5 ml/min, 20-µl sample. (From Ref. 129.)

Alkyl quaternaryammonium compounds have been analyzed by indirect optical detection [130]. Several inorganic anions have been determined by Cortes and Stevens [131] via indirect optical detection, and an amino column, with phthalic acid adjusted to various pHs as eluant. An ingenious approach reported by Andrasko [132] utilizes p-toluenesulfonate as eluant. Although this ion absorbs intensely around 210 nm, it is almost transparent at wavelengths greater than 240 nm. Consequently, dual-wavelength detection may be used to carry out both direct and indirect optical detection simultaneously. The pH of the column effluent may be monitored (see Section III.B) via introduction of a suitable acid/base indicator by a postcolumn reactor and optical monitoring of one of the forms of the indicator [133]. The full potential of such an approach remains unexplored.

One other notable aspect of indirect photometric detection is the possibility of using facile universal calibration, if the sample ion is essentially nonabsorbing at the monitoring wavelength. The signal

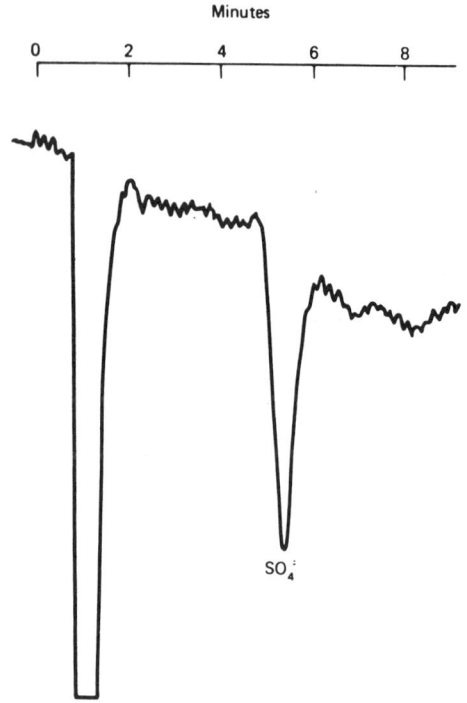

Figure 22. An example of the sensitivity of indirect optical detection. Sulfate peak due to a 100-μl injection of 1 μm sulfate. Eluant, 0.1 mM sodium sulfobenzoate, pH 8; flow rate, 1 ml/min; wavelength, 224 nm, 0.005 absorbance units, full-scale (AUFS). (From Ref. 129.)

that is produced originates from displacement of the eluant ion by an exactly equivalent amount of the sample ion in the sample band; the area under the dip (or peak, if the signal has been inverted) is directly related to the number of equivalents of sample ion eluted in that band, regardless of its specific identity. Note that for polyvalent ions, the ion must be considered in the form that it exists under the chromatographic conditions, i.e., phosphate should be considered as HPO_4^{-2} with a pH 8 eluant and not as PO_4^{-3}. If the sample ion displays significant absorption at the detection wavelength, the calibration factor should be modified as:

$$C_i = C_j \frac{\varepsilon_E}{\varepsilon_E - \varepsilon_i} \tag{23}$$

Approaches to Ionic Chromatography

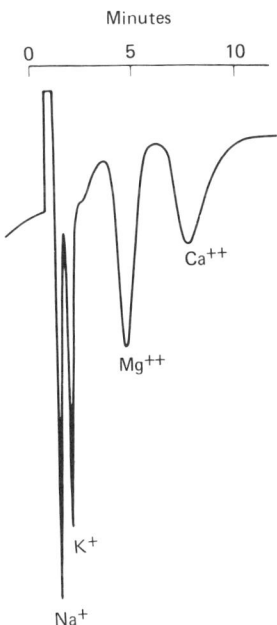

Figure 23. Separation of mono- and divalent cations by a split-column technique and indirect optical detection. First column: 2.8 × 250 mm surface-sulfonated styrene-DVB, 0.015 meq/g. Second column: 2.8 × 250 mm surface-sulfonated styrene-DVB, 0.087 meq/g. Eluant, 1.25 mM copper sulfate; detection at 216 nm. (From Ref. 129.)

where C_i is the calibration factor for the absorbing ion i (expressed in injected equivalents/unit dip area), and C_j is the calibration factor for a nonabsorbing ion j (e.g., sulfate); ε_E and ε_i are the equivalent absorptivities of the eluant ion E and the sample ion i, respectively (to obtain equivalent absorptivity, the molar absorptivity is divided by the charge on the ion). If ε_i is greater than ε_E (as is true for nitrite as a sample ion and nitrate as the eluant ion), a peak rather than a dip will be produced. Jenke [134] has recently presented data further verifying the validity and utility of universal calibration in indirect optical detection. Ultimately, universal calibration in indirect detection can be extended without any prior knowledge of absorptivities whatsoever, as shown by Wilson and Yeung [135].

The utility of indirect optical detection for anion SCIC with the commonly used phthalate eluant has also been demonstrated independently

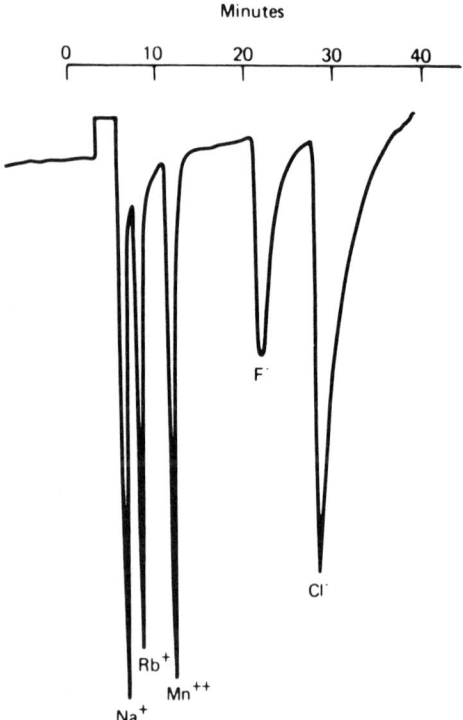

Figure 24. Joint determination of cation and anions by indirect optical detection. Columns: 4.6 × 250 mm Partisil SAX followed by 4.6 × 250 mm Partisil SCX. Eluant, 5 mM copper nitrate; detection at 241 nm. Sample: 20 µl NaF + RbCl + MnCl, 0.2 N each. (From Ref. 129.)

by Cochrane and Hillman [136]. Sub-parts-per-million LODs were claimed for a variety of organic and inorganic common ions.

The use of a fluorescent eluant ion for indirect fluorometric detection is clearly a potential avenue worthy of pursuit, but has not been explored in IC, except for the recent work of Mho and Yeung [714], to this author's knowledge. Anions derived from an aromatic amine, e.g., carboxylate and sulfonate derivatives of aniline and naphthylamine, are intensely fluorescent, exhibit a great range of eluting power, and as such, should be particularly interesting. It is likely, however, that attainable LODs are going to be limited by baseline noise, rather than by sensitivity.

The exploitation of optical properties such as absorbance and refractive index (RI) are interrelated. The optical design of most low-volume absorbance detectors are such that they routinely respond to changes in the RI of the column effluent. Haddad and Heckenberg [137] reported that for anion SCIC with phthalate eluents, indirect RI detection produces superior results, as compared to indirect absorbance or conductimetric detection (Figure 25). Using this approach, cation analysis was also shown to be feasible; an anilinium salt was found to be most suitable for separating alkali cations. Computer optimization of eluant concentration and pH for such separations has been described [138]. Buytenhaus has also reported his favorable experience with indirect RI detection [139]. It is also worthwhile to note in passing that with the introduction of eluants such as methanesulfonate [140], which display very low optical absorption down to 190 nm, direct optical detection below 200 nm may be attractive for some anions that do not absorb appreciably above 200 nm [141]. Jackson et al. [142] have used chloromethanesulfonate as eluant for the sensitive determination of nitrite and nitrate in cured meats by direct UV detection at 210-214 nm. Eek and Ferrer [143] found perchlorate to display significantly lower absorption in the low-UV wavelengths as compared to methanesulfonate, and therefore recommended the use of perchlorate as eluant for direct UV detection. Even with conventional carbonate or borate eluants in a suppressed system, direct UV detection may be a valuable tool, by itself or as an adjunct to conductivity detection [144]. It is of particular interest to note Williams's observation [144] that postcolumn cation exchange for H^+ not only decreases the background conductance of borate or carbonate eluants, it also decreases the background absorbance because the undissociated acids absorb less.

A novel indirect optical detection approach has been introduced by Downey and Hieftje [145] and termed *replacement ion chromatography*. For anion chromatography, a conventional carbonate-eluant suppressed system is followed by a short cation-exchanger column in K^+, or more commonly in Li^+, form. Elution of low-pK acids from the suppressor column is accompanied by a drop in pH. The free hydrogen ion in the effluent is translated into lithium by the lithium replacement column. The lithium concentration is monitored by flame photometry with a total consumption burner. Achievable LODs were not especially good with the carbonate eluant because of the significant background lithium bleed from the replacement column. With a hydroxide eluant, LODs of 15 and 35 ppb (250 μl injection) were achieved for F^- and Cl^-, respectively. If the background were truly pure water, the LODs would have been two orders of magnitude lower, according to the authors. Micromolar level impurities in the water and the reagents are believed to be the reason for the failure of the suppressed hydroxide eluant to reach calculated LODs. Indeed, when deionized water was used as the carrier, the LOD for directly injected Li^+ was 14 nM. Cation chromatography was conducted in a similar fashion via use of a nitric acid eluant, a cation exchanger separator, and an

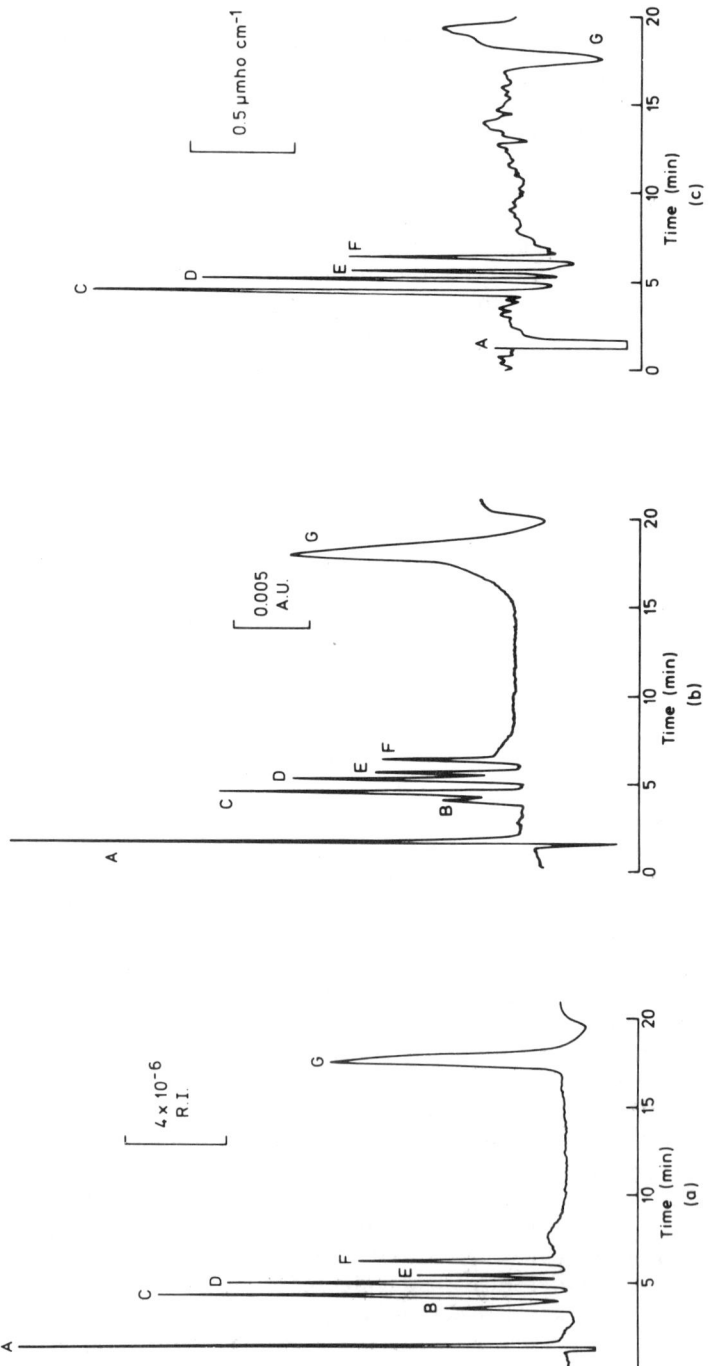

Figure 25. Separation of inorganic anions with (a) indirect refractive index (RI), (b) indirect UV absorbance, (c) indirect conductimetric detection. Vydac anion column; flow rate, 2 ml/min; eluant, 4 mM potassium hydrogen phthalate at pH 4.0; injection volume, 100 μl. (A) System peak, (B) 2 μg $H_2PO_4^-$, (C) 2 μg Cl^-, (D) 2 μg NO_2^-, (E) 2 μg Br^-, (F) 2 μg NO_3^-, (G) 2 μg SO_4^{2-}. (From Ref. 137.)

Approaches to Ionic Chromatography

OH⁻ form suppressor, followed by the lithium replacement column. Although a linear response over four orders of magnitude of injected concentration, and good LODs were cited as the preeminent advantages of replacement ion chromatography, practical applications may well be deterred by the complexity of the system.

IV. ION CHROMATOGRAPHY WITH ION INTERACTION REAGENTS

The term *ion interaction chromatography* denotes an approach in which eluants containing lipophilic counterions (ion interaction reagents, IIR) are used in conjunction with a nonpolar stationary phase (reverse phase) to separate ionic species. This approach substantially predates the introduction of suppressed or nonsuppressed conductimetric IC and has had the distinction of bearing a bewildering variety of names: soap chromatography [146], ion-pair chromatography [147], paired-ion chromatography [148], solvent-generated (Dynamic) ion exchange chromatography [149,150], detergent-based ion exchange chromatography [151], surfactant chromatography [152], hetaeric chromatography [153], ion exchange chromatography with adsorbed ion exchangers [154], chromatography on dynamically coated columns [155], and most recently ion interaction chromatography [156-158]. Although this technique has enjoyed widespread use in the past for the separation of ionic and ionizable organic solutes, including biomolecules, applications toward the separation of inorganic ions and smaller organic ions are of more recent origin.

Detailed quantitative treatment of the retention mechanism spanning the entire scope of applicability of ion interaction chromatography has not as yet been developed. The terms *dynamic ion exchange chromatography* and *ion-pair chromatography* exemplify the two extreme views of the retention mechanism believed to be involved in this type of separation, as detailed below.

A. The Dynamic Ion-Exchange Model

The dynamic ion-exchange model envisages as a first step, the adsorption of the lipophilic counterion on the stationary phase. For example, if tetrapropylammonium hydroxide (NPr_4OH) is being used as the IIR, NPr_4^+ ions will be adsorbed on the stationary phase by virtue of the lipophilic alkyl groups attached to this ion. The OH⁻ ions, in contrast, are present merely to satisfy charge balance requirements and are subject to replacement by other anions entering the system. The situation is quite analogous to a conventional anion-exchanger stationary phase that contains bonded quaternary ammonium anion exchanger functionalities, with the exchangeable ion being hydroxide. Indeed, the similarity is even greater when very highly lipophilic IIRs are used to render a nonpolar stationary phase into an ion exchanger, discussed in greater detail in Section IV.C. In general, the dynamic ion-exchange model correctly predicts the retention order of analyte ions; that is, their retention order is predicted to be the same as that observed with a conventional bonded ion-exchanger packing. Although it has been argued [158] that the observed retention orders with systems employing IIRs often display "anomalous" retention behavior (e.g., a singly charged analyte ion is occasionally retained more strongly than a doubly charged ion) as compared to what is expected strictly in terms of electrostatic forces, it should be recognized that even with conventional, bonded ion exchangers, forces other than electrostatic binding play a role, and may indeed dominate overall retention behavior; this is especially true with ion exchangers based on poly(styrene-divinylbenzene) (PSDVB) matrices: Perchlorate and trichloroacetate are often retained more strongly than sulfate; picrate, a singly charged anion, is retained so strongly that it cannot be eluted with any purely aqueous eluant. Lipophilic interactions clearly play a role in retention of ionic analytes, whether with bonded ion exchangers or IIR-modified stationary phases. The softer the ion (in terms of the hard-soft acid-base theory; see Ref. 159) and the stationary phase matrix, the greater will be the contribution of the lipophilic interaction to overall retention.

The dynamic ion-exchange model also correctly predicts the role of the eluant ion (the ion obligatorily present with the lipophilic counterion of the IIR, of similar charge to the analyte ion; e.g., OH^- when NPr_4OH is used as the IIR) in governing retention behavior of analyte ions. As the eluant ion is changed from OH^- to Br^- to I^- without changing the nature or concentration of the lipophilic ion of the IIR, the retention of any given analyte ion decreases in that order. Again, this is exactly analogous to the observed decrease in the retention of any given analyte anion to an anion exchange column when the eluant is switched from NaOH to NaBr to NaI; the eluting power of eluant ions follows the same order as their affinity for the ion exchange sites. The dynamic ion-exchange model does not, however, address the question whether the extent of adsorption of the IIR itself is influenced by the nature of the eluant ion.

Sorel and Hulshoff [160] have recently reviewed dynamic anion-exchange chromatography; the emphasis is largely on theory, and references to literature separations are limited to organic analytes.

Although the dynamic ion-exchange model is of significant utility for qualitative understanding of the retention behavior of analyte ions, there are serious shortcomings of the model toward a quantitative explanation of several aspects of ion interaction chromatography. If the dynamic ion-exchange model were strictly valid, total ionic concentration in the effluent from a column would always be the same and would be equal to the eluant ionic concentration, once equilibrium has been established, as is the case with a conventional ion-exchange column. The analogy between an anion exchanger column employing a

KOH eluant (and say, indirect conductimetric detection as in Ref. 77) and a PSDVB column with a NPr$_4$OH eluant will be complete, and the principle of universal calibration (which relies on the constancy of the effluent ionic concentration, see Section III.C) should be applicable for both cases. Because OH$^-$ displays an equivalent conductance higher than any other anion, we expect that the emergence of any injected anion will be accompanied by a dip in the background conductance (see Section III.A). Although this is indeed valid for the true ion-exchange system and constitutes the basis for a highly sensitive method [77], the elution of many strongly retained anions, e.g., perchlorate, is accompanied by a peak, rather than a dip, in the system with the nonpolar stationary phase and NPr$_4$OH as the IIR [161]. This is possible only if the effluent ionic concentration changes during elution of sample ions. When the eluant ion is especially hard (e.g., hydroxide or acetate), the total effluent ionic concentration invariably increases during the elution of softer analyte ions. Using benzyltributylammonium / 1-naphthylmethyltripropylammonium / 1-naphthylmethyltributylammonium acetates as IIR in conjunction with a C-18 silica column, Barber and Carr [162] monitored the concentration of the lipophilic cation in the eluant by its strong UV absorption. If the ionic concentration of the effluent were constant, no change in the background absorbance would be expected. With all the sample ions studied, elution of the analyte ion was indicated by a peak over the background absorbance, indicating that the concentration of the IIR counterion actually increased in the effluent during sample elution. Hackzell and Schill [163] have shown that for sample ions oppositely charged to the IIR, the IIR concentration in the effluent band containing the sample ion is less than the IIR concentration in the eluant, if the sample ion has a capacity factor less than that of the eluant ion. Conversely, when the sample ion is retained more than the eluant ion, i.e., when the sample peak elutes later than the system peak, the IIR concentration in the sample peak is greater than that in the eluant.

B. The Ion-Pair and the Ion Interaction Model

The ion-pair model envisages as a first step the formation of an ion pair between the lipophilic counterion and the sample ion. The ion pair is then thought to be adsorbed onto the stationary phase. According to this view, the retention of a sample ion is dependent on the relative affinity of the lipophilic counterion to form an ion pair with the sample ion rather than with the eluant ion. Further, the retention is dependent on the affinity of the resulting ion-pair for the stationary phase. For relatively small lipophilic counterions and sample ions in aqueous eluant systems, experimental evidence does not support the ion-pair model. An indication of ion-pair formation may be obtained from conductance studies because ion pairs do not contribute to electrical conductance [164]; no evidence of ion pair formation for any combination of a number of analyte ions and ion pairing reagents was found in methanol:water solvent systems [156]. The applicability of conductimetric detection methods for single-column IIR-based chromatography [6] also clearly indicates that pairs of ions, rather than ion-pairs are present in the column effluent.

The ion interaction model acknowledges that multiple forces are operative in such a system; i.e., an electrostatic force as well as an adsorptive (lipophilic) force. Imagine for example, a system containing a cationic IIR, e.g., NR_4^+. A double layer is formed on the stationary phase, the primary layer is positively charged because of adsorption of the cationic IIR, whereas negatively charged eluant ions are present in the secondary layer to maintain electrical neutrality. Beyond this is the bulk eluant, with a random but macroscopically even distribution of cations and anions. The retention of a negatively charged sample ion is thought to involve the following steps [158]:

1. Sample ion competes for access to the secondary layer.
2. Sample anion is electrostatically attracted to the primary layer.
3. If the sample anion is soft, its affinity for the primary layer is aided by lipophilic interactions with the stationary-phase matrix.
4. At this point, an extra negative charge is present in the primary layer. To maintain electrical neutrality, an additional NR_4^+ ion enters the primary layer.
5. The net result is that a *pair of ions*, not an *ion-pair*, has entered the primary layer because of a combination of electrostatic and lipophilic forces.
6. This sequence is reversed to bring about the desorption of a pair of ions (not necessarily the same pair, because a different anion may be desorbed).

The ion interaction model is supported by a variety of experimental evidence. The credit for first applying the concept of the electrical double layer to IIR-based systems should go to Cantwell and Puon [165]. In a lucid account of a brilliantly designed study, Cantwell and Puon [165] applied the Stern-Gouy-Chapman (SGC) theory of the electrical double layer to the adsorption of lipophilic organic ions on a nonpolar stationary phase, Amberlite XAD-2. A primary layer of lipophilic ions and a secondary layer of counterions satisfactorily explained observed microelectrophoretic behavior and adsorption isotherm data. The SGC theory also predicts increased retention of the lipophile (and consequently increased retention of the sample counterion) with increasing ionic strength of the eluant, the ionic strength being adjusted by some indifferent electrolyte. If a lipophilic ion is used as the sample, a linear relationship between the reciprocal of the capacity factor [166] and the reciprocal of the square root of the

Approaches to Ionic Chromatography

ionic strength of the eluant is expected and has been so found by Cantwell and Puon [165], as well as by Iskandarani and Pietrzyk [167]. If a lipophilic ion is used as the IIR in the eluant and the eluant ionic strength is altered by some indifferent electrolyte, the reciprocal of the capacity factor of the sample ion is directly proportional to the ionic strength [167,168]. The model is also supported by the quantitative aspects of the retention behavior of amines with alkylsulfonates as IIR [169]; the degree of retention of a sample ion has been shown to be directly proportional to the surface charge arising from the adsorbed IIR [170]. The SGC theory has also been applied to explain the retention of cisplatin in IIR-based systems [171]. The correlation of the Hammet substituent constants of a number of organic ions with retention parameters in ion interaction chromatography [172,173] also supports the ion interaction model, and quantitative treatment of the multiplicity of the forces involved has been initiated [174].

One note of caution is in order. The theory of ion interaction chromatography, referred to above, has been developed primarily for hydrophobic organic ions. Whether these considerations will strictly apply to hydrophilic inorganic ions is unknown.

C. Immobilized IIR Systems

Specific examples of some actual chromatographic systems are given below. A virtual cation exchanger column may be prepared by pumping an anionic IIR [e.g., 500 ml of a 0.25 mM solution of sodium 1-eicosylsulfate, $CH_3(CH_2)_{19}SO_3Na$ in acetonitrile:water (25:75)] through a C-18 silica column and then switching the eluant to water [155]. The IIR is so strongly adsorbed on the stationary phase that it is not appreciably removed by a purely aqueous eluant, for which it has little affinity. Cation exchange sites are generated by the replaceability of the sodium ion. Anion exchange properties can similarly be rendered to reverse phases by coating such stationary phases with dilute (0.67 mM) solutions of trioctylmethylammonium, tetraoctylammonium, or tridodecylmethylammonium iodides in 80:20 methanol:water [155]. Depending on the amount of solution passed through the column, the amount of the adsorbed IIR varies between 20 and 80 mg. Further details of this approach have been presented by the authors in a more recent paper [175]. There is essentially no difference between ionic separations on these types of modified stationary phases and conventional bonded ion-exchange supports. In a sense, this approach is equivalent to liquid/liquid chromatography (LLC) [176], which was the major separation mode in liquid chromatography prior to the development of bonded phase packings. There are advantages to such an approach:

1. For convenient renewal, the adsorbed IIR may be removed by washing the packed column in situ with a suitable solvent (typically methanol or acetonitrile) and recoated many times.
2. The same column can be rendered into a cation or an anion exchanger (many manufacturers do not recommend this, however).
3. The (virtual) ion exchange capacity can be varied from very high to very low by varying the amount of the IIR adsorbed on the column for facile separation of very weakly retained as well as very strongly retained analyte ions.

A typical separation employing this approach is shown in Figure 26.

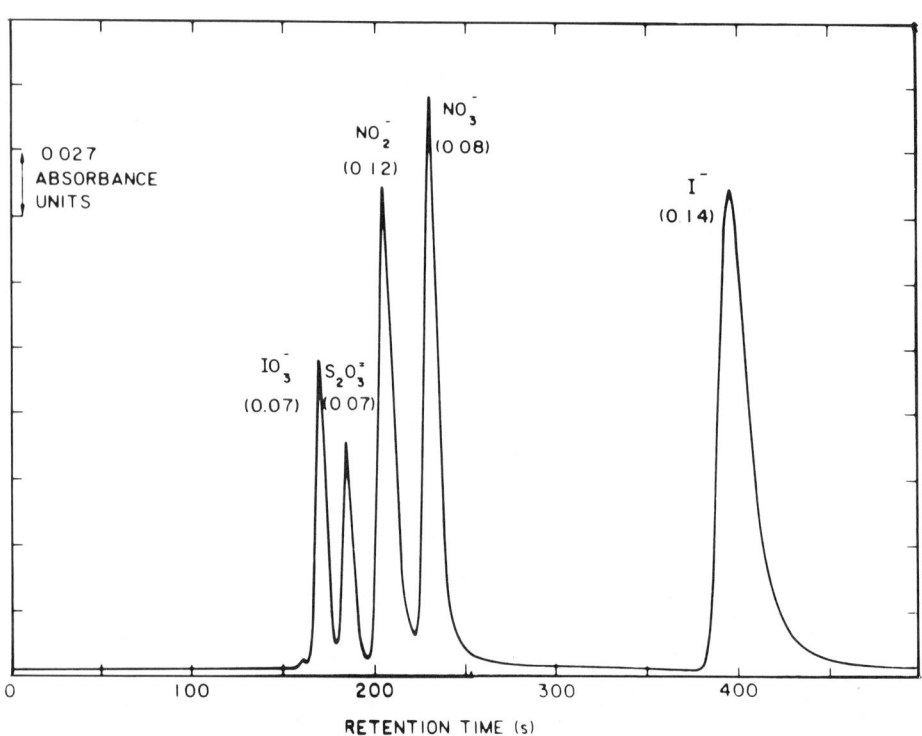

Figure 26. Separation of anions on a C-18 silica column previously equilibrated with 200 ml of 1 mM tridodecylammonium iodide. Coating solution in 75:25 methanol:water; flow rate, 1 ml/min. Numbers in parentheses are HETP values in millimeters. (From Ref. 155.)

D. Ion Interaction Chromatography: Dynamic IIR Systems

Tomlinson et al. [152] and Bidlingmeyer [177] have reviewed the basic principles and methods involved in ion interaction chromatography. Some specific applications in combination with different detection techniques are discussed below.

Various cationic drugs and biogenic substances were separated by Persson and Lagerstrom employing normal-phase LLC on 10-μm silica gel [178]. The stationary phase was coated with sodium perchlorate and perchloric acid, or methanesulfonate as the anionic IIR. The mobile phase was n-butanol/dichloromethane/n-hexane. Anionc substances were similarly separated with a stationary phase containing tetrabutylammonium ion as the cationic IIR. Excellent detection limits were attained for several compounds of interest via direct UV detection. Nakae et al. [179] have separated various cationic surfactants (alkylbenzyldimethylammonium and alkylpyridinium ions) on a 10-μm PSDVB column, using perchlorate as an anionic IIR and UV detection. Terweij-Groen et al. [151] have studied the chromatographic behavior of several organic acids on a 10-μm C-18 silica column, using cetyltrimethylammonium ion as the cationic IIR, with varied amounts of the IIR (0.1-2%, n-propanol (10-25%), NaBr (0-0.1 M), and phosphate (0-0.075 M) in the eluant at varied pH (1.8-7.5) and temperature (25°-45°C). Direct optical detection was used. The same IIR was also utilized to achieve separations of sulfa drugs, various anionic dyes used as food additives, and also salicylic acid and indomethacin in deproteinized blood plasma. In the same study, the use of dodecylsulfate as an anionic IIR was demonstrated for the analysis of quinine in lemon tonic, B vitamins in medicinal preparations, adrenaline in eye drops, and dipeptides.

Using 10-μm C-18 silica columns and hexylammonium ion as the cationic IIR, Van de Venne et al. [180] have studied the retention behavior of 15 carboxylic acids and nitrate, as a function of pH and methanol content and ionic strength of the eluant. The separation of several anionic dyes was studied by Tomlinson et al. [181], using various alkylbenzyldimethylammonium cations as the IIR in a 50:50 methanol:water eluant as a function of the IIR concentration with direct UV detection.

Excellent sensitivities were reported by Reeve [182] for the separation and detection of main-group inorganic anions on a 10-μm CN-silica column with direct UV detection and a methanol/water eluant containing cetyltrimethylammonium ion as the IIR and 0.1 M Na_2HPO_4 and 0.1 M KH_2PO_4 for buffering and elution. The detection limit ranged from 1.2 ng for $NaNO_2$ to 90 ng for KBr for a 10-μl injected sample.

In an exemplary study, Molnar et al. [6] described the high-performance liquid chromatography of ions via direct conductimetric or UV detection. With the use of aqueous 5 mM heptylsulfonate at pH 2 as the anionic IIR, Na^+, K^+, NH_4^+ and Ca^{2+} were separated on a C-18 silica column and detected by conductimetry. Anions such as F^-, Cl^-, SO_4^{2-}, NO_2^-, Br^-, CrO_4^{2-}, and NO_3^- were similarly separated and detected with the use of 2 mM aqueous tetrabutylammonium as the cationic IIR in a 0.05 M phosphate buffer at pH 6.7. Because of its high optical absorption, the detection of I^- by UV absorption was found to yield better sensitivity. The same study also compares the above results with separations attainable on silica-based ion exchange columns with or without suppressor columns.

Knox and Hartwick [170] studied the retention of cations, anions, and zwitterions as well as neutrals on a 5-μm C-18 silica column with an 80:20 water:methanol eluant containing a 20 mM phosphate buffer at pH 6. Alkyl sulfates containing 8-12 carbon chains were used at varied concentrations as the anionic IIR. The primary factor governing retention was found to be the surface charge arising from the adsorption of the IIR; ion pairing was found to be unimportant. Bartha and Vigh [183] also studied the retention of several cationic, anionic, and neutral solutes on a C-18 silica column; the IIR (tetrabutylammonium) concentration and methanol content (0-70%) of the aqueous eluant were varied. These authors have also carried out fundamental studies on the roles of various parameters of importance in IIR-based chromatography, e.g., the effect of the concentration of the hydrophilic counterion [184] and the role of the chain length of the IIR [185]. Rotsch and Pietrzyk [157] studied the retention of weak mono- and diprotic organic acids as well as ampholytes on a macroreticular PSDVB stationary phase (Amberlite XAD-2), employing various tetraalkylammonium salts as the cationic IIR, and studied the effect of pH and organic solvent (acetonitrile) content of the mobile phase. In subsequent work with a more efficient PSDVB stationary phase (10 μm, PRP-1, Hamilton) Iskandarani and Pietrzyk [167] separated various simple organic ions via direct UV, refractive index, or conductimetric detection. This was then extended to the separation and conductimetric detection of F^-, Cl^-, Br^-, I^-, CN^-, SCN^-, NO_2^-, NO_3^-, BrO_3^-, MnO_4^-, ClO_3^- IO_3^- $HCOO^-$, CH_3COO^-, S^{2-}, SO_4^{2-}, SO_3^{2-}, CO_3^{2-}, $C_2O_4^{2-}$, $Cr_2O_7^{2-}$, $CH_2(COO^-)_2$, CH_2CH_2-$(COO^-)_2$, and fumarate, with the use of tetrapentylammonium as the cationic IIR, with or without added NaF, at varied (17.5-35%) acetonitrile content of the aqueous eluant [168].

Whenever the eluant system is optically transparent relative to the analyte ion, direct optical detection is simple and attractive. Skelly [186] employed a number of C-18 and C-8 silica columns with 10 mM octylammonium ion as the cationic IIR (HCl or H_3PO_4 was added, chloride or phosphate functioning as the eluting ion) with or without the addition of 15% acetonitrile to the aqueous eluant. A large number of inorganic (IO_3^-, IO_4^-, Br^-, BrO_3^-, NO_2^-, NO_3^-, MnO_4^-, CNO^-,

$S_2O_8^{2-}$, I^-, SCN^-) and organic (glycine, acrylamide, glycolate, formate, acetate, lactate, acrylate, pyruvate, propionate, monochloroacetate, dichloroacetate, methacrylate, fumarate, 2,2-dichloropropionate, oxalate, maleate, trichloroacetate, itaconate) ions were separated and detected by UV detection at 205 nm. Detection at sub-part-per-million levels was routinely achieved for a 200-µl injected sample. A typical chromatogram is shown in Figure 27. The facile application of the methodology for ionic analysis of several "real-world" sample types (high-purity water, raw brine, nitration wastewater, silage samples) was demonstrated by Skelly. Cassidy and Elchuk [155] similarly employed UV detection at 215 nm for the separation and detection of IO_3^-, NO_2^-, NO_3^-, I^-, and $S_2O_3^{2-}$, via a 10-µm PSDVB column (PRP-1) with 1 mM trioctylmethylammonium or tetrabutylammonium hydroxide in the eluant, which contained 50% or 15% acetonitrile, respectively. The chromatographic efficiency was found to be poor with purely aqueous eluants, but increased dramatically with the incorporation of only a small amount (5-10%) of acetonitrile. Methanol was not found to be particularly effective for this purpose. The authors believe that the poor efficiency with aqueous eluants is caused by the inability of the purely aqueous eluant to wet the stationary phase. Interestingly, C-18 silica stationary phases were not found to suffer from a similar wetting problem. Excellent linearity of response (in terms of peak area counts) was found for the sample ions studied over a large concentration span.

Mullins and Kirkbright [187] have used 10 mM cetyltrimethylammonium chloride in aqueous acetonitrile, buffered typically at pH 6.8 with 10 mM phosphate, for the separation of nitrate, nitrite, iodide, bromide, and iodate on a C-18 silica column (employing UV detection). Retention behavior was shown to obey the solute partitioning equation for micellar chromatography originally derived by Armstrong and Nome [188].

Using a 10-µm CN-silica column and 0.1% hexadecyltrimethylammonium chloride with 0.1 M each of Na_2HPO_4 and KH_2PO_4 in an aqueous eluant containing 25% acetonitrile, de Kleijn reported [189] low-nanogram-level detection limits for 200-µl injected samples of IO_3^-, BrO_3^-, NO_2^-, NO_3^-, Br^- and I^- with UV detection at 204 nm. Using a 10-µm C-18 silica column, tetrabutylammonium ion as the IIR, and a phosphate buffer (pH 3-7), Vespalec et al. [190] studied the separation of several inorganic ions, employing direct optical detection at 210 nm. Aside from common main-group anions, e.g., NO_2^-, NO_3^-, Br^-, I^-, etc., the authors studied the retention of oxo metal anions, e.g., CrO_4^{2-}, MoO_4^{2-}, and WO_4^{2-} as a function of pH and ionic strength of the eluant. Interestingly, whereas the main-group anions are not significantly retained in the absence of an IIR on a C-18 surface, this was not the case for the oxo metal anions studied.

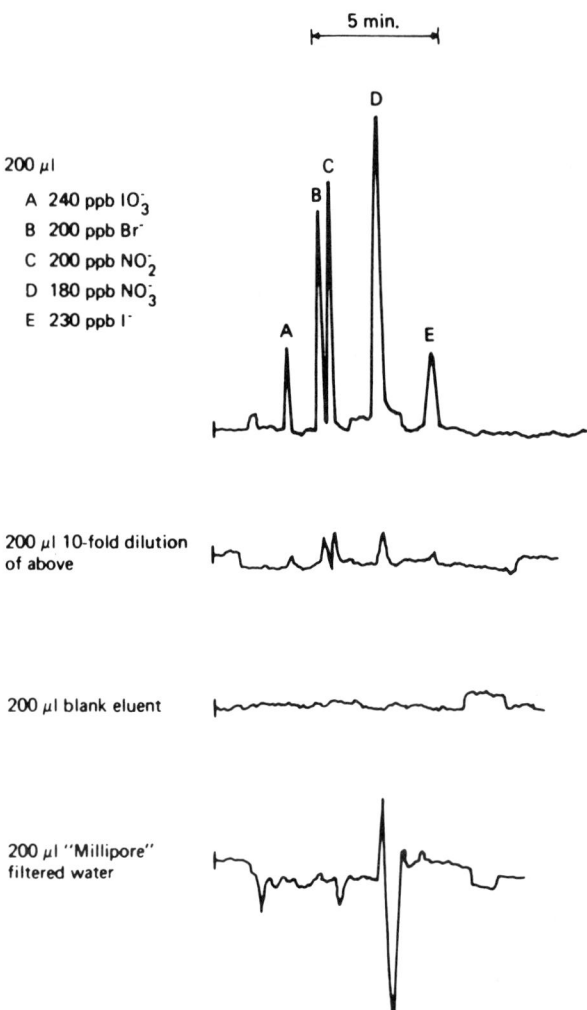

Figure 27. Determination of trace levels of inorganic anions in water on a Partisil-10 ODS-3 column. Eluant: 1 mM octylamine adjusted to pH 6.2 with H_3PO_4, 2 ml/min. UV detection at 205 nm. (From Ref. 186.)

For cases where the analyte ions have no useful optical absorption, indirect optical detection can be exploited in two different fashions in IIR-based chromatography. The first mode is analogous to indirect optical detection with bonded ion exchanger stationary phases in which highly absorbing eluant ions are used (Section III.C). Such high background negative signal techniques have not been extensively reported in the literature for obvious reasons. Attractive separations of many anions with good detection sensitivity may be obtained, however, by use of C-18 silica columns with purely aqueous eluants containing tetrabutylammonium iodide and potassium iodide for relatively weakly retained anions, and tetrabutylammonium phthalate and dipotassium phthalate for strongly retained anions [191]. A second approach to indirect optical detection in IIR-based chromatography is to use an optically absorbing IIR; no analogous situation exists in chromatography with bonded ion exchangers. Positive peaks are produced for sample ions that are retained more than the eluant ion and vice versa. This principle was applied by Denkert et al. [192] for C_6H_5-silica columns using aqueous solutions of naphthalene-2-sulfonate or 1-phenylethyl-2-picolinium ions as anionic and cationic IIRs, respectively, along with (sodium) acetate and phosphate as buffering agents in the eluant. Separation and sensitive fixed-wavelength (254 nm) UV detection are described for a large number of organic ions and ionizable species, e.g., sulfonates, sulfates, carboxylates, amino acids, dipeptides, and alkylammonium salts. Using a surface-modified silica stationary phase (Lichrosorb diol), the same authors described sensitive chromatographic analysis of various organic and inorganic anions (halides, alkylsulfates, alkylsulfonates, carboxylates, and sulfonamides); the support was coated with N,N-dimethylprotriptyline (MPT) phosphate (pH 6.2), and elution was done with a chloroform:n-propanol (9:1) mobile phase 90% saturated with MPT; and UV detection was employed at 254 nm [193]. An especially notable work on indirect optical detection of inorganic ions in IIR-based chromatographic systems was reported by Barber and Carr [162]. Benzyltributylammonium, 1-naphthylmethyltripropylammonium, and 1-naphthylmethyltributylammonium in an acetate buffer were chosen as the optically absorbing cationic IIR. Because acetate is a particularly hard anion, elution of sample ions such as Cl^-, Br^-, NO_2^-, NO_3^-, SO_4^{2-}, etc. simultaneously increased the concentration of the cationic IIR in the effluent band, enabling sensitive optical detection in UV. It was found useful to add an anionic IIR, e.g., hexanesulfonate, to the eluant to decrease retention of the sample ions. A typical chromatographic system employed a C-18 silica column and a precolumn (25 + 5 cm) with an aqueous eluant containing 4 mM naphthylmethyltributylammonium hydroxide, 10 mM acetate buffer (pH 4.75), and 0.25 mM sodium hexanesulfonate pumped at 1-2 ml/min. The background absorbance at the monitoring wavelength of 316 nm was 1.1 AU. The system capacity factor was 2.36. Reported capacity factors for F^-, IO_3^-, Cl^-, BrO_3^-, NO_3^-, Br^-, NO_2^-, ClO_3^-, tartrate, SO_4^{2-}, SO_3^{2-}, citrate, and I^- were 0.60, 1.82, 3.48, 3.83, 5.02, 7.84, 11.5, 14.7, 15.3, 16.4, 16.9, 18.9, and 55.9, respectively. Capacity factors for ClO_4^-, BrO_4^-, and IO_4^- were greater than 60. For a 50-µl injection volume, stated detection limits for Cl^-, Br^-, NO_2^-, and NO_3^- ranged from 0.5 to 1.0 nmol. More recent papers by these authors on these systems describe in detail factors that control retention and selectivity [194] and factors that govern sensitivity and detectability [195]. With the specific UV detector used in the study, it was found that attainable LODs deteriorate above background absorbance levels of 1 absorbance unit (AU). Helboe [196] has similarly added p-toluenesulfonate to the eluant (methanol:water, 55:45) for the separation and detection of alkyltrimethylammonium ions. Interestingly, CN- or C_6H_5-bonded silica columns yielded better peak shapes than C-8 or C-18 silica columns, whereas p-toluenesulfonate yielded better peak shapes than naphthalenesulfonate as the IIR. Dreux et al. [197] have studied the selectivities of IIR-based anion chromatography systems, employing indirect optical detection as a function of the nature of the IIR and the eluant ion.

Dasgupta [45] has studied the utility of tetraalkylammonium hydroxides as cationic IIRs for the separation of common inorganic and organic ions. A purely aqueous eluant containing the IIR in hydroxide form was used with a PSDVB column (because silica-based columns, although more efficient, are not stable to alkaline eluants) and a filament-filled helical membrane suppressor (Section II.E, filament-filled Helical Membrane Suppressors). Capacity factors for 33 anionic solutes with various concentrations of NMe_4OH, NEt_4OH, and NPr_4OH were determined and are shown in Table 9. The utilization of IIR-containing eluants with an ion exchange resin-packed suppressor column has been termed mobile-phase ion chromatography (MPIC is a trade name assigned to Dionex Corporation) and is the subject of a U.S. patent [44].

In an exemplary paper (the sole flaw is that the title has been visited by the Printer's Devil), Cassidy and Elchuk [198] showed how high resolution and sensitivity may be obtained for single-column conductimetric anion chromatography with the use of a PSDVB column and a cationic IIR. Both dynamic and adsorbed IIR systems were used. In both cases, the eluant contained a small portion of acetonitrile (7:93). Excellent efficiencies were exhibited by the commercial 10-µm PRP-1 column and were marginally better for the dynamic system. The dynamic system used 0.5-0.75 mM tetrabutylammonium salicylate as eluant; a typical separation of common anions at sub-part-per-million levels is shown in Figure 28. In the "permanently" sorbed IIR mode, 10-80 mg of cetylpyridinium salicylate was adsorbed on the column (150 × 4.1 mm). A salicylate-containing eluant, typically 0.5 mM tetramethylammonium salicylate, was used with

Table 9. Capacity Factors (k') for Various Sample Anions on a PRP-1 Column

	IIR			
Solute	NEt$_4$OH, 1 mM	NEt$_4$OH, 5 mM	NEt$_4$OH, 10 mM	NPr$_4$OH, 1 mM
IO_3^-	0.12	0.40	0.44	3.16
F^-	0.12	0.40	0.44	3.19
$NH_2SO_3^-$	0.14	0.42	0.46	3.46
CH_3COO^-	0.19	0.44	0.49	3.56
$HCOO^-$	0.21	0.46	0.52	3.60
Cl^-	0.16	0.46	0.52	3.81
NO_2^-	0.19	0.53	0.60	4.81
BrO_3^-	0.22	0.56	0.69	5.20
Br^-	0.22	0.56	0.69	5.20
N_3^-	0.28	0.65	0.84	6.02
NO_3^-	0.28	0.66	0.86	6.76
ClO_3^-	0.33	0.79	1.28	9.37
I^-	0.42	1.23	2.56	13.2
ClO_4^-	1.00	2.67	3.79	60.2
SCN^-	1.19	3.16	4.42	49.9
Sulfanilate	1.33	3.30	4.12	30.9
Cl_3COO^-	10.1	28.8		
Picrate	∞	∞		
SO_3^{2-}	0.19	0.79	1.07	28.8
Tartrate	0.19	0.84	2.16	37.0
SO_4^{2-}	0.23	0.88	1.12	30.9
$C_2O_4^{2-}$	0.23	0.88	1.14	37.8
CrO_4^{2-}	0.19	0.88	1.19	
$S_2O_3^{2-}$	0.16	0.93	1.23	
Maleate	0.23	0.91	1.23	41.3
HPO_4^{2-}	0.16	1.02	1.67	
$S_2O_6^{2-}$	0.21	1.14	1.60	
Phthalate	0.37	2.46	3.74	
$S_2O_8^{2-}$	0.33	2.77	4.81	
Citrate	0.23	1.81	2.67	
$Fe(CN)_6^{3-}$	0.30	4.88	8.91	
$Fe(CN)_6^{4-}$	0.26	4.63	8.44	

Source: Ref. 45.

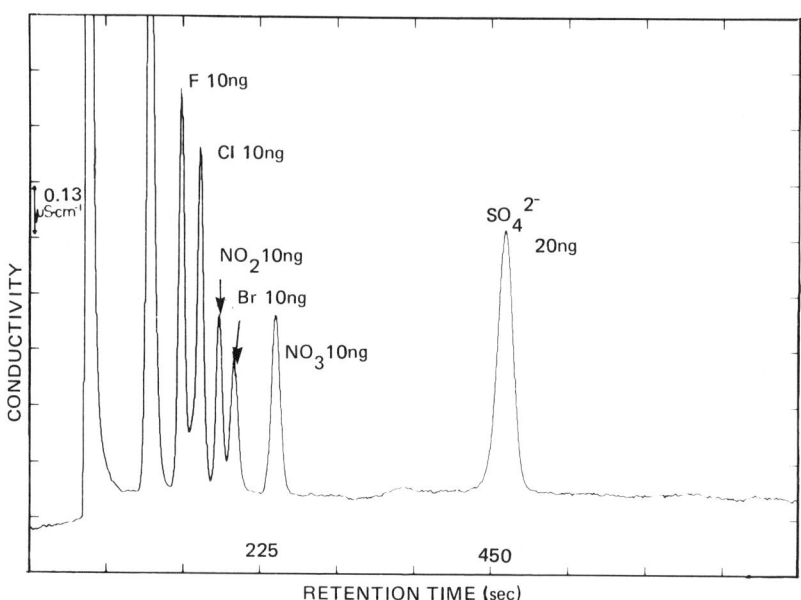

Figure 28. Separation of anions with a tetrabutylammonium salicylate eluant and conductimetric detection: 4.1 × 150 mm PRP-1 column, 0.75 mM TBA-Sal in 7:93 acetonitrile:water, pH ca. 6.3. Injection volume, 20 µl. (From Ref. 198.)

the "coated" column. No loss of the adsorbed modifier was observed, even after prolonged use. With 22 mg of the cetylpyridinium salicylate adsorbed on the column, and 0.5 mM tetramethylammonium salicylate as eluant, capacity factors for F^-, HCO_3^-, Cl^-, BrO_3^-, NO_2^-, Br^-, NO_3^-, ClO_3^-, SO_4^{2-}, and $S_2O_3^{2-}$ were reported to be 2.0, 2.4, 2.5, 2.8, 3.0, 3.4, 4.2, 5.8, 12.3, and 15.6, respectively. The system was easily amenable to trace enrichment; part-per-billion levels of anions were enriched on a C-18 silica cartridge and then switched in-line. The resulting chromatogram is shown in Figure 29. The exact retention times of sample ions were found to be influenced by the precise sample concentration injected, increasing by up to 10% at high sample concentrations. Similar observations have been made on other IIR-based systems at the present author's laboratory. Although Cassidy and Elchuk did not rationalize this observation, it is probable that this increased retention is a consequence of increased ionic strength, as predicted by the Stern-Gouy-Chapman theory of double layers. Obviously the effect is more pronounced if the eluant

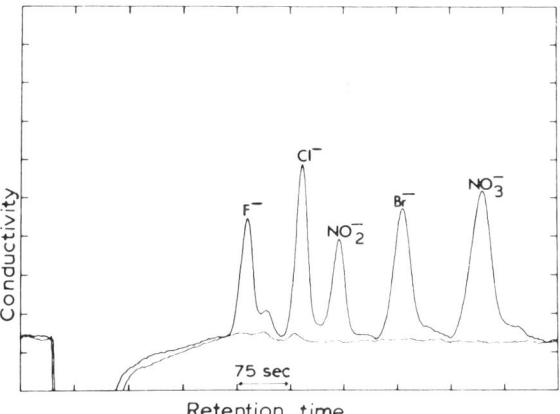

Figure 29. Separation of anions after trace enrichment. Eluant, 0.5 mM tetramethylammonium salicylate; flow rate, 1 ml/min; PRP-1 column coated with ca. 45 mg of cetylpyridinium chloride; 2-ml sample containing 100 ng each of F^-, Cl^-, and NO_2^-, and 300 ng each of Br^- and NO_3^- enriched on a coated C-18 silica column and then switched in-line. (From Ref. 198.)

ionic strength is low and the total amount of adsorbed IIR is high. A more recent paper by Cassidy and Elchuk [175] discusses in detail the use of tetraethylammonium and tetramethylammonium salicylate as eluants with a number of reverse phase columns; long-term stability, column efficiencies, attainable LODs with careful temperature control, and comparability with suppressed chromatography systems were addressed.

Cassidy et al. have also described excellent separation of lanthanides and Th and U, using reverse phase columns and octanesulfonate as IIR [199]. A complexing agent, α-hydroxyisobutyric acid, was incorporated in the eluant, and both isocratic and gradient elution modes were investigated. A postcolumn reaction with Arsenazo III was used for detection.

Direct electrochemical detection is often attractive for electroactive analytes because of the extraordinary sensitivity of this technique for applicable cases. Kagedal et al. [200] used a C-18 silica column with a 85:15 water:methanol eluant. The aqueous phase contained NaH_2PO_4 (2 mM), Na_2EDTA (0.1 mM), and tetrabutylammonium hydrogen sulfate (17.6 mM), adjusted to pH 6 with NaOH. A mercury-based working electrode operating at 0.0 V versus the saturated calomel electrode (SCE) was used for electrochemical detection; micromolar levels of thiosulfate in urine samples were easily detected.

Wheals [201] has recently carried out a thorough study on the use of multiple columns and detectors for anion analysis. The recommended IIR-based eluant contained 1.75 g citric acid and 0.125 g cetyltrimethylammonium bromide in 2.5 L methanol:water (30:70); it was applicable to both C-18 silica and silica-based anion exchange columns, employing UV, electrochemical, conductivity, or refractive index (RI) detection. Retention data for the different stationary phases, response behavior, and sensitivities with various detectors were reported for a large number of common organic and inorganic anions with the use of the above eluant. A facile method of preparation of the anion exchanger phases was reported; the separation efficiencies obtained on these "homemade" stationary phases are among the very best reported in the literature to date. Results with several "real" samples clearly established the attractiveness of the approach.

Another recent study, of Schmuckler et al. [202] employs tetrabutylammonium salicylate as eluant in a C-18 silica column with conductimetric detection. Six common inorganic anions were easily separable and detectable with reasonable sensitivity. An unusual finding was that pretreating the column with a cationic dye (Methyl Green) improves resolution, reduces retention times, and minimizes effects of eluant pH variation. The method was applied to anionic analysis of fruit juices. Cetyltrimethylammonium ion has been used as the IIR by Andrasko [132] as well, to carry out anionic separation on a C-18 silica column.

Chiu [203] has found it possible to separate and quantitate ethyl sulfate from inorganic sulfate or phosphate by use of tetrabutylammonium as IIR in a methanol:water (30:70) eluant and RI detection. Nitrite and nitrate were separated by Kok et al. [204] on a C-18 silica column with the use of tetramethylammonium phosphate as eluant. Chromatography of trivalent chromium complexes has been studied by Lethbridge and Reeve [205] on C-18 silica, using either cationic (tetrabutylammonium) or anionic (lauryl sulfate) IIRs. Inorganic divalent platinum complexes have been similarly separated by Riley et al. [206], using trimethylammonium ion as IIR. To separate platinum antineoplastic agents, cetyltrimethylammonium ion was used as the IIR [207]. One particularly interesting example is the IIR-based (tetrabutylammonium) chromatographic determination of borate as the boron-chromotropic acid complex, which is injected preformed [208]. The sample is treated with excess IIR prior to injection.

Atomic plasma spectrometric detection has been used in IIR-based separations. The separation of anionic arsenicals including dimethylarsinate, methylarsenate, arsanilate, arsenate, and phenylarsonate was accomplished by Gast et al. [209], using a C-18 silica column, cetyltrimethylammonium as a cationic IIR, and UV/plasma spectrometry for detection. Flame photometric phosphorus-selective detection was used by Chester et al. [210] to determine inorganic phosphates

Approaches to Ionic Chromatography

which were separated on a PSDVB column with tetramethylammonium formate as eluant. Flame emission-based phosphorus- and sulfur-selective detectors are highly sensitive and well developed [211]. Octadecyl silica stationary phases were used by Krull et al. [212, 213] with alkanesulfonic acids as anionic IIR for cation separation. These authors also successfully used tributylphosphate, an uncharged IIR, in the eluant to separate Zn, Cd, and Hg on a reverse phase column [214]. Although the authors view this as liquid-liquid partition chromatography, it is well described as IIR based.

In this context, it is useful to note that many advantages of an IIR-based system can be realized without any IIR on a amino-bonded phase because the effective retention can be varied by varying eluant pH [215]. Excellent separations were reported by Cortes [215] and extended later to indirect optical detection methods by Cortes and Stevens [131]. This approach was used earlier for the measurement of bromide and nitrate in foods [216]. The long-term stability of amino-bonded silica columns with pure aqueous eluants is not known, however.

A novel concept in the use of IIRs in the eluant involves introducing the IIR as programed pulses by a loop valve preceding the sample injection valve [217]. The approach is easily adapted toward optimizing the amount of IIR necessary for a given separation. Berry [218] has recently reported how pulsing with a weak eluant can be highly useful in resolving unseparated constituents. Doubtless, these approaches are of great potential interest to the aspiring ion chromatographer.

E. Chromogenic, Electroactive Chelating Ligands as IIR: Ion Interaction Chromatography of Transition Metals

One of the truly interesting aspects of dynamic ion interaction chromatography is the use of a lipophilic chelating ligand in the eluant. The complexes formed may be amenable to direct optical or electrochemical detection. Berthod et al. [219] used C-8 silica columns with varying amounts of oxine (8-hydroxyquinoline) (typically 3-5 mM) in a phosphate buffered (typically pH 8.5) mobile phase (typically 60:40 methanol:water) to separate a variety of transition metal ions, including Cd^{2+}, Cu^{2+}, Mn^{2+}, Co^{2+}, Zn^{2+}, Ni^{2+}, Hg^{2+}, and Fe^{3+}. A variety of detection methods were used:

1. UV/VIS detection (for all metals other than Fe^{3+}. The absorption maximum for the oxinate complex is at 385 nm; for Fe^{3+}, there are two absorption peaks, at 470 and 580 nm).
2. Detection by atomic absorption spectroscopy, using a continuous flow-through system (at 325 nm for Cu, 241 nm for Co, and 248 nm for Fe).

3. Electrochemical detection on a glassy carbon electrode at -0.7 V versus an Ag/AgCl reference electrode for Fe^{3+}, Cu^{2+}, and Hg^{2+}.

For 20-µl injection volumes, determination in the nanogram-microgram range was easily accomplished; very short (5 cm) columns could be used for a number of separations. Hoffmann and Schwedt [220] subsequently extended this approach for the separation and detection of Cr^{3+}, Mn^{3+}, Mn^{2+}, Co^{2+}, Cu^{2+}, Zn^{2+}, and Al^{3+} with the use of a 7-µm C-8 silica column and a mobile phase containing 1 mM oxine in 60:40 methanol:water. The water in the mobile phase was prepared with 20% borate buffer, pH 9.0. Detection was generally accomplished at 394 nm. Mn^{3+} and Mn^{2+} were determined in natural waters and soil samples; slightly different chromatographic conditions and detection parameters were used for this analysis.

Bond and Wallace [221] pioneered the use of dithiocarbamates in the eluant for direct injection of the metal ion leading to on-column complex formation, chromatographic separation, and then detection by electrochemical means. The first study [221] involved Cu^{2+}, which could be detected down to 1 ng on a variety of electrodes. A C-18 silica column was used with a 70:30 acetonitrile:water eluant containing the ligand, 0.02 M acetate buffer, and 0.2 M $NaNO_3$ as supporting electrolyte. Excellent agreement was achieved with results from parallel determinations by atomic absorption spectroscopy for real samples. In a subsequent study [222], this approach was extended to the simultaneous determination of Cu^{2+} and Ni^{2+}. Interestingly, Cr^{3+} and Co^{2+} could be analyzed if the complex was preformed, but not if the metal ion was directly injected onto the column. Automated determination of Ni^{2+} and Cu^{2+} via on-column complex formation was next reported [223]. A more recent effort by these workers [224] has combined spectrophotometric detection with electrochemical detection (superior for Ni^{2+} and Cu^{2+}); automated analysis of six metals (Pb^{2+}, Cd^{2+}, Hg^{2+}, Co^{2+}, Ni^{2+}, and Cu^{2+}) has been made possible. Nanogram detection limits were achieved using photometric detection and a 10-µl injected sample. Addition of a ion exchange-based column prior to the photometric detector to remove excess dithiocarbamate ligand improved the detection limits for Ni^{2+}, Cu^{2+}, Co^{2+}, and Hg^{2+}, but was found to be deleterious for the detection of Pb^{2+} and Cd^{2+}. The incorporation of this column also shortened attainable unattended operating period of the instrument. Smith and Yankey [225] have also examined the utility of an 80:20 methanol:water eluant containing 0.1% ammonium or 0.05% sodium diethyldithiocarbamate for the separation of Cu^{2+}, Ni^{2+}, Co^{2+}, Pb^{2+}, and Fe^{3+} on a C-18 silica column via variable-wavelength (320-440 nm) photometric detection. Capacity factors for Cd^{2+}, Pb^{2+}, Ni^{2+}, Co^{2+}, Fe^{2+}, and Cu^{2+} were reported to be 1.56, 3.10, 3.78, 3.87, 4.00, and 5.96, respectively.

Low-nanogram detection limits (10-μl injection) were obtained, with good repeatability and linearity of response for each metal.

O'Laughlin [226] has shown the utility of 1,10-phenanthroline (o-phenanthroline) as a chromogenic ligand in the eluant, with perchlorate being present as the anionic IIR, for the separation and detection of Fe^{2+}, Ru^{3+}, Ni^{2+}, Zn^{2+}, Co^{2+}, Cd^{2+}, and Cu^{2+}. Silica-based ion exchanger columns as well as CN-silica and PRP-1 columns were used with an acetonitrile:water-based eluant. Low-nanogram levels were reported to be detectable. Earlier, O'Laughlin and Hanson [227] had successfully separated 1,10-phenanthroline chelates of Fe^{2+}, Ni^{2+}, and Ru^{3+} on CN-silica columns with methanesulfonate or heptanesulfonate as IIR. The same IIRs were employed by Valenty and Behnken [228] for the separation of Ru^{3+}-bipyridyl on a C-18 silica column with the use of a THF:water eluant. Bipyridyl chelates of Fe^{2+} and Ni^{2+} were separated on a CN-silica column by Mangia and Lugari [229], who used SCN^- as the IIR and UV or atomic absorption spectrometry (AAS) of discrete fractions for detection.

Using tetrabutylammonium or cetyltrimethylammonium as IIR, Hoshino et al. [230] separated 4-[2-pyridyl(azo)]resorcinol (PAR) chelates of V^{5+}, Cr^{3+}, Fe^{2+}, Fe^{3+}, Co^{2+}, Ni^{2+}, and Pd^{2+} on a C-18 silica column, with a methanol:water eluant and detection at 254 nm; low LODs were attainable. In a subsequent paper [231], these authors separated diethyldithiocarbamate chelates of Co^{2+}, Ni^{2+}, Cu^{2+}, V^{5+} and the oxine chelate of Al^{3+}, employing the nonionic surfactant Triton X-100 in the eluant to solubilize the insoluble metal chelates. Molybdenum and tungsten have been determined as their oxoanions on a C-18 silica column with the use of tetrabutylammonium as the IIR and tiron as the chromogenic agent, the latter being also incorporated in the eluant [232].

It is not necessary for IIR-based chromatography of transition metals to employ chromogenic or electroactive ligands in the eluant. An IIR-based eluant system can be used to accomplish the separation, and the eluted metal ion can then be detected by suitable postcolumn chemistry. Cassidy and Elchuk [155] have shown how several transition metal ions can be separated on a C-18 silica column on which a large IIR (e.g., 1-eicosylsulfate) is immobilized, or on a C-18 silica column in equilibrium with an eluant containing hexanesulfonate. The incorporation of a chelating ligand, e.g., tartrate, is necessary to accomplish elution. Detection was accomplished by postcolumn reaction with pyridyl(azo)resorcinol (PAR), and low parts-per-million levels of several metal ions were easily detected. Similar approaches to the analysis of lanthanides have been reported by the same workers [199]. Recently, Pohl and Riviello [233] have utilized nonpolar stationary phases in combination with octanesulfonate as the IIR in the eluant. The eluant also contained citrate and tartrate for elution and separation of several transition metal ions. Detection was accomplished by postcolumn reaction with PAR. These authors found, however, that the selectivity in these systems is not as great as that attainable on a conventional ion-exchange column using chelating eluants.

Some metal complexes are nonlabile; this property permits their separation without dissociation during chromatography. Such complexes may display sufficient optical absorption by themselves to perform direct optical detection. Buckingham et al. [234] have separated diastereoisomers of cobalt-ethylenediamine-amino acid complexes, using p-toluenesulfonate or hexanesulfonate as IIR. Lam-Thanh et al. [235] separated L-methionine dipeptide complexes of Pd(II), employing cetyltrimethylammonium or trioctylmethylammonium as IIR. Similarly, Hoshino and Yotsuyanagi [712] have reported sensitive analysis of the preformed PAR-chelate of Cr^{3+}, using hexamethylbutanediammonium ion as the IIR on a C-18 silica column.

Finally, an extreme example is presented by the work of Tsuda et al. [236]: Ni^{2+}, Cu^{2+}, and Co^{2+} were separated as their chloro complexes with methyl(tri-n-octyl)ammonium as IIR in 7.5 N HCl, and detection at 210-220 nm. An all glass/PTFE microbore (0.5 × 138 mm) chromatographic setup was used to deal with the highly corrosive eluant. Similarly, Pb^{2+}, Sb^{3+}, and Sn^{4+} were separated by a step gradient from 0.02 N to 0.25 N $HClO_4$, both in 8 N HCl.

V. CHROMATOGRAPHIC ANALYSIS OF METAL SPECIES

The original emphasis of IC as introduced by Small et al. was on anionic analysis; a similar trend had persisted until fairly recently. This state of affairs may be attributed to two separate factors: (a) "Suppressed" conductimetric IC employing an OH^--form suppressor column was applicable only for cations that form soluble hydroxides, i.e., alkali, alkaline earth, and ammonium or related cations; (b) sensitive atomic spectrometric methods have been in vogue for determinations of heavy metal ions.

Chromatographic analysis of heavy metal ions, although never very widely practiced, has been in the domain of active scientific endeavor for more than three decades. One reason for these continued research efforts is that atomic spectrometric techniques have their own shortcomings, the most serious of which is the inability to distinguish between the different chemical forms and/or oxidation states (e.g., Cr^{3+} vs. CrO_4^{2-}; Fe^{2+} vs. Fe^{3+}) of the same element. Further, atomic absorption spectrometry is not readily adaptable to sensitive simultaneous multielement analysis. On the other hand, emission spectrometers, including direct current plasma (DCP) and the more recently introduced induction coupled plasma (ICP) spectrometers, which can perform sensitive, multielement analysis rapidly and routinely, require substantial initial capital investment. (To be fair, it is prudent to warn a prospective buyer that available commercial IC equipment capable of performing routine trace-level metal ion analysis is far from cheap. Additionally, sample throughput rate is significantly lower than that attainable with an ICP spectrometer.)

A second reason for the interest in the chromatography of metal ions has been for the purposes of separation, rather than of determination. This is particularly true for the nuclear industry, where metal ion chromatography has had, and continues to have, a stronghold; much effort has been devoted to the separation of lanthanides, actinides, and *trans*-actinides by ion exchange. More often than not, the ions to be separated are radioactive, with characteristic radiative emissions, which may provide a means for unambiguous detection.

In recent years, there has been a resurgent interest in the determination of trace metals by chromatography by electrochemical and spectrophotometric detection, often after a suitable postcolumn reaction. Remarkable developments have taken place in postcolumn reactor design. This progress has been an important factor in the increased attractiveness and the perceived potential of chromatographic analysis of heavy metals.

Schwedt has repeatedly reviewed the chromatographic analysis of metal species [237-239]. The latest review [239] is a book which surveys the application of all types of chromatographic methods (gas chromatography, liquid chromatography, paper/thin-layer chromatography) for trace inorganic analysis; this review is recommended to the reader interested in the general area. Chromatographic analysis of metal species has been reviewed in recent years by Willeford and Veening [240], Cassidy [241], Krull [242], and Nickless [243]. The first of these [240] is devoted to the separation and analysis of organometallics and metal chelates. The major emphasis is on the separation of similar complexes of the same metal, although the last section does address multielement separations. The review by Cassidy [241] addresses the interest of the analytical chemist to a greater degree; most of this review is devoted to simultaneous multielement analysis. It is extremely thorough, with clear delineation of the principles involved in each case. In addition, the obvious insight of an acknowledged master of the field makes this review a must reading for the metal ion chromatographer. Krull's review [242] is also very detailed and excels in its coverage on atomic absorption/emission spectrometers interfaced as a detector to a liquid chromatograph. The most recent review in the area is by Nickless [243], a second-generation chromatographer. Perusal of this work is well worth the reader's time. Except for the summarized account of a few salient points, the contents of these reviews are not repeated here. The biennial reviews on column liquid chromatography in *Analytical Chemistry* by Majors et al.

[5,244] also contain information pertinent to the chromatographic separation of metallic species.

A. Novel Suppressor and Eluant Chemistries: Metal Ion Chromatography with Conductimetric Detection

The limitation of suppressed conductimetric detection employing an OH^--form suppressor, which is due to the formation of insoluble metal hydroxides, may be obviated by clever chemistry. Nordmeyer et al. [245] used eluants containing $BaCl_2$, $Ba(NO_3)_2$, or $Pb(NO_3)_2$ and an anion exchange resin in the SO_4^{2-} form in the suppressor column to precipitate $BaSO_4$ or $PbSO_4$ in the suppressor and thus reduce the eluant background. The metal ions (including Mg^{2+}, Ca^{2+}, Sr^{2+}, Mn^{2+}, Fe^{2+}, Co^{2+}, Ni^{2+}, Cu^{2+}, Zn^{2+}, and Cd^{2+}) elute as the corresponding sulfates. The inclusion of yet another column containing a cation exchange resin in the H^+ form *after* the SO_4^{2-}-form suppressor served to translate the metal sulfate signal to an equivalent amount of sulfuric acid and increased the sensitivity of conductimetric detection by five fold. The chemistry of the system is less straightforward than the above account suggests, however. The authors suggest that the complexity arises primarily because the sulfate bound to the suppressor resin is really bound as HSO_4^- at eluant pH levels useful in the system. The consequent protic equilibrium for the dissociation of HSO_4^- causes a decrease in H^+ concentration as the metal ion elutes from the sulfate-form suppressor, although any change in the pH of the effluent accompanying sample ion elution is not detectable after the H^+-form postsuppressor. However, the system response is dependent on the pH of the eluant, increasing with increasing eluant pH up to a pH of 5.2-5.8 [which is approximately the pH of the unmodified 1 mM $Ba(NO_3)_2$ or $BaCl_2$ eluant] for all metal ions studied except Cu^{2+}, which exhibited maximum response with a pH 4 eluant. Copper was anomalous also in producing unusually broad peaks, compared to the other metal ions. Low-micromolar detection limits were attained for the other metals studied, and good agreement between results for Ca^{2+} and Mg^{2+} were demonstrated for pond water, soil, and blood samples with results from analyses by atomic absorption spectroscopy (AAS). There are two important shortcomings of this approach. The suppressor bed is not easily regenerated. Although some schemes were devised to accomplish this, it was found easier to simply replace the resin with fresh material each time. The second problem concerns a broad H^+-wave, which is produced at a high retention time (arising from ion exclusion of H^+ after the separator column) whenever the sample has a different pH from the eluant.

In subsequent work [246], the same authors found that a $Pb(NO_3)_2$ eluant with an IO_3^--form suppressor [like $PbSO_4$ or $BaSO_4$, $Pb(IO_3)_2$ is removed by precipitation] was somewhat more attractive because

Ba^{2+} could be analyzed in this system and Cu^{2+} did not exhibit any anomalous behavior. The incorporation of an OH^--form rather than the H^+-form postsuppressor improved detection limits for alkaline earth metals through removal of excess H^+ (resulting in less baseline noise). The transition metals are precipitated on the OH^--form postsuppressor, and any possible interference from such metals (especially in the determination of Ca^{2+}) is thus removed. If the focus is to be on alkali/alkaline earth metals only, the advantages, if any, of this more complex system--as compared to the simpler diamine salt eluant, and single, easily regenerable OH^--form suppressor systems, originally introduced by Small et al. [1]--have not been clearly delineated. With the phenylenediamine eluant, undesirable oxidation products reportedly accumulate on the columns [75, 247]; however, a balanced account of the advantages/disadvantages of each approach is yet to appear in the literature. It should also be noted that the selectivity among various transition metals observed in the sulfate/iodate-suppressed systems [245, 246] is quite poor.

It is also worth noting in this context the apparent absence of any attempts in the literature, other than the original work of Small et al., to use Ag^+ as an eluting ion with a halide-form suppressor. Even though Ag^+ is singly charged, it is soft and displays very high affinity for cation exchange sites (greater than any other monovalent cation; see for example, Ref. 248). Precipitating Ag^+ as the halide is well known not only in the context of IC: Using an HCl eluant and an Ag^+-form suppressor (the converse situation) is a principal mode of operation of ion exclusion chromatography. Further, removing precipitated AgCl is much easier than removing $PbSO_4$, $BaSO_4$, or $Pb(IO_3)_2$ because of the facile dissolution of AgCl in ammoniacal solutions. The fact that a number of transition metals form chloro complexes and that chloride is more easily replaced than sulfate should be additional incentives.

Another related approach involving ingenious chemistry utilizes $Zn(NO_3)_2$ along with HNO_3 as eluant for the analysis of alkaline earths [247]. The usual OH^--form suppressor precipitates $Zn(OH)_2$, an amphoteric hydroxide that is solubilized as the ZnO_2^{2-} anion during suppressor regeneration with strong NaOH. Aluminum was also investigated for this application instead of zinc, because of the similar amphoteric behavior of $Al(OH)_3$. However, it was found to be less suitable than zinc in terms of attainable resolution of analyte ions on available commercial columns.

Sevenich and Fritz [249] have pursued the single-column approach for conductivity detection of divalent as well as trivalent ions, extending Fritz et al.'s previous work on the separation of alkaline earth metal ions on a cation exchange column using ethylenediamine salts in the eluant [75]. In the present case, additional complexing agents, namely tartrate and α-hydroxyisobutyrate, were incorporated in the

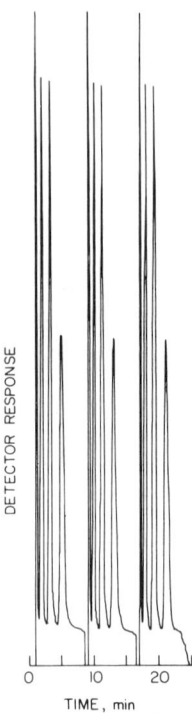

Figure 30. Rapid separation of Mg^{2+}, Ca^{2+}, and Sr^{2+} for successive injections of the same sample containing 0.5 mM of each metal. Eluant: 4 mM ethylenediamine, 3 mM tartaric acid, pH 4.5. (From Ref. 249.)

eluant to tailor elution behavior. The chromatographic behavior of a large number of metal ions, including Ba^{2+}, Ca^{2+}, Cd^{2+}, Ce^{3+}, Co^{2+}, Dy^{3+}, Er^{3+}, Eu^{3+}, Fe^{2+}, Gd^{3+}, Ho^{3+}, La^{3+}, Lu^{3+}, Mg^{2+}, Mn^{2+}, Nd^{3+}, Ni^{2+}, Pb^{2+}, Pr^{3+}, Sm^{3+}, Sr^{2+}, Tb^{3+}, Tm^{3+}, Yb^{3+}, and Zn^{2+}, was studied. Well-formed peaks were found for all of these ions. Fe^{3+}, Al^{3+}, and Cu^{2+} could be masked with EDTA so that they did not interfere in the determination of other metals of interest. Figures 30 and 31 show some representative capabilities of this type of a chromatographic system.

B. Electrochemical Detection

Brunt [250], and more recently Kissinger [251] have reviewed the principles and scope of electrochemical detection under continuous-

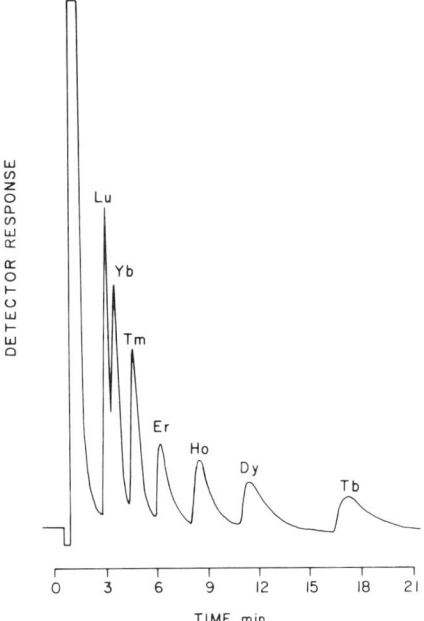

Figure 31. Separation of the seven heaviest lanthanides via 4 mM ethylenediamine, 3 mM α-hydroxyisobutyrate (pH 4.5) as eluant. Metal concentrations: 10 ppm each. (From Ref. 249.)

flow conditions. Concise, well-written reviews on the same topic have been published earlier by Kissinger [252] and Rucki [253]. A very simple and sensitive carbon fiber-based electrochemical detector has been described by Stulik et al. [254].

Aside from their studies on eluants containing the ligand for the electrochemical detection of metal ions, as discussed in Section IV.F, Bond and Wallace [221-224] also carried out studies with preformed metal complexes to determine the utility of electrochemical detection. Takata et al. have studied coulometric detection of metal ions since the mid 1960s [255]. Among other developments, they devised indirect postcolumn detection methods [256-260,262] for metals not amenable to direct electrochemical detection, by reactions such as:

$$HgL + M^{2+} + 2e^- = Hg + ML$$

$$CuL + M^{2+} + 2e^- = Cu + ML$$

where the ligand (L) is a chelating ligand with high affinity for the metal, e.g., diethylenetriaminepentaacetic acid (DTPA). In a similar vein, Takata and Muto [261] have reported on the anion exchange separation and electrochemical detection of silicate, phosphate, and germanate after postcolumn reaction with molybdate. An electrocatalytic detection method for Co^{2+} has been devised by Takata et al. [263]. An electrocatalytic detection method for Sb^{3+} in liquid chromatography (LC) effluent has been developed by Taylor and Johnson [264]. A review, describing the coulometric detection of trace metals in seawater, has also been published by Takata and Muto [265].

Cassidy and Elchuk [266] determined hexavalent uranium in groundwater and urine via a bonded-phase cation exchanger, on-line preconcentration, and coulometric detection; α-hydroxyisobutyric acid was used as eluant. With this system, 0.5-50 ppb U in ground water and 25-400 ppb U in artificial urine were measured. Using the same complexing eluant and a pH gradient, the rare earths and Sc and Y were determinable with high sensitivity via coulometric detection; the total separation time required was under 75 min. Chromatographic determination of pertechnate, TcO_4^-, via reductive electrochemical detection was reported by Lewis et al. [267]. Krull et al. [268] reported electrochemical detection of platinum chemotherapeutic agents. Electrochemical detection of Pt complexes with cetyltrimethylammonium ion as IIR has also been reported [269].

In the United States, MacCrehan et al. [270-273] have pushed the frontiers of electrochemical (especially differential-pulse) detection for organometallic species, notably those of Hg, Pb, and Sn in synthetic as well as real samples. A variety of high-efficiency stationary phases (C-18 silica, $-NH_2$, bonded cation exchange) were used successfully with saline, partially aqueous eluants to attain excellent detection limits. The LOD of methyl mercury in tuna or shark meat was 2 ng/g [270].

C. Atomic Spectrometric Detection Methods

Atomic absorption spectrometry, because of its high specificity, sensitivity, and relatively low cost, is particularly attractive as a detector when small amounts of different molecular forms bearing the same element are to be separated. Interfacing a flame atomic absorption spectrometer to a continuously flowing liquid stream is relatively simple and is an established technique [276]. Furnace AA, although more sensitive than flame AA, is not easily coupled as a truly continuous detector unless the analyte can be introduced as a gas (e.g., where hydrides are conveniently generated; see below). Although Krull [242] has perhaps overstated the case in remarking that AAS is surely the most commonly used form of detection in chromatographic analysis of metal species, the excellence of his review on this area certainly makes his enthusiasm understandable. For the reader interested in atomic spectrometric detection coupled to an LC system, this work is a must reading. Other reviews on atomic spectrometric detection coupled to liquid chromatography include those by Fuwa et al. [274] and Sun [275].

The earliest report of LC-AA is likely that of Jones and Manahan [276], who have published extensively on the technique since [277-279]. These studies were primarily concerned with the trace determination of chelating ligands such as ethylene glycol bis(β-aminoethyl ether)N,N'-tetraacetic acid (EGTA), nitrilotriacetic acid (NTA), ethylenedinitrolotetraacetic acid (EDTA), and $trans$(1,2)-cyclohexylenedinitrolotetraacetic acid (CDTA), which were isolated as their copper chelates and then chromatographed while the copper in the effluent was monitored. One paper [277] reported on the separation of organochromium species. Among other early reports is the work of Fanasaka et al. [280], who reported on a analytical method for organomercurials; in the work of Cassidy et al. [281], organosilicon compounds were chromatographically separated and continuously detected by flame AA. The work of Botre et al. [282], also from this early period, reported on analysis of environmentally important organolead species. Detection of chromium species in an LC effluent by continuous-flame AA was reported by Fernandez [283]; this article is also one of the better reviews of the early efforts in this area. Although not strictly concerned with chromatographic separation, a recent paper by Cantwell et al. [284] reported on the AA detection of metal ions eluting from an ion exchange column. In agreement with previous reports [285], the AA signal was found to be strongly dependent on the flow rate (nebulization rate). Beginning at small flow rates, the AA signal increases linearly with increasing flow rate, and finally becomes invariant at high flow rates. Thus, depending on the region of operation, the detector may behave either as mass sensitive or as concentration sensitive.

Collecting discrete fractions of the eluate and analyzing such by AA is less technologically sophisticated and convenient but always a viable approach. Van Loon et al. [286] separated and detected chromium species by such an approach. Grabinski [287] separated inorganic As(III), As(V), monomethyl arsonate (MMA), and dimethyl arsinate (DMA) in spiked environmental samples with attractive LODs. In both of the above efforts [286,287], an unusual combination of both cation and anion exchange resins were used for the stationary phase. Platinum complexes in blood plasma and urine of cancer patients treated with platinum chemotherapeutic agents have been separated by anion exchange or IIR-based chromatography and detected by AA in discrete fractions of the eluate [288,289]. The chromatographic results from the two different systems have been compared [290]. Tills and Alloway [291] have studied the occurrence and distribution of leachable Cd in polluted soils via size exclusion and reverse phase chromatography in conjunction with UV and discrete AA detection.

Burns et al. [292] were apparently the first to report the use of on-line hydride generation with subsequent continuous detection of the hydrides in quartz-tube-furnace AA for organotin compounds. This technique yielded impressive LODs, as small as 1.8 pg Sn; whereas direct coupling of the column effluent to flame AA produced quite poor LODs. C-18 silica columns were used with a pentane/acetone eluant. Using a similar approach, Ricci et al. [293] reduced the arsenic in the column effluent to arsine by on-line reaction with acidic persulfate and sodium borohydride. The generated arsine was swept out with argon, and the gas stream, after an appropriate gas-liquid separation stage, was directed into the heated quartz cell of an AA spectrometer. The experimental setup is schematically shown in Figure 32. Dionex anion separator columns were used with borate and/or carbonate eluants. Low parts-per-billion detection limits were attained for arsenite, arsenate, MMA, DMA, and p-aminophenyl arsonate. One other notable feature of this work is the application of gradient elution. With this type of detection system, gradient elution is facile because the eluant background is essentially zero. A chromatogram of trace arsenicals under gradient elution conditions is shown in Figure 33.

Unlike the above approaches of coupling quartz furnace AA to a gaseous effluent, a truly continuous coupling of graphite-furnace AA (GFAA) to a LC system has not as yet been realized because of the inherently temporally discrete operation of GFAA. Couplings that produce virtual analogue chromatograms as a series of histograms is possible, however, and have been found attractive for a variety of applications. The fact that histograms are produced is of no real concern; indeed, all digital data-acquisition devices really do the same thing. The main point of concern is that unlike a digital data-acquisition device, the temporal resolution is, at the time of this writing, so poor that the prospect of GFAA is essentially nonexistent for microbore or high-speed LC, and even of limited value for highly efficient separations on conventional columns. There is a limit to which the thermal capacitance of a furnace AA system can be reduced; barring a major scientific breakthrough, the present author is not enthusiastic about the future of GFAA as a LC detector.

The type of semicontinuous LC-GFAA coupling referred to above can be relatively easily accomplished with a carousel-type autosampler, as originally described by Brinckman et al. [294] and utilized by other workers [295]. More recently, an interesting flow injection interface has been exploited to accomplish the coupling [296].

In recent years, Brinckman, Fish, and co-workers have extensively reported on the application of LC-AA for analysis of environmental organometallic compounds. The occurrence and determination

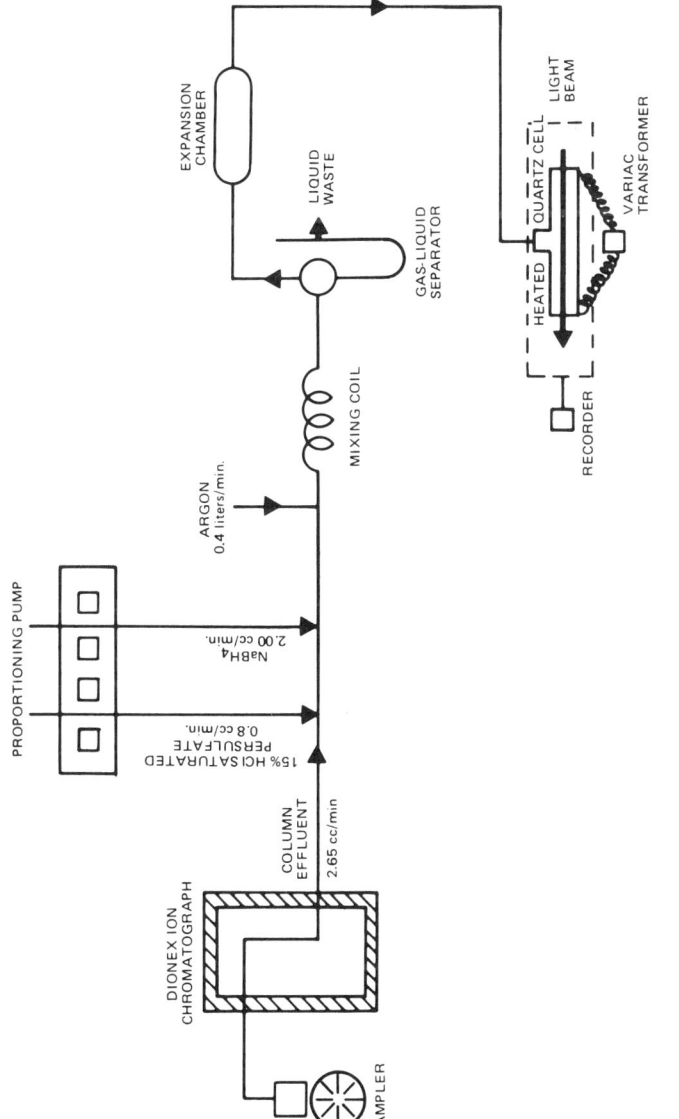

Figure 32. Coupling of ion chromatograph to automated arsine generator and AAS detector. (From Ref. 293.)

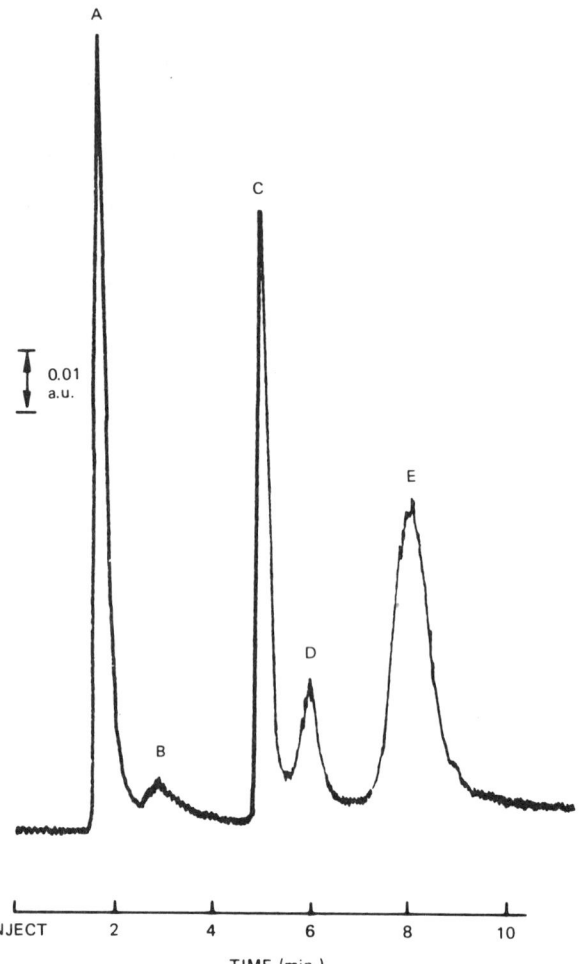

Figure 33. IC-ASS chromatogram of trace inorganic and organic arsenic species separated by gradient elution. (A) Dimethylarsonate, (B) AsO_3^{3-}, (C) monomethylarsonate, (D) p-aminophenylarsonate, (E) AsO_4^{3-}. All but peak E (60 ng/ml) represent injected concentrations of 20 ng/ml. (From Ref. 293.)

of organoarsenicals [297-301] and organotin [294,302-306] compounds by LC-AA have been addressed in a number of papers. In a number of studies, size exclusion columns were used for separation; a recent example involves the effort toward molecular weight characterization of vanadyl porphyrins in heavy crude petroleum [307]. Jewett and Brinckman [308] have authored a review specifically on the use of element-specific detectors.

As for applications, discrete batch-mode analysis has been used as frequently as the semicontinuously coupled systems. Aside from examples already cited above, methods have been developed for the separation and GFAA detection of amino acid complexes of copper in human serum as well as other complexes of copper (EDTA, triethylenetetramine, Ref. 295), organophosphates and phosphites [309], protein and ligand-bound calcium and magnesium in biological matrices [296], cis-platinum antitumor drugs in original preparations and metabolites [206,288-290], organomercurials [294], inorganic and organic lead compounds [294,310,311], inorganic and organic selenium compounds [312], and inorganic and organic arsenic compounds [313,314] including organoarsenical pesticide residues [315, 316].

Although a continuum source such as a deuterium lamp can be used in AAS to facilitate rapid multielement capability, detection limits deteriorate substantially. Tunable UV lasers are now on the horizon and may eventually shift the focus from emission to absorption, for simultaneous multielement detection.

Compared to AAS, as a detector atomic fluorescence spectroscopy (AFS) has been used much less thus far. Even though simultaneous multielement analyses are significantly easier to perform by AFS, attainable LODs by GFAA are frequently substantially better. However, continuous coupling with a flame AFS instrument is easily achieved. Van Loon et al. [317] first discussed the application of AFS as a LC detector. EDTA, triethylenetetramine, and glycine chelates of Cu, Ni, and Zn were separated on a bonded-phase cation exchanger and detected by concurrent multielement AFS. The work has since been described elsewhere as well [318,319]. Siemer et al. [320] have also used a LC-AFS system, monitoring the iron fluorescence for an analysis of acetylation products of ferrocene.

Because of their multielement capability, plasma emission spectrometers have significant potential in chromatographic detection in cases where different metals, each of which are present in different molecular forms of interest, are to be separated and determined. In contrast, simple flame emission spectrometry can yield good sensitivities only for a few metals and has been used as a LC detector [321]. Krull and Jordan [322] have reviewed the potential of interfacing plasma emission spectrometers to chromatographic systems. Krull's more recent review [242] also provides excellent coverage of this area. The interfacing of liquid chromatographs to atomic emission spectrometers and the potential capabilities of such systems have been demonstrated in depth by Gast et al. [209] in a paper which remains outstanding to date. Although a dummy column was used rather than a real chromatographic system, the paper by Fraley et al. [323] is a good source

Table 10. Limits of Detection (LOD) Values for the ICP under Time-Resolved Continuous Aspiration Conditions and for Sample Introduction as a Chromatographic Peak

Element	LOD at continuous aspiration (μg/L)	MDC as a chromatographic peak (μg/L)[a]
Al	2.2×10^2	1.2×10^2
As	(did not give meaningful results)	
B	39	64
Ba	1	1.5
Ca	2.1	9.2
Cd	46	89
Co	17	21
Cr	19	20
Cu	3.3	6.8
Fe	15	28
K	7.5×10^2	6.3×10^2
Li	4.2	6.4
Mg	15	1.8×10^3
Mn	1	1.3
Mo	26	62
Na	5.7	82
Ni	34	43
P	1.6×10^2	2.9×10^2
Pb	2.1	2.3×10^2
Sb	1.3×10^2	4.1×10^2
Se	4.3×10^2	2.8×10^2
Sr	1	7.2
Ti	0.7	2.1
V	10	11
Zn	14	19

[a]Concentrations in this column are concentrations at the peak maximum of the smallest detectable peak for that element, i.e., a peak with an amplitude equal to twice the noise level.
Source: Ref. 323.

of detection limits that may be expected from a liquid chromatograph-induction coupled plasma emission spectrometer (LC-ICP) system. Some results of their work is reproduced here in Table 10. In subsequent work, Fraley et al. [324] described the use of LC-ICP for the analysis of NTA and EDTA chelates of Cu, Zn, Ca, and Mg. Morita et al. have reported on the application of LC-ICP for minotoring Fe, Mn, Cu, Zn, P, and C in various protein and enzyme samples [325], for determining organoarsenicals in biological samples [326], and for monitoring phosphorus in inorganic phosphates as well as various nucleotides [327]. Taylor et al. have utilized LC-ICP for determining organically bound metals, notably Mg, Ca, Ti, Fe, Cu, Zn, Hg, Cd, in solvent-refined coal and processing solvent [328, 328]; several organoiron, organocobalt, and organocopper compounds [329]; and organoiron and organosilicon compounds; as well as a mixture of metal complexes involving some twenty different metals [330]. These papers represent a number of unusual features: (a) use of size exclusion columns for chromatographic separation, (b) use of highly volatile and combustible solvents for chromatography, and (c) the consequent excellent LODs achieved by the overall system. Krull et al. [212-214] have reported on the use of LC-ICP for determining chromium in different oxidation states [213], and also monovalent, divalent, and trivalent metals [212], employing IIR-based separation methods in both cases. They have also used aqueous mobile phases containing tributylphosphate to separate Cd, Zn, and Hg on a reverse phase column with a croww-flow nebulizer system for coupling the LC effluent to the ICP. The work of Gast et al. [209] was cited earlier as exemplary in this field; these workers have demonstrated high-efficiency separations of various organoiron, organomolybdenum, arsenic, and mercury compounds with ICP detection. A recent paper by Yoshida and Haraguchi [331] employed ICP detection for lanthanides which were separated on a cation exchange column; the LODs varied from 1 to 300 ppb for a 100-μl injected sample. ICP has been successfully used as a detector with microbore systems [332]. Tetraalkyl lead compounds have been separated on a C-2 silica column with a butanol-ethanol-water eluant and ICP detection [333]. Arsenic and selenium compounds have been separated by anion exchange and detected by ICP spectrometry [334].

Uden et al. have described the use of a liquid chromatography-direct current plasma emission spectrometer (LC-DCP) system for analyzing mixtures of β-diketone complexes of Co^{2+} and Cr^{3+} [336], as well as diethyldithiocarbamate chelates of Cu^{2+}, Ni^{2+}, and Co^{3+} [337]. The system has been described in detail [338]. Krull et al. [335] have separated and detected trivalent and hexavalent chromium species via IIR-based reverse phase chromatography and DCP detection [335]. The elution order was reversed as the IIR was changed from tetrabutylammonium to camphorsulfonate.

In his review, Krull [242] has commented extensively on the disparity of LODs attainable by direct ICP and by LC-ICP, as well as on the lack of and the necessity for simultaneous presentation of such data. The detection limits from the LC-ICP system are frequently two to three orders of magnitude worse than those attainable by direct ICP. Krull suggests that sample dilution of at least one to two orders of magnitude is a necessary evil of the chromatographic system; a self-respecting chromatographer should take issue with this assumption. Not only are the actual dilutions with an efficient gradient separation much smaller, but it is also actually possible to have greater analyte peak concentration in the effluent band than in the sample, especially with large sample volumes or concentrator columns. The work of Gast et al. [209] is a seminal paper at least partly because these authors found it more feasible than did other authors to carry out the desired chromatographic separations with high efficiency and with consequent satisfactory LODs.

D. UV/Visible Spectrophotometric Detection

As in any other type of chromatographic separation, direct absorptiometric detection is attractive whenever it is applicable. An excellent example is the chromatographic separation and detection of mercury alkanethiolates [339]. Many other organometallics are similarly amenable to direct UV detection. With the use of EDTA or similar chelating ligands in the eluant, direct UV detection of many metal ions that themselves have no useful optical absorption, e.g., Ca^{2+} or Mg^{2+}, is possible because the complexes absorb strongly. Unfortunately, for simple metal ions themselves, there is rarely sufficient optical absorption as such to permit sensitive detection, mercury being a notable exception. In the few cases in which sufficient UV absorption is present, the eluant of choice may not be optically transparent in that region. Fritz et al. [340-344] have separated metallic chloro complexes on an anion exchange column, using strong HCl as eluant and monitoring the absorption of the chloro complexes for detection. Similar studies have been done by others with HCl in acetone as eluant [345]. Detailed spectral data were reported for a very large number of metal ions in 6 M HCl [346]. The homemade "forced-flow" chromatographic system used by Fritz et al. for these studies is not a high-pressure system by current standards. Material compatibility will clearly be a serious problem with the use of most existing commercial instrumentation and columns, precluding such an approach.

By and large, the method of choice has been to introduce a suitable chromogenic ligand through a postcolumn reactor to form a highly colored metal complex and then to monitor the resulting optical absorption. These approaches are discussed in Section VI.B. Although it has rarely been attempted, using the chromogenic ligand directly in the eluant obviates the need for a postcolumn reactor. Zenki [347] separated and detected the alkaline earth metals on a cation exchange column, using a 0.7 M sulfosalicylic acid eluant containing 20 µM chlorophosphonazo III as the chromogen, adjusted to pH 7.5 with NH_3. Of the many bisazochromotropic acid dyes available (arsenazo I, arsenazo III, carboxyarsenazo, sulfonazo III, carboxynitrazo, chlorophosphonazo III, etc.), the last was chosen because it reacted with all the alkaline earths to produce highly absorbing complexes. The method was used to determine low parts-per-million levels of calcium in environmental water samples; excellent agreement with results from AAS analyses was demonstrated. The recent work of Bond and Wallace [224] has already been cited with respect to electrochemical detection. In the same study, the authors also employed spectrophotometric detection with 0.1 mM diethyldithiocarbamate or pyrrolidinecarbodithioate in the eluant. Sub- to low-nanogram LODs were attained for all six metals studied (Pb^{2+}, Cd^{2+}, Hg^{2+}, Cu^{2+}, Ni^{2+}, and Co^{2+}).

E. Other Detection Systems

Fluorescence detection, especially after postcolumn reaction, should be applicable for a variety of metal ions. The fact that metals often quench the fluorescence of other molecules may also be utilized for fluorescence detection, although it is admittedly less desirable than low-background, high-signal methods. Only a few rare earths, uranyl (UO_2^{2+}), and, to a smaller extent, beryllium ions fluoresce in solution; the present author is unaware of any report in the literature of applications of these properties. The best known examples of metal complexes that fluoresce are those of oxine (8-hydroxyquinoline); however, this fluorescence is apparently limited to the precipitated complex; no significant fluorescence is observed in solution. Morin (2, 3, 4, 5, 7-pentahydroxyflavone) also forms fluorescent complexes with several metals. Although these complexes, strictly speaking, are insoluble, frequently they remain in colloidal solution to permit continuous-flow detection.

In an entirely different and novel approach, Beckett and Nelson [348] derivatized EDTA to form 4-aminophenyl-EDTA. The amino group of this compound is capable of being rapidly derivatized with a suitable fluorogenic reagent, e.g., fluorescamine, to form a fluorescent product. The authors preformed the chelates of Zn, Cd, and Pb with 4-aminophenyl-EDTA, separated the individual complexes by anion exchange chromatography on a bonded stationary phase, and used fluorescence detection after postcolumn reaction with fluorescamine with an attained LOD of 0.92, 0.71, and 0.38 pmol, respectively. The unreacted 4-aminophenyl-EDTA is not eluted within 60 min under the chromatographic conditions (0.2 M NaOAc, pH 6.50), whereas the desired separation requires approximately 20 min. Further development and application of fluorescence detection of metal ions is clearly dependent on the appearance of new chemistry.

Scintillation-based flow-through radioactivity detectors [349,350] incorporating postcolumn reactors to introduce the scintillation cocktail are now available from a number of manufacturers. Although these detectors are primarily intended for biological applications involving detection of low-energy emitters; e.g., for the detection of tagged or tritium-labeled compounds, there is potential for such detectors in the nuclear industry. For γ-emitters, γ-spectrometry provides an attractive detection method. The dead volume of the necessary flow cell can be sufficiently minimized for successful capillary-column use [351-353] for both cationic and anionic analysis. Radioactivity detection with more conventional columns has been widely applied for alkali and alkaline earth metals [354-357], and especially for lanthanides and actinides [358-370], which are of particular importance in nuclear science and technology. More recently, the analysis of radiopharmaceuticals bearing the metastable technetium isotope (Tc-99) has become important because of its increasing use in nuclear imaging for medical diagnostics [371]. Scintillation detectors have been employed [372], and specially designed cadmium telluride detectors have been developed as well [373,374].

Universal detectors such as a differential refractometer have been successfully applied for chromatography of metal species. Yamamoto et al. [375] and Suzuki et al. [376,377] applied UV and/or RI detection for the separation of metal β-diketone chelates. With the suitable choice of a nonpolar eluant, complexes of very similar size were surprisingly separable by size exclusion on a PSDVB gel column. Suzuki and Saitoh [378] have since reported on the elution behavior of metal acetylacetonates for a great range of solvent polarity ranging from chloroform to methanol on the same stationary phase. The behavior of other β-diketone complexes as well as the effect of employing other size exclusion stationary phases has also been studied [379]. The behavior of mixed-ligand beryllium-β-diketone complexes on size exclusion-type stationary phases has also been investigated [380]. These authors have reported earlier [381] on the separation of fluorinated and nonfluorinated β-diketone chelates of metal ions. Two recent papers [382,383] provide clues to the actual basis of separation of such similarly sized molecules on size-exclusion stationary phases. Attempts have been made to accomplish reverse phase separation of β-diketone complexes of various metals [384]; decomposition of the metal chelate can be prevented by incorporating excess ligand in the eluant.

Differential refractometry has also been used, in combination with UV detection, by Gasparrini et al. [385] for the separation and detection of mixed substituted hydrazone complexes of Pd on a diol stationary phase with n-hexane/dichloromethane as eluant.

Creber and Wan [386] have demonstrated, in a preliminary report, the successful use of an electron spin resonance (ESR) spectrometer as an LC detector to establish the formation of a rhenium complex.

Approaches to Ionic Chromatography

Although Krull [242] has discoursed at length on the potential of a mass spectrometer as a detector for chromatography of metal species, many fundamental developments are necessary before such a method can become routine. The fact that high salt and buffer concentrations are often used with metal ion chromatography poses a formidable obstacle to facile application of mass spectrometric detection. However, interfaces between atomic plasma sources and mass spectrometers have recently been designed and may prove to be the best solution for those able to afford an LC-ICP-MS.

F. Metal Ion Chromatography with Chelating Ligand Eluant Systems: Considerations on Instrumentation

The use of any powerful chelating ligand in the eluant requires certain precautionary measures to prevent significant leaching of metals from system components during operation. Prior deoxygenation of the eluant (e.g., by purging with N_2 or He) is highly recommended. Some manufacturers (e.g., Dionex) currently market columns that are made of an inert synthetic polymer. Glass-lined steel columns that withstand much higher pressures are available empty or filled with a variety of high-performance adsorbents, from a number of manufacturers (Chrompack, EM Science, Scientific Glass Engineering). Glass-lined stainless steel tubing is also produced by the last-mentioned manufacturer. For long-term use, such inert column material is highly desirable for metal ion chromatography [387]. A number of inert (nonmetallic or titanium-coated) pumping systems of moderate pressure capability (1000-4000 psi) are available (Dionex Corp., Eldex Laboratories, LKB Inc., Perkin Elmer Corp., Toyo Soda/Bio-Rad). Detectors are usually supplied with metallic connecting tubes; however, in most cases, manufacturers will supply them with PTFE connecting tubes on request; it is also relatively easy to replace the metallic parts in-house. The limited contact time and exposed metallic surface area in a typical detector rarely presents a serious problem. The situation is different with the pump; a chelating eluant should not be allowed to remain sitting in a pump with metallic contact parts for appreciable periods of time. If it *is* left overnight somehow, the column must be disconnected from the pump before eluant flow through the column is resumed. These considerations assume increased importance if trace-level analysis is to be performed, directly or after enrichment [388].

G. Separation Modes

Ion Exchange

The use of an ion exchange column is by far the most common approach to chromatographic separation of ionic metallic species. Separations have been carried out on conventional crosslinked ion exchangers and bonded-phase ion exchangers as well as macroreticular ion exchangers, with applications spanning from alkali, alkaline earth, transition, and posttransition metals, including the highly radioactive and unstable man-made elements. The review of Cassidy [241] covers this area in detail, but the interested reader should also consult the treatises on ion exchange and ion exchangers by Dorfner [389], Helfferich [59], and Kunin [248]. General treatment of ion exchange chromatography has been dealt with by Kraus [390] and Scott [391]; many details on specific separations are to be found in the works of De [392], Inczedy [393], Rieman and Walton [394], Samuelson [395], and Schubert [396]. The work of De [392] provides data on ion exchange properties of many metal ions as well as on the related topic of solvent extraction; a series of volumes edited by Marinsky and Marcus [397] report on the advances of the same. The collection of ion exchange chromatographic data reported by Strelow in Volume 5 of this series [397] is particularly noteworthy; the same volume also contains a comprehensive account of the use of pellicular ion exchangers by Horvath. The 1982 biennial review of column liquid chromatography by Majors et al. [244] in *Analytical Chemistry* also contains more than thirty citations on ion exchange separation of metal ions, principally lanthanides and actinides. The most recent review on ion exchange chromatography and ion exchangers is by Walton [398]; it is also likely the most useful.

Fritz et al. have studied the application of low-capacity macroreticular cation exchangers for ion exchange separation [399,400]; this work was of crucial importance for the further refinement of metal ion chromatographic methods. Cassidy and Elchuk [401] have also employed ion exchange-based separation methods for the analysis of Mn, Fe, Co, Ni, Cu, Zn, and Pb (with optical detection after postcolumn reaction) in a variety of matrices of interest to the nuclear industry. These authors were the first to demonstrate [402] the truly impressive separation capabilities of bonded ion exchangers and small-size ion exchange resins for the separation of lanthanides. The studies of Takata et al. [256-263] employing electrochemical detection and ion exchanger stationary phases have been mentioned earlier in Section V.B.

In more recent years, impressive contributions to high-efficiency separation of metal ions, coupled with high sensitivity detection, have originated from the Dionex Corporation. Slingsby and Riviello [403] demonstrated high-efficiency separation of transition metal ions with a citrate/tartrate complexing eluant (12 mM citric acid, 40 mM tartaric acid, adjusted to pH 4.1 with LiOH) on a Dionex cation exchange column. Optical detection at 520 nm was used for quantitation after postcolumn reaction with PAR. An example of low- to sub-parts-per-million level separation of seven transition metal

Approaches to Ionic Chromatography

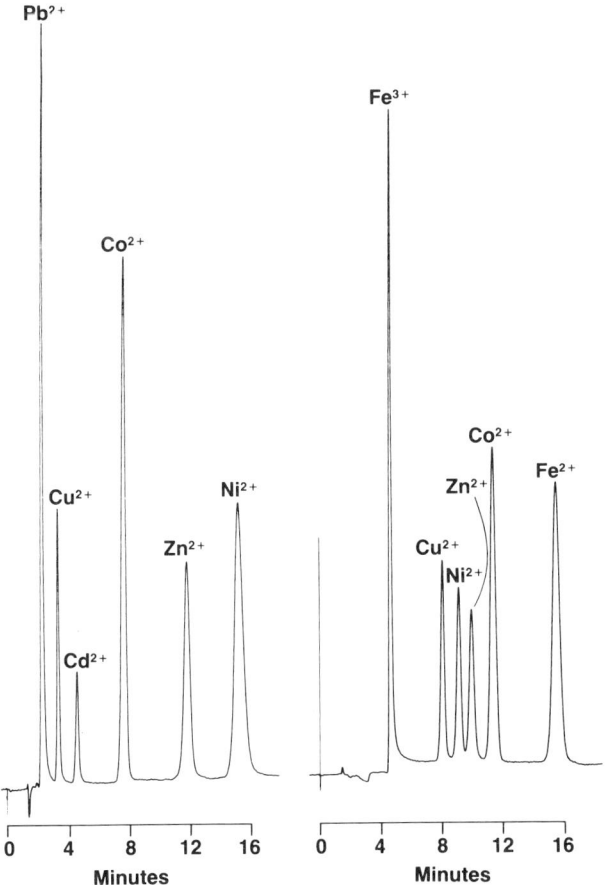

Figure 34. Separation of transition metal ions on a Dionex CS-5 column and detection after postcolumn reaction with PAR in a membrane reactor. (Left) Eluant: 50 mM oxalic acid + 95 mM LiOH, pH 4.8, 1 ml/min. Sample: 50 μl; 4 ppm Pb^{2+}, 0.5 ppm Cu^{2+}, 4 ppm Cd^{2+}, 2 ppm Co^{2+}, 2 ppm Zn^{2+}, 4 ppm Ni^{2+}. (Right) Eluant: 6 mM pyridine-2,6-dicarboxylic acid, 8.6 mM LiOH (pH 4.8), 1 ml/min. Sample: 50 μl; 1 ppm Fe^{2+}, 1 ppm Cu^{2+}, 3 ppm Ni^{2+}, 4 ppm Zn^{2+}, 2 ppm Co^{2+}, 3 ppm Fe^{3+}. (Courtesy of J. M. Riviello.)

ions is shown in Figure 34 for a 50-μl injected sample. Via a suitable complexing eluant such as oxalate (25 mM oxalic acid adjusted to pH 4.8 with NaOH), transition metal ions can also be separated on an anion exchange system. Clearly, selectivity differences between the cation and anion exchange systems allow the chromatographer unusual flexibility to tailor the analytical protocol to a given sample. The same paper [403] also demonstrates efficient and sensitive chromatographic determination of alkali metal ions with a membrane-based suppressor and conductimetric detection. Suppressed IIR (hexanesulfonic acid)-based analysis of alkylamines on a reverse phase column was reported. Separation of alkaline earth metal ions with a m-phenylene diamine eluant was reported by these authors, as well as by Rocklin [404].

Riviello and Pohl [233] have reported on their extensive studies of the various parameters that affect efficiency and selectivity in separations of metal ions, while working with commercial Dionex systems. There are essentially three principal modes of separation: (a) cation exchange on pellicular cation-exchange packing, (b) anion exchange on Latex-agglomerated anion exchange packing (0.1-μm anion-exchange Latex beads coated on 15-μm surface-sulfonated cation-exchange substrate), and (c) IIR-based separation. The last mode can employ either a cationic or an anionic IIR, as long as a suitable complexing agent is incorporated in the eluant. In the cation exchange mode, a solution of 40 mM tartaric acid + 12 mM citric acid (pH adjusted to 4.3 with LiOH) was found to be the best choice for separating Ni^{2+}, Zn^{2+}, Co^{2+}, Pb^{2+}, Fe^{2+}, Cd^{2+}, and Mn^{2+}. The metal ions are cited according to their elution order: Ni^{2+} elutes first, Mn^{2+} last. The citrate/tartrate eluant is particularly good for the analysis of strongly retained metal ions such as Cd^{2+} and Mn^{2+}. A solution of 10 mM oxalic acid + 7.5 mM citric acid (pH adjusted to 4.2 with LiOH) was better suited for the separation of Fe^{2+}, Cu^{2+}, Ni^{2+}, Zn^{2+}, Co^{2+}, Pb^{2+}, and Fe^{3+}. If the pH of the eluant is increased, the complexing ability of the ligand in the eluant increases, resulting in decreased retention of the metal ions. For example, log k' values for Fe^{2+}, Pb^{2+}, and Zn^{2+} were found to decrease linearly with increasing pH (adjusted with LiOH) for a 20 mM oxalate eluant, within the pH range 3.6 to 4.2. The metal ion chosen to be associated with the base used to adjust the pH can also profoundly affect the retention. As these ions compete for cation exchange binding sites, all retention times decrease, without any change in retention order, as the base chosen for pH adjustment is changed from LiOH to NaOH to KOH. For the citrate/tartrate eluant, selectivity is dependent on the precise choice of eluant pH, which affects the ratio of the respective complexing ligand anions because of the differences in pK values of the two ligands. For example, Pb^{2+} is eluted after Zn^{2+} with a 35 mM tartaric acid + 15 mM citric acid eluant at pH 4.4, whereas the situation is reversed with the eluant at pH 4.6. Similarly, the retention order is affected by changing the tartrate/citrate ratio at constant pH. Eluants with more modest complexing ability, e.g., 20 mM ammonium sulfate + 10 mM sulfuric acid are also useful in some cases, particularly for the separation of Al^{3+} and Fe^{3+}. The detection of Al^{3+} is better accomplished with tiron (1,2-dihydroxybenzene-3,5-disulfonic acid) as the postcolumn reagent whereas detection of most other metal ions is satisfactorily accomplished with PAR [pyridyl(azo)resorcinol]. It should be noted that accurate determination of Fe^{2+} in complexing media is impossible if oxygen is present. Oxygen diffuses rapidly through PTFE and many other plastics. For obvious reasons, trace analysis of Fe is impossible in stainless steel-based LC systems.

In the anion exchange mode, 25 mM oxalic acid at pH 4.8 (with NaOH) is an attractive eluant. In this mode, both cation and anion exchange sites are accessible. As a result, the effect of pH with an oxalate-based eluant is not predictable a priori. For example, log k' values for Cu^{2+} decrease monotonically (but nonlinearly), whereas those for Zn^{2+} and Co^{2+} remain virtually unchanged, as the pH of the 25 mM oxalate eluant is changed from 3 to 5.5. Similarly, the effect of the monovalent counterion present in the base used for pH adjustment (Li^+, Na^+, K^+, or NBu_4^+) is somewhat unpredictable. The degree of crosslinking of the substrate cation-exchange resin (on which the Latex anion-exchange beads are agglomerated) also significantly affects the selectivity of the chromatographic system, presumably by controlling access to the subsurface cation-exchange sites.

In the IIR mode, a cationic IIR (e.g., NBu_4^+) can be used with the oxalate eluant, or an anionic IIR (e.g., octanesulfonate) can be used with the tartrate/citrate eluant in conjunction with a reverse phase column. Utilization of this technique to separate Au and Fe as their anionic cyanide complexes with suppressed conductivity detection has been described [84]. IIR-based metal ion separation methods have been discussed in more detail earlier in Section IV.F.

Employing Latex-agglomerated anion-exchange columns is described above, Rocklin [405] determined a variety of metal ions with the use of an eluant containing 0.05 M HCl, 0.30 M $NaClO_4$. UV detection, at 215 nm, for the chloro complexes of the metal ions was attractive for most of the metals studied; pulsed amperometric detection was found to be attractive for Au. Levels lower than 1 ppb were detectable with preconcentration techniques. Perchlorate is employed in the eluant because of the very high affinity of the metal-chloro complexes for the ion exchange binding sites.

Normal Phase Chromatography

Although ion exchange is the dominant mode of separation utilized for metallic species, normal phase chromatography (polar stationary phase, e.g., hydrated silica; nonpolar eluant, e.g., hexane) has been widely used for the separation of organometallic compounds and metal complexes with organic ligands.

Separation of organoiron compounds has been extensively studied [320,209,406-412], as has been that of organochromium species [406, 409,413-416]. Separation of various organomanganese [406], organoruthenium [417], organohenium [386], organocobalt [409,412,416, 418], organomercury [280,419], organolead [294], and mixed-metal species containing Ge-Hg-Pt-Ge, Sn-Hg-Pt-Ge, and Ge-Cd-Pt-Ge moieties [420] have been successfully carried out by normal phase chromatography. Homo- and heteroclusters containing Fe, Ru, Os, and Ni have been separated both by normal and reverse phase chromatography [421].

A variety of metal complexes have been separated by normal phase chromatography. Complexes of substituted and unsubstituted dithiocarbamates [336,338,422-433], and dithizone [387,424,434,435] have been extensively studied. Urinary Pt has been converted to the dithiocarbamate chelate and then chromatographed on a CN-silica column, using a heptane-isopropanol eluant [436]. In many cases, on-column dissociation has been reported to be a problem. Xanthate complexes of As^{3+}, Sb^{3+}, Bi^{3+}, Se^{2+}, Te^{2+}, and Ni^{2+} were studied by Eggers and Russell [437]; serious decomposition was observed with polar stationary phases or eluants. Tollinche and Risby [438] carried out extensive studies on the separation of β-diketone chelates via a variety of stationary phases; normal phase chromatography was judged to be the best mode.

Galushko et al. [439] have studied the separation of preformed pyridyl(azo)naphthol chelates of Co, Fe, and Ni, using normal phase chromatography. The early report of Huber et al. [440] on the separation of metal β-diketonates involved liquid-liquid partition in a strict sense, but it may also be regarded as a normal phase separation. Other studies on normal phase separation of metal β-diketonates have been cited earlier [337,338]. Other examples of metal complexes separated on normal phase LC include hydrazones and semicarbazones [385,422,434,411], β-ketoamines [442], oxinates [443], thiooxinates and 1-hydroxy-2-pyridinethionates [444], and porphyrin IX [445]. General criteria for chromatography of metal complexes have been outlined [446]. An especially interesting example is the separation of four diastereomers of tris[(+)-3-acetyl-camphorato]chromium(III) [447].

Separation on Reverse Phase and Other Bonded Phases

In recent years, the general practice of liquid chromatography has come to be dominated by the use reverse (nonpolar) and other bonded phases. The shift from normal phase to these modes is also noticeable for the analysis of metal species. IIR-based chromatography (Section IV) is carried out with reverse phase supports almost without exception and has already been discussed.

Among organometallics, examples include the separation of organoiron [448-452], organocobalt ([450,453-456]; Ref. 456 is a review solely devoted to the liquid chromatographic separation of coordination complexes of Co^{3+}), organomolybdenum [209,454], organoruthenium [450,457], organoiridium [450], organophosphorus [309], organosilicon [281], organotin [292,294,303,458], organoarsenic [294, 313,314], organolead [282,294,310,311], organomercury [270-273,294, 209,419], and metallic clusters containing Fe/Ru, Fe/Co, Fe/Ru/Os, and Co/Ru atoms [459].

Among reverse phase separation of metal complexes, notable are dithiocarbamates [222,460-466] and β-ketoamines [338,467,468]. Separation of Cu, Zn, Fe, and Co diethyldithiocarbamate chelates on a microbore column (0.5 × 120 mm) coupled with ICP detection has been reported [332]. Shih and Carr [469] reported exceptionally stable chelates involving n-butyl-2-naphthyldithiocarbamate, which can be readily subjected to reverse phase chromatography. Ni, Cu, Hg, Co, and Fe chelates are preformed by pretreatment with an excess of the labile Zn complex and then separated on a C-18 silica column, using a methanol/water eluant. Roston [470] has shown the utility of precolumn chelation of Cu, Co, Ni, and Fe with PAR; in combination with UV or amperometric detection, this is a very attractive and sensitive technique for the analysis of these metals. Dithizonate of Co has been successfully separated from those of Ni, Zn, Pb, and Cd on a C-18 silica column with a methanol/water eluant [471]. Conditions for baseline separation of acetylacetone and benzoylacetone chelates of Mn, Be, Cd, Cr, Rh, Ru, Ir, Pd, and Pt on C-18 silica columns have been reported by Gierira and Carr [472]. Other examples include the separation of crown ether complexes of mercury halides [473] and the determination of Se after reaction with 2,3-diaminonaphthalene [474] or o-dianisidine/ 4-chloro-o-dianisidine [359]. Weber and Schwedt [475] have investigated a number of bonded phases for the determination of Zn, Cu, Fe, Co, and Ni in tea and coffee; the range varied from 20 ppb Co in tea to 15 ppm Zn in coffee. Direct analysis of organically bound Cu, Zn, and Fe in such beverages has also been carried out [476]. Different separation modes were also explored by this group for the speciation of manganese in biological growth media [477]. Schwedt and Budde [478] have described the best conditions for the reverse phase separation of pyridyl(azo)naphthol chelates of various metals.

A variety of stationary phases (-CN, PSDVB gel, C-18) have been studied by Jessen et al. [479] for the separation of organotin halides. Gast and Kraak [480] similarly studied -CN, -diol, and C-18 silica for the separation of bis(tetraazadiene) complexes of Mn, Co, Fe, Ni, Mo, and Cu; mixed metal clusters containing Co/Mn; as well as other organometallic compounds containing Ru, Fe, and Mo. Igarashi et al. [481] separated porphyrin complexes of Cu, Zn, and Pd by reverse phase chromatography, using dodecylsulfate as IIR.

Armstrong et al. [482] have recently introduced a highly unusual stationary phase consisting of cyclodextrins (CD) bonded to silica. Cyclodextrins are bacterially formed, cyclic oligomers of dextrose that are naturally chiral. The cyclodextrin molecules are shaped much like open-ended wastebaskets with a hydrophilic exterior and a hydrophobic interior. The size of the cavity is dependent on the number of dextrose units in the oligomer, and cyclodextrins are designated as α-, β-, γ-, σ-, etc. in the order of increasing cavity size. The cavity of α-CD accommodates a molecule of the size of benzene, whereas β-CD accommodates a molecule of the size of naphthalene--to give some idea of the actual size of the cavities. Because cyclodextrin molecules are inherently chiral, they lend themselves well to the otherwise monumentally difficult separation of optical enantiomers. Using hydrolytically stable cyclodextrin-bonded phases, Armstrong et al. separated a variety of optical isomers of substituted ferrocenes, osmocenes, and ruthenocenes with methanol/water eluant systems [482].

Size Exclusion Chromatography

Although in a number of cases metal species have been separated on stationary phases designed for size exclusion (also called *steric exclusion* or *gel permeation*), it is clear that the operative physical separation mode is likely not based on size discrimination, e.g., for the separation of metal β-diketonates [375-380] or simple organolead compounds [310]. However, with element-specific atomic spectrometric detection, a true size exclusion mechanism has been successfully exploited by Brinkmann, Fish, et al. [302,307]; see also other works by these authors cited in Section V.C), and by Renoe et al. [296] and Morita et al. [325] for size (or MW)-differentiated elemental speciation in environmental, biological, and synthetic polymeric samples.

H. Stationary Phases

Chelating Supports

Complexing Supports for Chromatography and Preconcentration--Appropriate chelating ligands can be bonded to or otherwise immobilized on a support matrix, thus providing a complexing stationary phase. Such material can be used for chromatography and/or trace enrichment of element(s) of interest. In the following, the citations refer to either one of these uses. Modeling of equilibria involved with chelating resins has been carried out by Szabadka et al. [483].

The use of a resin with chelating iminodiacetic acid [$-N(CH_2COO^-)_2$] functionalities was introduced commercially several decades ago as the

Dowex-Al resin; this type of resin is still available under the trade name Chelex-100. The affinity of such a stationary phase for various metal ions is well characterized quantitatively [484]. More recent examples include the preconcentration of heavy metals in snow, and limestone and biological materials on the calcium-form resin [485]; the same form was also particularly useful in trace enrichment of Cd, Co, Ni, Cu, and Zn from natural waters [486]. A large number of metals (Ba, Ca, Cd, Ce, Co, Cr, Cu, Fe, La, Mg, Mn, Sc, U, V, and Zn) have been preconcentrated on Chelex-100 from seawater [487]. It has been found possible to preconcentrate tungstate and molybdate on Chelex-100 [488]. Chromatography on Chelex-100 may be carried out in the presence of nonaqueous solvents as well; glycine has been used as the eluting ligand in an acetone:water medium [489]. Some analytical measurement techniques (e.g., x-ray fluorescence) are particularly facilitated if enrichment is carried out on a membrane filter by loading the filter with the resin and passing large volumes of sample through it [490].

Resins similar to Chelex-100 include propylenediaminetetraacetic acid bonded to macroporous XAD, introduced and characterized for chromatography by Moyers and Fritz [491]. Gimpel and Unger [492] have bonded monomeric and polymeric iminodiacetic acid functionalities to silica supports. Supports with chelating diamine functionalities [493-496] as well as primary and secondary amines bonded to silica have been described [493]; of course, commercial amino columns bonded to silica have been commercially available for some time now. Among other chelant-based stationary phases containing N-donor atoms are α-hydroxyoximes [497], amides [498,499], and hydroxamic acids [500], all introduced by Fritz et al.

Ghosh and Das [501] bonded 1-nitroso-2-naphthol to PSDVB resin; the same support has been used by Lundgren and Schilt [502] for attaching ferroin-type reagents; this latter resin was found useful for the preconcentration of Fe, Co, Ni, Cu, and Zn from seawater and various reagent-grade chemicals. Dimethylglyoxime immobilized on polyurethane foam was successfully used for the preconcentration and separation of Ni [503]. One viable approach to incorporation of the ligand into the stationary phase matrix is to introduce it during the synthesis (polymerization) of the stationary phase itself. Poly-(acrylamidoxime) [504,505] and styrene-3(5)-methylpyrazol [506] copolymer supports made in this fashion have been described.

Among supports containing oxygen donor atoms, β-diketone-containing bonded phases have been described [494,507].

Oxine is the most notable among immobilized ligands containing both N and O donor atoms; one commercial resin containing bonded oxine is available (Spheron Oxine 1000). There have been a large number of reports on oxine-containing stationary phases. Parrish and Stevenson [508] and Parrish [509] incorporated oxine into a formaldehyde-resorcinol resin. Parrish [510] has also described chromatography on macroporous supports coated with a liquid ion exchanger containing oxine; detailed description of the chromatographic setup has been published [509]. More recently, he has given an account of the synthesis and characteristics of various types of chelating resins from oxine [511]. Several other reports [512-518] concern oxine-containing stationary phases. Oxine has been bonded to silica [512] and to controlled-pore glass [513,514], incorporated in glycomonomethacrylate-glycodimethacrylate resin [505], and immobilized on glass [516] and carboxymethylcellulose [517], the last being used for preconcentration of Pd, Zn, Cd, Mn from seawater. Chromatographic characterization of the oxine-bonded CPG support is particularly extensive [514]. It is interesting to note that the addition of oxine to a water sample, followed by adsorption of the metal oxinates on activated carbon, was found to be a particularly attractive method of preconcentration of trace metals for neutron activation analysis [519]. Preformed oxine complexes have been concentrated on C-18 silica [520,521]. A PSDVB stationary phase containing bonded 8-aminoquinoline (the N analogue of oxine) has been reported [522].

Among chelating supports containing sulfur donor atoms, dithiocarbamate-bonded silica supports have been described [493,494,496]. Fritz et al. [523,524] have synthesized and chromatographically characterized thioglycolate-bonded stationary phases. King and Fritz [525] have described the utility of bis(2-hydroxyethyl)dithiocarbamate (which forms soluble complexes) in preforming metal complexes for preconcentration on XAD-4 resin. Dithizone immobilized on carboxymethylcellulose has been used for preconcentration [517], and plasticized foam containing dithizone was found useful for the trace enrichment of silver [526].

Cellulose has been used as the support for immobilizing salicylic acid, pyrocatechine-3,5-disulfonic acid (tiron), 1-(2-hydroxyphenylazo)-2-naphthene (hyphan), and PAR, and used for trace enrichment purposes by Forster and Lieser [527,528]. Sarri and Seitz [529] have developed a calcein-bound cellulose support. One particularly unusual support is graphite, modified by adsorbed Eriochrome Black T [530]. This work describes a number of novel features, the most notable being the controlled liberation of the metal ions (Fe, Ni, Co, Mg, and Ca) adsorbed and enriched on the stationary phase, by applying a potential to the conductive bed. Other unusual methods of preconcentration include (a) the reactive uptake of oxidant anions, e.g., permanganate, chromate, and vanadate, on Fe^{2+}-form cation exchange resin [531]; (b) enrichment of gold on tributylphosphate-impregnated polyurethane foam [532]; and (c) the uptake of gallium by tributylphosphate-coated polyfluorocarbon [533].

Another novel and particularly facile technique of creating complexing stationary phases, of special utility in trace enrichment, is the binding of terfunctional reagents on ion exchange resins. The desired threefold properties of a terfunctional reagent are:

1. The ability to sequester selectively the desired analyte that is to be preconcentrated.
2. The possession of an ionic functionality separate from (1) above, which allows the ionized reagent to be bound to an ion exchange resin by electrostatic forces.
3. The presence of a strong lipophilic interaction of the terfunctional reagent molecule with the matrix of the ion exchange resin, such that the binding in (2) is substantially augmented by adsorptive forces.

The binding of the terfunctional reagent to the ion exchange resin through (2) and (3) must be sufficiently strong such that there is no appreciable leaching of the reagent from the support with the usual eluants. Grote et al. [534] have used formazan sulfonate as a terfunctional reagent for binding on silica-based quaternary ammonium functionalized packing. Tanaka et al. [535] have described their work in this area in exemplary detail. A PSDVB-based anion exchange resin (Amberlite IRA-400) was shaken with the aqueous solution of the terfunctional reagent (30°C, 1 hr), filtered, washed (water, methanol), and air dried. Terfunctional reagents examined included tetraphenylporphinetrisulfonic acid (TPPS, 0.05), dithizone (0.1), arsenazo III (0.3), sulfonazo III (0.4), zincon (1.0), azothiopyrinesulfonic acid (ATPS, 1.3), and dithizonesulfonic acid (1.9), the values in parentheses indicating the exchange capacity of the treated resin for the desired metal in millimoles per gram. Except for arsenazo III, which released 10% and 1% of the reagent upon treatment with 1 M NaCl, and 0.1 M HCl, respectively, the other resins did not release the reagents to any perceptible extent. Reagents such as p-mercaptobenzenesulfonic acid, thiosalicylic acid, and oxinesulfonic acid produced large exchange capacities but were leached off relatively easily. The release of a metal ion taken up by such a stationary phase is typified by the example of Hg(II) collected on a ATPS-loaded resin. With a very strong eluant, e.g., 8 M HNO_3, both the terfunctional reagent and Hg(II) were liberated from the resin. With a more moderate and selective eluant, e.g., 10% thiourea in 0.1 M $HClO_4$, Hg(II) alone was released, the ATPS-loaded resin being regenerated in the process. Clearly, whenever practicable, the second type of approach is preferable. The dithizonesulfonate-loaded resin was found to be remarkably efficient for silver uptake, even from spent photographic fixer solutions, which contain silver as its highly stable thiosulfate complex.

This approach involving the use of terfunctional reagents is not limited to the enrichment of cations; fluoride was selectively enriched,

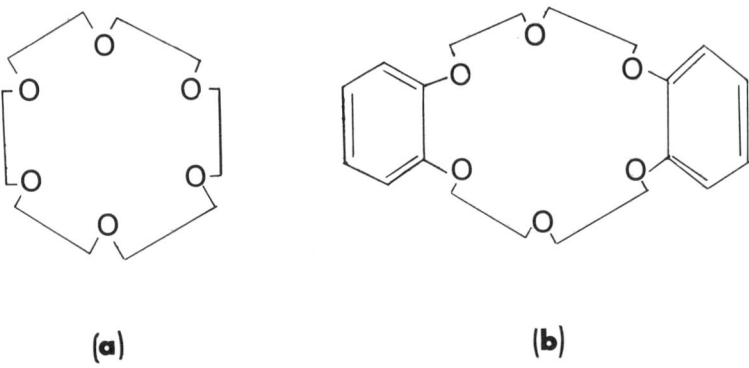

Figure 35. (a) Structure of 18-crown-6; (b) structure of dibenzo-18-crown-6.

even from a 1 M NaCl solution, on a lanthanum-alizarin complexone-loaded resin, and removal was facile with any alkaline eluant.

Crown Ethers in Chromatography--Since the serendipitous synthesis of the first crown ether (also called cyclic polyethers; the IUPAC recommended but rarely used designation is *coronands*) was reported by Pedersen in 1967 [536,537], the interest and research activity in crown ether chemistry have increased exponentially. Crown ethers contain a cavity composed of oxygen atoms, typically joined by $-CH_2-CH_2-$ bridges. A common name, e.g., that of the first crown ether synthesized, 18-crown-6, implies that the ring contains 18 members, including the six oxygen atoms (Figure 35a). Other substituents present on the ring (if more than one, they are invariably symmetrically placed) precede the name, e.g., dibenzo-18-crown-6 (or DB-18-C-6 in the usual abbreviation), represents the compound in Figure 35b. Because oxygen is a hard donor atom, hard cations, such as those of alkali and alkaline earth metals are particularly well complexed by crown ethers. Crown ethers in which one or more O atoms are replaced by S or N atoms are appropriately called thia or aza crowns. Although these may show different selectivities in complexation as compared to simple crown ethers, the oxygen atoms are generally more numerous and dominantly determine the hardness of the overall molecule as a base. The characterization of such compounds thus far has been limited. Cryptands and spherands, as opposed to coronands, contain two or more interconnected rings, respectively, producing a truly three-dimensional cavity. Although these are, in general, more powerful complexing agents than crown ethers, complexation/decomplexation rates are too slow to be of utility in efficient chromatographic separations.

It is generally believed that the selectivity of a given crown ether toward a given metal ion is largely determined by the relative size of the metal ion and the crown ether cavity; the metal ion that most closely fits the cavity is most strongly bound. Thus 12-C-4, 15-C-5, and 18-C-6, respectively, prefer Li^+, Na^+, and K^+. Formation of sandwich (or club sandwich) complexes, in which a metal ion is interposed between two crown ether molecules (or two metal ions are stacked between three planar crown ether cavities), is not uncommon when the metal ion is too large to fit well into the crown ether cavity. Before the simplistic rationalization presented above is overextended, it is wise to point out that the cavities of large, flexible crown ethers can easily conform sterically to a range of metal ion sizes, and the majority of these show a preference for K^+ [538].

The particular promise of crown ethers in ionic chromatography is the potential use of completely nonionic eluants (namely, pure water) for ionic separation on crown ether-based stationary phases. If feasible, such an approach easily permits sensitive conductimetric detection or detection with any type of bulk property detector including a differential refractometer. A critical examination of the chromatographic results obtained to date, however, indicates that the initial promise is yet to be fulfilled and that efficient separations are indeed yet to be realized.

Although chromatography has been successfully carried out with pure methanol or methanol/water mixtures, the use of pure water as eluant for meaningful separation of metal ions remains a relatively elusive goal. Affinity of crown ethers for metal ions increases with decreasing solvent polarity; water itself is a hard base and competes in aqueous media with the crown ether for metal ion complexation. As a result, complexation constants in pure water are too low, and crown ether stationary phases do not generally produce sufficient retention of metal ions with hard counterions (e.g., chloride) in purely aqueous eluant systems to be useful for chromatographic separation.

A second important aspect concerns the influence of the associated anion on the retention of a given cation by a crown ether stationary phase. This effect can be very large and may well overshadow the influence of the metal ion itself on the observed retention. In experiments conducted by the author, DB-18-C-6 was coated on a commercial 5-μm C-18 silica column from an aqueous acetonitrile solution, and then chromatography was carried out with pure water. DB-18-C-6 is sufficiently insoluble in water; therefore, operation for several hours is possible before retention characteristics change perceptibly. Although it was not possible to resolve NaCl and KCl on this stationary phase, KCl and KSCN could be easily resolved and KI and KSCN could be partially resolved [539]. Indeed, reasonably efficient applications of pure water as eluant with crown ether-based stationary phases have primarily been reported for the separation of anions, all introduced into the column as the salt of the same cation. Blasius et al. [540,541] have reported the separation of $F^-/Cl^-/Br^-/I^-/SCN^-$ (both as Na and K salts); SO_4^{2-}/Cl^-, SO_4^{2-}/SO_3^{2-}, HPO_4^{2-}/Cl^-, $Cl^-/NO_2^-/NO_3^-$, $Cl^-/Fe(SCN)_6^{3-}$, $SO_4^{2-}/MoO_4^{2-}/CrO_4^{2-}$ (all as Na salts); and $IO_3^-/BrO_3^-/ClO_3^-/ClO_4^-$ (as K salts) on polymeric crown ether-containing stationary phases with water elution. In a similar vein, Igawa et al. [542] reported the separation of $F^-/Cl^-/Br^-/I^-$ as Na salts on a polyamide crown stationary phase and somewhat better separation of the same mixture on a 10-μm silica support coated with the polyamide crown. The second stationary phase was sufficiently more efficient to separate the more demanding mixture of $SO_4^{2-}/F^-/Cl^-/Br^-/NO_3^-/I^-/SCN^-$ as their Na salts. Nakajima et al. [543] separated $Cl^-/Br^-/I^-$ as their K salts on a silica column containing bonded bis(benzo-15-C-5) functionalities. It may well be that crown ethers will be better suited for anionic than cationic separations.

It is obvious that to obtain meaningful chromatograms, all the cations in the sample need to be converted to the same form. Direct loading into the injection valve through a small column packed with H^+-, or better, Li^+-form cation exchange resin seems to be the best way. Conversion to the lithium salt rather than the corresponding acid is preferable both because Li^+ is the most weakly held of all cations and because weak acids may be retained in the H^+-form resin through molecular exclusion. Unfortunately, no efforts have been made in this area; the crown ethers incorporated in various stationary phases studied thus far have all been too large for the effective complexation of Li^+. The question remains that if the counterions are not all converted to the same form, in what forms, in reference to the cation-anion combinations possible, does the injected mixture eventually elute? In a rather extreme case studied by Blasius et al. [540], injection of equimolar amounts of Li^+, Rb^+, Cl^-, and SCN^- on a stationary phase containing the DB-18-C-6 moiety resulted in almost complete association of the softer anion with the preferentially retained cation; i.e., elution took place as LiCl followed by RbSCN. Similar results were obtained when NH_4^+ was substituted for Li^+ [544]. In a less extreme example studied by Igawa et al. [542], the injection of equimolar amounts of Li^+, K^+, Cl^-, and Br^- on a stationary phase also incorporating the DB-18-C-6 moiety resulted in the elution of all four possible compounds, LiCl, KCl, LiBr, and KBr, in that order. This situation is likely to be more typical and also exemplifies the importance of the anion in determining retention behavior. Blasius et al. [540] have also studied the retention behavior of a mixture of LiCl, NaH_2VO_4, and $Na_6V_{10}O_{28}$ on a stationary phase containing DB-18-C-6 as a function of the pH of the aqueous eluant. At pH 9.0, LiCl was completely resolved from the oxovanadium anions, which were only partially resolved and eluted as Na_2HVO_4 and $Na_6V_{10}O_{28}$. At pH 5, LiCl was partially resolved from NaH_2VO_4, which was well resolved from $Na_5HV_{10}O_{28}$. At pH 3, $H_2VO_4^-$ was transformed to cationic VO_2^+, which eluted as a single peak with LiCl, whereas $Na_5HV_{10}O_{28}$ was well separated from this. At pH 0.5, the polyvanadate was also hydrolyzed to the vanadyl cation.

Examples of cation separation with pure water elution are not, except as noted below, especially impressive. These include the separation of HCl/NaCl and NaOH/KOH on a stationary phase involving DB-18-C-6 [540], the separation of KCl/RbCl/CsCl and NaI/KI/RbI/CsI on stationary phases involving DB-24-C-8 [541], and the separation of Cs^+ from 18 other metals (which all elute together preceding Cs^+) on a stationary phase containing DB-24-C-8 in a simulated nuclear waste sample in 1 M HNO_3 [545]. Although their initial report [546] was not especially promising, substantially more efficient separations of five alkali halides (chlorides, bromides, and iodides) on poly(benzo-15-C-5) bonded to silica were reported by Nakajima et al. [543].

The incorporation of a soft anion in the aqueous eluant greatly increases the retention and generally improves the separation factor. Detection however becomes a difficult problem, except when the sample metals are radionuclides or are so tagged. Thus Smulek and Lada [355-357] have separated radioisotopes of Na, K, Rb, and Cs on polyether coated columns, with $NaClO_4$ or NaSCN as eluants. Similar separations have been reported by Shen et al. [547] as well.

The third and perhaps the most disconcerting aspect of crown ether-based chromatography concerns the rate of mass transfer in such systems, which in turn affects the attainable column efficiency at virtually all but unacceptably low column flow rates. In his review [241], Cassidy astutely suggested a need for more intensive studies on the mass transfer characteristics of crown ether-containing stationary phases because the large peak widths for the late-eluting species in the large majority of studies then available suggested low column efficiencies. In the present author's studies [539], for a 4.6 × 250 mm DB-18-C-6 coated 5-μm C-18 silica column, chromatographic efficiency continued to increase all the way to the lowest flow rate studied, 100 μl/min. For conventional 3- to 4-mm i.d. columns, flow rates used have been as low as 29 μl/min [542] or 33 μl/min [545]; these are clearly unacceptable to the practicing analyst with a finite amount of time at his or her disposal. It is unfortunate that a number of investigators have swept the problem under the rug by presenting the chromatograms plotted in units of retention volume rather than retention time. Poor mass transfer rates must somehow be involved with the presence of the crown ethers themselves; certainly the poor column efficiency (≤ 650 plates/m) calculated for SCN^- in Figure 36 [542] is not characteristic of the 10-μm spherical porous silica used as support.

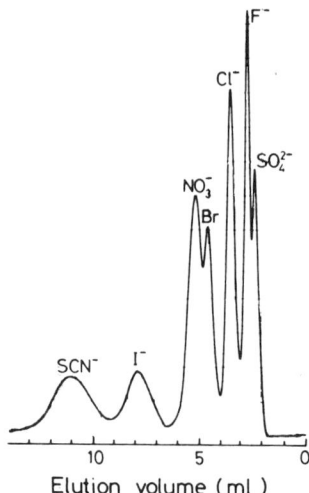

Figure 36. Chromatogram for seven anions on a silica stationary phase coated with polyamide crown. Column, 4 × 250 mm; flow rate, 60 µl/min; mobile phase, water; 20-µl sample, 0.1 M in each ion. (From Ref. 542.)

To be sure, the design of the stationary phase is important. For stationary phases in which the crown ether is part of the polymer matrix (as opposed to being bonded to the support), slow diffusive transport of the solute in and out of the stationary phase may well be at least partly responsible for the poor overall efficiency. The most efficient separations reported on a crown ether stationary phase thus far are attributed to Nakajima et al. [543]; furthermore, these authors conducted their studies at a more conventional flow rate (1.0 ml/min). Poly(benzo-15-C-5) or bis(benzo-15-C-5) was bonded to 10-µm irregular silica and for the chromatogram shown in Figure 37, 2600 plates/m may be calculated for the KCl peak. It is also, however, clear that specific chemistry plays a role. In the same study, a highly asymmetric peak and poor column efficiency (800 plates/m) were observed for the last eluting peak (Ba), in the separation of alkaline earth chlorides with water:methanol (50:50) elution.

Several independent investigators, using different techniques, have reported nonchromatographic studies on the rates of complexation/decomplexation of metal ions by various crown ethers in a number of different solvents (including methanol). The conclusion of all these studies [548-553] is essentially the same; complexation occurs at virtually diffusion controlled rates. It is the unqualified

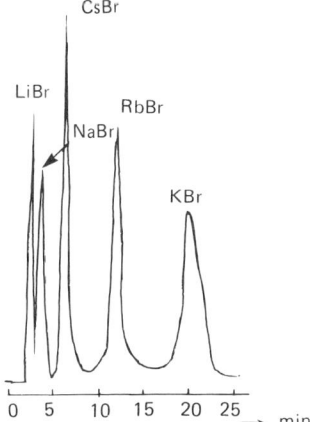

Figure 37. Separation of alkali metal bromides on poly(benzo)crown-coated silica. Column: 4 × 300 mm; eluant, water. Sample: 5-30 μg of each salt. (From Ref. 543.)

opinion of this author that the problem in chromatography with crown ether-based supports partly originates from the close proximity of the crown ether moiety to the nonpolar support matrix, in the bulk of the stationary phases made and studied thus far. This configuration tends to force a hard cation into the neighborhood of a lipophilic surface, a situation not particularly attractive for the cation. The actual problem may also be conformational; it may be difficult for the crown ether to properly conform so as to enclose the metal ion when it is in close proximity to the support. Therefore, the crown moiety should be bonded to the stationary phase by a hydrolytically stable 6- to 10-atom spacer. Further, the crown itself should *not* contain hydrophobic substituents (e.g., benzo-, dibenzo-, etc.), which are likely to cause the hydrophobic end to reattach itself adsorptively to the nonpolar support. If possible, the crown moiety should have one or more polar groups (e.g., $-CH_2OH$) attached, such that there is greater tendency for the polar end group and thence the crown cavity to jut out into the polar solvent environment. It is interesting to note that in the stationary phases introduced by Nakajima et al. [543,546], the atom spaces between the crown and the support are greater than those in any other crown-containing stationary phase thus far.

Separations on crown ether-based stationary phases using methanol or methanol:water eluants are more varied than those using pure water. One interesting application involves the determination of the water of hydration of a salt, e.d., $LiCl \cdot xH_2O$, $CrCl_3 \cdot 6H_2O$, etc., by elution with pure methanol. Because the metal ion is desolvated as it is complexed by the crown ether, the salt itself and the associated water elute separately [545]. Other examples include the separation of alkali halides; alkaline earth halides [540,543]; Zn, Hg, and Cd chlorides [541]; alkylammonium chlorides [554]; LiOH/NaOH; and $HCO_3^-/OAc^-/Cl^-/NO_3^-/SCN^-$, $HCO_3^-/CN^-/Cl^-/Br^-$, NO_2^-/NO_3^- (all as Na salts [555]). King and Heckley [556] achieved a partial separation of Pr and Er as nitrates with an acetone:hexane eluant on a pure crystalline DB-18-C-6 bed as stationary phase. Heumann and Schiefer [557] have reported 3.3% enrichment of Ca-48 over its natural abundance, using a commercial poly(benzocryptand) as stationary phase with a methanol:chloroform eluant containing a small amount of water. Blasius et al. have also described the separation of various nonionic organic compounds on crown ether stationary phases in a number of their publications.

The types of crown ether-containing stationary phases are varied. The pure crystalline monomeric ether [556] is of little more than academic interest. It is relatively easy to fabricate columns coated with hydrophobic crown ethers either by in situ [539] or prior coating [355-357,542,547,558]. Blasius et al. [540,541,544,545,554,555] have described at length their manifold approaches to making crown ether-containing stationary phases, which include polymeric matrices incorporating the crown ether obtained by condensation polymerization or vinyl copolymerization, PSDVB-based bonded phases incorporating amino or ether bridges, and silica-based bonded phases incorporating methoxy or propylamino bridges. The silica-based stationary phase described by Nakajima et al. [543,546] also contains propylamino bridging groups. An unusual stationary phase is formed by adding a dichloromethane solution of a crown ether to a solution of phosphomolybdic acid in 2 M HNO_3 [559]. The heteropolyanion and the crown form an insoluble adduct, which exhibits extraordinary affinities for alkali cations. The precipitate with DB-18-C-6 is so strongly retentive that Na^+ or K^+ cannot be removed from this material even with 5 M HNO_3. Na^+ can be quantitatively separated from K^+ by eluting the former with 0.05 M HNO_3 and the latter with 0.5 M HNO_3 on a column consisting of the DB-24-C-8 precipitate. Strangely, however, the phosphomolybdate adduct of several other crowns and a cryptand did not exhibit any ion exchange capacity.

Acylic polyethers have also been bonded to glass, silica, or PSDVB supports and used for chromatographic separations [560].

In part because of their cost, crown ether solutions as mobile phases have received limited attention. Delphin and Horwitz [561] studied both the cis-*syn*-cis and cis-*anti*-cis isomers of dicyclohexano-18-C-6 in 1.0 M HCl, 80% methanol solution for the separation of alkali cations on a sulfonated PSDVB column. The separation obtained with the second isomer was far superior to that obtained with the first. Slight separation of Na-22 and Na-24 was also noted in the same study.

Crown ethers have been used in the mobile phase in the author's laboratory in yet another fashion. A water/acetonitrile eluant containing dicyclohexano-18-C-6 and an indicator dye, bromocresol green, in the acid (HA) form was pumped through a C-18 silica column until equilibration was attained. Both the indicator and the crown ether adsorb on the stationary phase. As an alkali metal salt is injected, it binds to the stationary phase as follows:

$$MX_{(m)} + Crown + HA_{(s)} \rightarrow M(Crown)^{+}(A^{-})_{(s)} + HX_{(m)}$$

where the subscripts m and s denote the mobile and the stationary phase, respectively. When the metal-crown indicator aggregate eventually elutes from the column, the base form of the indicator (A^-) is obligatorily present with the metal ion to maintain electroneutrality. If an optical detector is set to monitor the base form of the indicator (which should be chosen such that there is very little overlap between the absorption characteristics of the acid form and the base form), a low-background, high-sensitivity method of detection is formulated. Presence of the lipophilic indicator anion greatly aids the retention of the metal ion and improves the separation factors. All but the most lipophilic anions originally associated with the sample are eluted in the solvent front, because the dye anion is typically much more lipophilic. Consequently, prior anion exchange is unnecessary; there are no compounding effects from the variation of the sample anion. Although the alkali metal ions could be separated and detected with good sensitivity, the observed column efficiency was poor [562].

The role of crown ethers in the mobile phase has attracted greater attention in the chromatography of organic solutes [563,564].

Because cyclodextrins form inclusion complexes (Section V.G, Separation on Reverse Phase and Other Bonded Phases) that are somewhat analogous to crown-metal complexes, Armstrong et al. [565] studied the utility of cyclodextrin-bonded phases for ionic separations. Although differences in retention times were noted for various halide ions with methanol:water eluant systems, further studies [566] have shown that the retention is dependent on both the injected solute concentration and the associated cation. It is not clear at the present time whether the retention is dominantly controlled by the cation or the anion. Observed column efficiencies were poor.

Ion Exchanger Stationary Phases

A brief account of the important types of stationary phases is in order as the concluding portion of this subsection. Bonded silica-based ion exchangers have been commercially available for some time but are of limited utility in many applications because of the inherent pH limitations of a silica support (pH 2-8). Excellent efficiencies, may, however, be obtainable on such columns [567], and for this reason silica-based ion exchange columns have enjoyed wide use in SCIC despite their limitations [568]. Preparation and properties of silica-based, weak [570] and strong [569-571] cation exchangers and strong anion exchangers have been described. Sulfonic acid-type cation exchangers may also be conveniently prepared by peroxide oxidation of thiolated substrates [572]. It is interesting to note that a perfluorosulfonate polymer (Nafion)-coated C-18 silica stationary phase has been described [573]. Conventional ion exchange resins, even in microparticulate form, are also not especially desirable because slow mass transfer rates in the stationary phase matrix limit attainable column efficiency. For cation exchangers, a surface-sulfonated PSDVB resin yields good separation efficiencies and was the type of stationary phase used in the celebrated work of Small et al. [1]. This type of resin was introduced by Small in 1961 [574]. Some interesting x-ray electron micrographs showing the surface sulfur distribution have been published [575]. Fritz et al. have studied in great detail the utility of macroporous supports that are partially sulfonated; the paper by Fritz and Story [576] describes the selectivity behavior of such material. A paper by Gibson and Bailey gives the interested reader valuable information on the sulfonation of polystyrene surfaces [577]. Sulfoalkylated PSDVB resins have been described by Doerscher et al. [578]. Caude and Rosset have compared the chromatographic performance of silica-based and PSDVB-based cation exchangers [579]. It is also pertinent to draw attention at this point to a paper by Mackey, which describes the cation exchange behavior of supposedly nonionic stationary phases [580]. Polyaromatic resins (XAD-2, Chromosorb 101 and 102, Porapak P, PS, Q, and QS, Benson BN-X4, and BN-X10) and well end-capped octadecyl silica display cation exchange capacities in the 0.1- to 0.5-μeq/g range. Somewhat higher cation-exchange capacity has been reported more recently for octadecyl silica [581]. Exchange capacities of activated carbon and polyacrylic ester resins XAD-7 and XAD-8 were reported to be larger by one to two orders of magnitude [580]. Underivatized silica columns have been shown to contain sufficient cation-exchange sites such that many cationic solutes can be well resolved on them with suitable eluants [582]. Kel-F, a rigid chlorofluorocarbon, has been modified by reacting first with phenyllithium and then sulfonating [583].

"Spheron" is a glycolmethacrylate macroreticular gel [584] that has been rendered into diethylaminoethyl anion exchangers [585], and carboxylate- [586], phosphate- [587] and sulfate- and sulfonate- [588] based cation exchangers.

Anion chromatography has thus far been of greater interest, and, because many metals are separated as their anionic complexes, these types of packings are described here as well. Commercial anion exchanger columns based on matrices other than PSDVB are available, and include polymethacrylate (Toyo Soda) and "polyvinyl aromatic" (Interaction Chemicals). Such new polymeric IC columns are the subject of an article by Lee [589], which also describes the macroreticular microparticulate spherical PSDVB-based anion exchange columns introduced by Hamilton Company. Gjerde and Fritz [590] have discussed the effect of capacity on the behavior of macroporous anion-exchanger PSDVB resins. For the ion chromatographer interested in synthesizing his or her own stationary phase, the more recent papers by Barron and Fritz, which describe the effects of the structure of the quaternary ammonium functional group and exchange capacity on the selectivity of the resin toward monovalent [591] and divalent [592] ions, are compulsory reading. A particularly interesting aspect is that as the exchanger group becomes more hydrophobic (e.g., in going from triethylammonium to trihexylammonium), the relative importance of the electrostatic binding of the anion decreases. The preparation of these supports has been described in detail [593]. Modifying a purely aqueous eluant with an organic solvent, e.g., methanol, can also markedly affect selectivity [594]. It has been shown by computer simulation that low-capacity columns are essential in IC whether the system is suppressed or nonsuppressed [595]. Use of pellicular anion exchangers has also been shown to be of utility in IC applications [596].

In actual chromatographic work, the surface agglomerated anion-exchange packing has proven to be quite efficient. The basic type of packing involves agglomerating quaternary ammonium-functionalized microparticles on larger, surface-sulfonated cation exchanger resin. Electrostatic forces hold the agglomerate together; essentially, most of the cation exchange sites are covered by the anionic surface agglomerate (but not all, as shown by the successful cation exchange chromatography on this type of anion exchange columns by Wimberley [597]), and this type of column was also used first in the original work of Small et al. Important advances were made later by Stevens and Langhorst [33] within the domain of the same basic approach, but improving the column efficiency by a large factor. Dionex scientists have continued to expand on this theme; although much of this important information is not in the public domain, some glimpses appear in the recent review of Johnson and Haak [84] and the presentation of Riviello and Pohl [233]. Aside from the size of the surface-sulfonated substrate resin and that of the agglomerated microparticle (as small as 200 Å), column efficiency and especially selectivity are greatly influenced by the degree of crosslinking of the substrate and the exact nature of the quaternary ammonium functionality of the microparticle.

Carboxycellulose cation exchangers have been reported in the literature [598,599]; no details are available on chromatographic efficiency. Cellulose-based ion exchanger phases are commercially available. Exchangers consisting of activated carbon filled with hydrous Zr oxide or mixed oxides of Zr and P have been proposed as high-temperature ion exchangers [600].

VI. IMPROVING DETECTABILITY

A near obsessive determination to surpass the records of previously attained limits of detection is a passion unique to that special breed of people known as analytical chemists.

First of all, it is important to recognize that chromatographic factors, especially width of late-eluting peaks, may be of great importance in determining detectability, all other conditions being the same. Gradient elution techniques can not only improve separation efficiency and reduce analysis time, but also significantly improve detectabilities of late-eluting species. For this reason, gradient elution IC is considered in this section.

Aside from chromatographic factors, detection limits can be reduced by one of two basic approaches: innovating detection techniques to improve sensitivity, and use of a large sample volume, typically accompanied by preconcentration.

Plasma spectrometric detection of metals and some nonmetallic elements, electrochemical detection of many electroactive species, and wavelength-discriminated (or fluorescence lifetime-discriminated, as yet unexplored in chromatographic systems) fluorescence detection of natively fluorescent species are examples of detection methods that are both highly sensitive and specific. (This is not to imply, however, that any relationship exists between specificity and sensitivity of a detection method.) In favorable cases, such detection methods allow routine analysis at parts-per-billion levels with conventional sample volumes. Because Chapter 4 of this volume is dedicated exclusively to the application of various detectors, further discussion of this aspect is omitted here. However, conventional optical detectors (variable-wavelength UV/VIS absorption or fluorescence detectors of modest capabilities) may often be used with very high sensitivity after the modest (or even nonexistent) chromophore or fluorophore in the eluting sample is converted, by addition of a suitable reagent, to an intensely absorbing or fluorescing product. This technique of postcolumn reactions and the design of various postcolumn reactors are discussed in this section.

Using a large sample volume is the other alternative for detecting low concentrations of sample ions. Direct injection of sample volumes as large as 10 ml with a sample loop of appropriate size is occasionally feasible if the desired species of interest is strongly retained by the stationary phase, particularly if gradient elution is practiced. In most other cases, the band broadening from direct injection of large sample

Approaches to Ionic Chromatography

volumes is too large to be acceptable. Prior to chromatography, enrichment of the desired constituent(s) directly on the column or preferably on a small precolumn, often referred to as the *concentrator column*, is the generally practiced route. Very large sample volumes, exceeding tens of liters, can be preconcentrated this way, permitting detection of species present in the original sample at low parts-per-trillion levels. This and some other preconcentration techniques are discussed below.

A. Trace Enrichment

Evaporation, precipitation, and extraction are the usual procedures of preconcentration employed in classical chemistry. Evaporative concentration is extremely tedious and error prone. Recoveries from precipitation and subsequent dissolution processes are often unknown. For these reasons, these two techniques have little use as a preconcentration technique for chromatographic analysis.

Extractive Preconcentration

Extraction methods however, are often of utility. A recent study of Reuter and Schwedt [601] focused on the relative merits of nine separate preconcentration methods (including extraction as PAR-chelate ion pair, coprecipitation with $Mg(OH)_2$, extraction with benzoyl-N-phenylhydroxylamine, and uptake by Chelex-100) for the preconcentration of vanadium from environmental waters, with chromatographic analysis being the final step. Based on the high preconcentration factor, low susceptibility to interferences relative to the matrix, and selectivity, extractive preconcentration with benzoyl-N-phenylhydroxylamine was judged to be the best approach. A number of authoritative volumes, including the monograph by De [392], Kolthoff and Elving's classic series of volumes on the analytical chemistry of the elements [602], and the recently published handbook on organic analytical reagents [603] cover the topic of extractive preconcentration, especially for metal ions, in detail; further discussion is omitted here except for two important recent developments.

The first of these concerns the extractive preconcentration of alkali and alkaline earth metals by extraction with suitable crown ether-based chelants. Although the general literature on crown ether-based extraction systems is quite extensive, of special interest to the analytical chemist in the present context are the reports of Bartsch et al. [604-606] which describe the use of crown ether carboxylic acids for extractive preconcentration. In these studies, the crown ether itself contained an ionizable acid group, and thus complications due to variations in the anionic composition of the sample being subjected to extraction were avoided. With simple nonfunctionalized crown ethers, extractability of alkali and alkaline earth metals into organic solvents is enhanced in the presence of soft anions such as perchlorate, thiocyanate, or picrate. Furthermore, excellent preconcentration factors and often selective extraction of a specific metal, depending on the choice of the particular crown ether carboxylic acid used, were possible with crown ether carboxylic acid extraction systems. The rates of transport across liquid-liquid membranes for such systems have been described by these workers [607,608]. The use of phosphonic acid crown derivatives [609] and acyclic polyether carboxylic and phosphonic acids [610] has also been reported to be of particular utility in selective extractive preconcentration.

The second item of interest involves the use of surfactant phases above their cloud point for extractive preconcentration. Although surfactants (ionic or nonionic) dissolve in water to form micellar solutions, above a temperature called the cloud point (CP, characteristic of the surfactant) the surfactant phase separates from the bulk aqueous phase and behaves like conventional liquid-liquid extraction systems. This type of extractive preconcentration system may be of special utility in reverse phase chromatography. The important applications in this area have been carried out by Watanabe et al.; their work prior to 1980 is well summarized in Ref. 611. The thiazolylazonaphthol (TAN) chelate of Ni was extracted into Triton X-100 at elevated temperature (CP 70°C) [612], and the procedure could be used, in presence of suitable masking agents, for the determination of Ni in soil samples [613]. A similar extractive preconcentration of the xylidyl Blue I - zephiramine complex of Mg into Triton X-100 was sufficiently successful to determine the concentration of the metal in Antarctic snow [614]. The pyridylazonaphthol (PAN) chelate of Zn was preconcentrated from environmental waters by polyoxyethylene nonyl phenyl ether with 7.5 average ethylene oxide units (PONPE-7.5), at room temperature (cloud point 0°C) [615]. PONPE-7.5 was also used to extract the ion-pair of the cationic copper complex of $\alpha, \beta, \gamma, \delta$-tetrakis(1-methylpyridinium-4-yl)porphin and dodecylbenzenesulfonate [616] and several nonionic complexes of zinc [617]. In all cases, the surfactant phase is heavier and easily separated by centrifugation. The theory of such extraction processes has been presented [618].

Enrichment on Stationary Phases

Several unusual chelating supports have been described previously (Sec. V. H, Chelating Supports: Complexing Supports for Chromatography and Preconcentration) for the preconcentration of metal ions. The common practice, however, is to use a short column filled with the same packing material (typically an ion exchanger) as the analytical column for preconcentration. There are no special barriers against using a different packing material for the trace enrichment column, however, as long as it is compatible with the eventual indended elution (if one bears in mind the species the eluant will leach off from this column onto the analytical column). The capacity of the enrichment column and the affinity of the desired species to be preconcentrated are important factors that must be taken into account when one is practicing enrichment of trace species from large volumes of high ionic strength samples, e.g., seawater. The loss of the enriched species by elution with the sample itself is a potential problem. Scott [619] has succintly stated the requirement: The eluant strength of the mobile phase must be substantially higher than that of the solvent in which the sample is dissolved. On the basis of this principle, he has illustrated how a large sample volume can be injected directly on column without attendant band dispersion. The system involved a reverse phase packing with hexanesulfonate and citrate in the eluant, with the first, functioning as an anionic IIR and serving to retain the cations; and the second, serving as the chelating eluting species. If the sample was dissolved in the eluant, large injection volumes produced proportionately broad bands in the eluate. However, when the sample (Cu, Zn, Co, etc.) was dissolved in water only, it eluted without undue broadening, up to an injected volume of 2 ml. In the absence of the eluting species (citrate), the cations in the sample were strongly retained by the adsorbed hexanesulfonate at the head of the column. Similarly, Heckenberg and Haddad [621] have achieved the determination of a number of anions at the parts-per-billion level by using directly a 2-ml sample volume and indirect optical detection. Jupille et al. [622] and Buchholz et al. [97] have earlier shown the applicability of large-volume sample injections in SCIC.

More than a decade ago, Kirkland [620] elaborated on the use of precolumn and on-column enrichment techniques with the aid of small-dead-volume, high-pressure switching valves. Two of the principal modes of operation suggested by Kirkland are as follows: In one mode, a single column is used with two interconnected switching valves. One valve is placed between the pump and the injector, and the second is placed between the column and the detector. A large sample volume is injected and the flow proceeds in normal fashion, injector to column to detector. When the species of interest have eluted, strongly retained undesired species are removed by backflushing the column. This is accomplished by switching the valves so that the pumped eluant is directed by the interconnecting tube to the second valve, which directs it into the back end of the column, reversing the flow direction. The effluent from the column returns to the first valve and is directed to waste. The merit of this procedure is doubtful, especially because many manufacturers recommend only a single flow-direction through the column to preserve packing integrity and thus column efficiency. Second, it may be a strongly retained species that is of interest. Third, some undesired species may well be irreversibly retained.

The second mode of operation suggested by Kirkland involves a separate precolumn. A switching valve is placed between the precolumn and the analytical column. The valve is also connected to the outlet of the analytical column. A three-port union connects the outlet of the analytical column and the tube from the valve to the detector. A large sample volume is injected with an injection valve. Initially, the flow at the outlet of the precolumn is switched by the valve to bypass the analytical column and directly enters the detector. After early-eluting species have eluted from the precolumn, the precolumn effluent is switched to the analytical column for better resolution of species of interest.

Plumbing Configurations for On-Column and Precolumn Enrichment

Available hardware, especially different types of multiport high-pressure valves, are now much more extensive, permitting a variety of approaches. An excellent recent review by Harvey and Stearns [623] specifically addresses various plumbing configurations of utility in trace enrichment and is especially recommended. Although this review primarily focuses on the use of 10-port switching valves (Valco), a combination of two conventional 6-port injection valves can always be substituted for a 10-port valve.

Sample volumes of 10 ml or less are preferably injected directly with a loop injector because exact reproducibility of injected volumes are easily attained. However, sample loops constructed entirely of 1.5-mm (1/16 in.) o.d. tubing are likely to be inconveniently long. For example, the largest practical internal diameter for 1.5-mm o.d. tubing is 1.25 mm (0.05 in.), and close to 8 m will be required to make a 10-ml loop. Even if one considers the potential utility of multiple injections, reducing unions to connect larger-bore tubing should be considered in constructing large-volume loops. It is clearly impractical to use loop injection with sample volumes much larger than 10 ml. A metering pump can be used to pump the desired sample volume through the column or precolumn. Because of the high back pressures associated with typical analytical columns, the metering pump must be capable of delivery at high pressures if enrichment is carried out directly on the analytical column. The use of the eluant pump to load the sample is not recommended. A microprocessor-controlled valving arrangement to utilize a single pump for both sample preconcentration and eluant delivery has been described by Haddad and Heckenberg [624], however. The premise that an inexpensive pump leads to considerable volumetric inaccuracies in loading the sample is questionable for the amounts of sample typically used for preconcentration. Once the cost of valves and necessary control equipment to practice this single-pump technique is considered, it

Approaches to Ionic Chromatography

is not clear to this author why this approach should be considered especially meritorious.

It must be noted that direct enrichment on the analytical column is risky, unless the sample is known to be free of particles and irreversibly retained species. Irreversible retention of unwanted material may permanently damage expensive analytical columns, and unless sample solutions are carefully membrane-filtered, particulate matter in the sample may result in column blockage with attendant high back pressures. Analysis of residual ions in high-purity water [625] is a relatively rare example for which direct on-column enrichment is permissible.

The most common mode of using a precolumn for preconcentration purposes is to substitute it for the loop in the sample injection valve. For most 6-port injection valves (e.g., Rheodyne series 7000), the pump/column and syringe/waste ports are interchangeable and matter little for conventional loop injection. When one uses an enrichment column, after the desired volume of the sample has been put through from the syringe port with the valve in the load position, it is preferable that the enrichment column be backflushed onto the analytical column when the valve is switched to the inject position. This means that the flow direction through the enrichment column is reversed between the load and the inject modes. The pump/column connections should be made accordingly, and they will no longer be interchangeable. Because precolumns are generally short and often are packed with larger pellicular packing, as compared to smaller-diameter packings employed in the analytical column, the back pressure may be small enough to enrich sample volumes as much as 100 ml manually with a syringe. However, sample loading with a noncontaminating metering pump is preferable and is a necessity with higher back pressures or very large sample volumes. Additionally, chromatographic reproducibility is often improved if the sample passes through the enrichment column at a constant rate, which cannot be assured with manual syringe loading. Certain enrichment procedures (e.g., reactive adsorption described in Sec. V.H., Chelating Supports: Complexing Supports; Ref. 531) may also be sufficiently slow such that too fast a sample loading rate through the enrichment column may result in subquantitative capture of the desired species.

For sample sizes of 10 ml and below, the reproducibility of loop injection can be combined with precolumn enrichment. Two 6-port injection valves and two pumps are needed. Pump 1 is connected to valve 1, which contains the large sample loop. The column port of valve 1 is connected to the syringe port of valve 2, which contains the enrichment column in place of the loop. Pump 2 and the analytical column are, respectively, connected to the pump/column ports of valve 2. Both valves 1 and 2 are initially kept in the load mode. After the loop of valve 1 is loaded manually, pump 1 (which need not

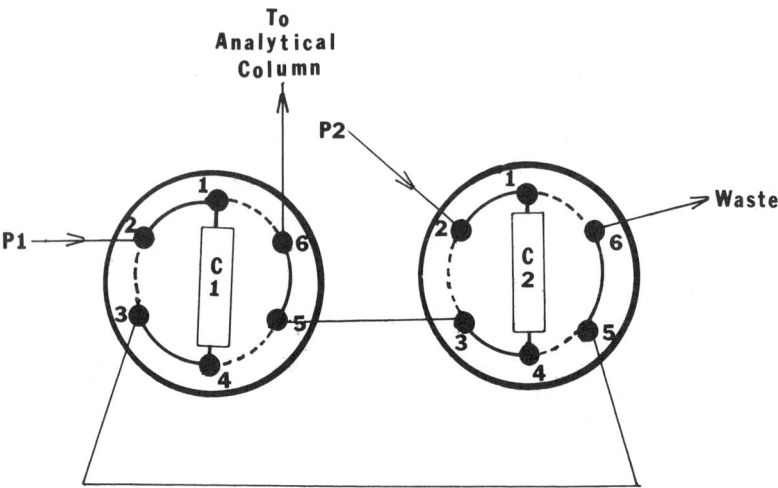

Figure 38. Plumbing configuration for trace enrichment with two six-port valves, showing one enrichment column being backflushed to the analytical column while the other is being loaded with sample.

necessarily be a high-pressure pump) is started to pump deionized water. Valve 1 is now switched to the inject mode. After a sufficient amount of water has been pumped to transfer completely the contents of the loop to the enrichment column (and thence to waste), pump 1 is stopped. Pump 2 is then started, to pump the eluant. Bringing valve 2 into the inject mode now backflushes the enrichment column onto the analytical column. A single 10-port valve can replace the two 6-port valve combination, as described by Harvey and Stearns [623,626].

Occasionally, it may be useful to use two different types of enrichment columns for selective enrichment of two different types of species. Two 6-port valves and two pumps may be used in a somewhat unorthodox fashion to perform enrichment on one column while the second column is being backflushed onto the analytical column. Of course, two identical enrichment columns may also be used with this arrangement. This reduces sample loading times which, for large sample volumes, could well be significantly greater than the analysis time. The configuration is shown in Figure 38. These valves are conventional 6-port injection valves, where ports 1-6 are typically designated loop, pump, column, loop, syringe, and waste. According to the plumbing configuration shown in Figure 38, with both valves in position A (solid connecting lines operative), metering pump P1 performs trace enrichment on column C1, while eluant pump P2 backflushes column C2 on to the analytical column. When both valves are simultaneously switched to position B (dotted connecting lines operative), the roles of the two columns are reversed. Simultaneous switching of both valves can be accomplished by a common mechanical link or by electropneumatic switching. Alternatively, a single 10-port valve can be used, as detailed by Harvey and Stearns [623]. These authors have also described a plumbing configuration wherein two separate enrichment columns are connected serially and can be separately backflushed to the analytical column. The plumbing is complex and requires one 10-port and one 6-port (or presumably three 6-port) valve(s). The particular merit of this configuration in any actual application is not clear to the present author. A related, but somewhat different, application of a two-column method of preconcentration of trace metal ions in natural waters has been described by Wan et al. [628].

Occasionally, the desired analysis is difficult to carry out in a single step; chromatography, collection of the desired fraction, followed by rechromatography under a different condition are carried out [627]. Such situations can often benefit from judicious plumbing and switching arrangements so that the entire sequence is automated. An interesting sample pretreatment procedure has been described by Siemer [629] for determination of halides in a complex matrix. The halides in the sample are taken up on a Ag^+-form cation exchanger and later eluted by aqueous NH_3. Prior to chromatography, the silver in the eluate is removed by a Jones reductor. Roberts et al. [630] have described the utility of sample preconcentration in SCIC for the analysis of anions in boiler feedwater and steam condensate.

Applications

Any discussion of trace enrichment in chromatographic analysis, especially for metals, is incomplete without homage to the works of Takata et al. and Cassidy and Elchuk. The work of Takata et al. [260], which includes the use of sample volumes as large as 140 L for the simultaneous chromatographic determination of Ca, Mg, Mn, Co, Ni, Zn, and Cu, present at levels as small as 10 parts per trillion (picogram per milliliter), is summarized in Cassidy's review [241]. Further, Cassidy and Elchuk have authored a two-part exposé of their experience with trace enrichment methods for the chromatographic determination of metal ions [388,631]. Some highlights of these investigations are noted below. In the bulk of their studies, a citrate-based eluant was used with cation exchange columns. The same packing was used in the enrichment column, and detection was accomplished through postcolumn reaction with PAR, described in greater detail in Section VI.B. Samples and low-level standards were preserved in 1 mM citrate, which helped stabilize these very dilute solutions of metal ions. Loss of the metal ion was not detectable within experimental error for a citrate-stabilized sample containing 10 pg/ml Co^{2+} stored in a polyethylene bottle. However for large sample volumes, larger concentrations of citrate caused incomplete capture of the metal ions on the enrichment column. With 1 mM citrate as preservative, recovery was found to be quantitative. For natural water samples, significant differences were observed for acidified and unacidified samples. Acidified samples yielded higher concentrations, which suggested the presence of acid-labile metal complexes. Although not especially important for analysis at the nanogram-per-milliliter level, electrolytic purification of all solutions at a mercury pool cathode was found necessary for analysis at picogram-per-milliliter levels. With reagents thus purified, low picogram-per-milliliter levels of several metals (Mn, Co, Ni, Cu, Zn, Pb) were detectable, the LOD for Co^{2+} being 0.5 pg/ml for a 2-L sample volume. In terms of trace metal content, deionized water was found to be significantly superior to distilled water. The extent of trace metal contamination due to leaching of metal ions from metallic surfaces of the chromatographic system exposed to the flow stream was studied by repeatedly recycling 10 ml of deionized water through the enrichment system for a sample volume equivalent to 2 L. Figure 39 is shown here as a representative example of these studies.

Zolotov et al. [708] have shown that Mo, Cr, and W can be determined by conductimetric IC as the oxoanions, after alkaline oxidation with alkaline peroxide and preconcentration.

Dialytic Preconcentration

Finally in this section, dialytic preconcentration methods needs to be mentioned. An automatic dialyzer injection system has been developed by Nordmeyer and Hansen [632] for the preconcentration/purification of ions and molecules and applied to the analysis of Ca in biological samples. DiNuzio and Jubara [633] employed a Donnan dialysis preconcentration system specifically for preconcentration of ions prior to chromatographic analysis. Cox and Litwinski [35,634] reported a similar Donnan preconcentration system in which preconcentration factors of more than 50 were attainable in 30 min. Cox and DiNuzio [635] have earlier reported cation concentration by Donnan dialysis.

Postcolumn Reactors

Postcolumn detection schemes are designed to improve detectability of eluting species that otherwise elude sensitive detection. Probably the oldest as well as most widely applied exploitation of such a scheme involves the detection of amino acids through postcolumn introduction of ninhydrin. Although amino acids can be regarded as ionic analytes and indeed are most commonly separated by ion exchange chromatography, they are hardly typical ionic analytes, the principal focus of

Approaches to Ionic Chromatography

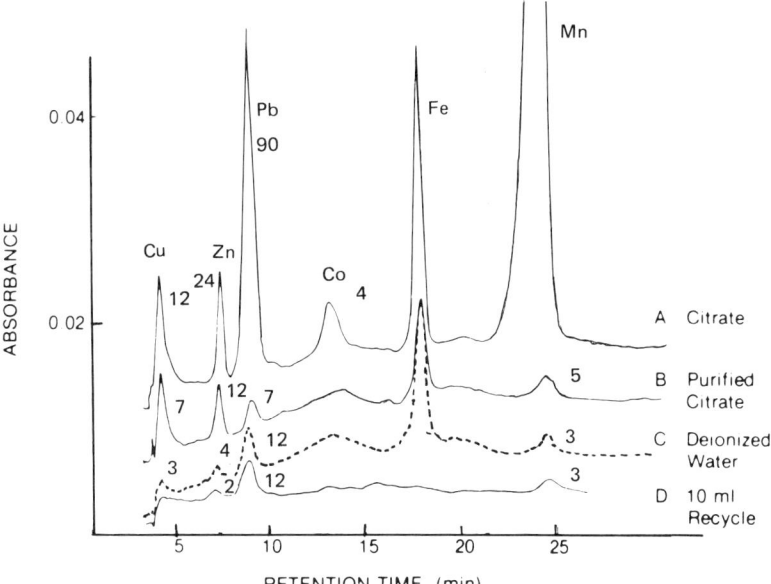

Figure 39. Contamination problems at the picogram-per-milliliter level. Samples A to C were 2 liters each, and D was 10 ml of deionized water recirculated through the system to give an equivalent volume of 2 liters. The concentrations shown are in picograms per milliliter; the citrate solution was 0.5 mM. (From Ref. 388.)

this volume. However, in a sense it is fortunate that this initial use of postcolumn detection involved the use of the ninhydrin-amino acid reaction because this reaction is particularly difficult to deal with in a postcolumn reaction (PCR) scheme. The impetus to develop better postcolumn chemistry was great because of the importance of amino acid analysis in a variety of biological disciplines. Much faster reactions with o-phthalaldehyde and fluorescamine, which form fluorescent reaction products with amino acids, were eventually devised. Various chromogenic and fluorogenic reagents, for detecting other classes of compounds, were newly synthesized or rediscovered along the way. As columns became more efficient, designing PCR systems that introduce minimum extraneous dispersion (so as to preserve the efficiency of the separation attained by the column) and minimum mixing noise (so as to attain maximum possible sensitivity) became necessary.

Unusual PCR Systems

Not all PCR systems involve the introduction of a reagent per se. Exposing the column effluent to intense radiant energy often converts the analyte to a fluorescent or electroactive product [636,637]. However, thus far the applications have been limited to organic solutes. A photoconductivity detection system which photodecomposes organic solutes to produce ionic decomposition products that are detected by their electrical conductivity is commercially available [638]. In these systems, the reaction coil is typically wrapped around an intense radiant source. An analogous concept is the use of thermal energy to accomplish a desired reaction. Thermal conversion of catecholamines into fluorescent products has been reported [639]. Ivey [640] has thermally decarboxylated acetoacetic acid, used as an eluant for anion chromatography, by heating the column effluent to convert the eluant into acetone and CO_2. It is important to note that the purpose of the PCR system is to enhance analyte detectability; whether this is accomplished by converting the analyte into a more detectable product, or by converting the eluant to some product that exhibits a substantially lower detector response, is immaterial. In this sense, the ion chromatography system introduced by Small et al. [1] containing a suppressor column can be regarded as a solid-phase PCR which does both of the above. Solid-phase PCRs have been used otherwise as well; again most applications thus far have involved organic analytes. Some examples include the reduction of carbonyl compounds by solid-phase polymeric borohydride resin [641] and the release of easily detectable moieties, bound to a polymeric support, by the eluting analyte [642,643].

PCRs with Reagent Introduction

With the introduction of a liquid reagent, there are several possible scenarios. The most common of these is one in which the reagent and the analyte react together to form essentially one product, which is easily detectable. This is the case for the ninhydrin reaction with amino acids, or for the detection of metal ions via formation of a suitable complex (e.g., with PAR). A second possible case is that the analyte converts the reagent to or liberates from the reagent, some easily detectable product. Examples of this type include (a) the reaction of Ce^{4+} as postcolumn reagent with reducing ions (e.g., nitrite) to produce fluorescent Ce^{3+} [644]; (b) the reaction of nonfluorescent Pd-calcein complex with sulfur-containing compounds to liberate free calcein, which fluoresces strongly [645]; and (c) the displacement of Cu or Hg from a labile complex of the metal, being used as the postcolumn reagent, by other metal ions eluting from the column followed by electrochemical detection of the liberated Cu or or Hg [173-178] (discussed in detail in Section V.B). Another possible type of PCR employs catalytic reactions [646]. Postcolumn

introduction of a reagent to reduce the background conductivity of
the eluant has been explored by Cassidy and Elchuk [155] for conductimetric IIR-based chromatography. In yet another type of PCR
applications, no real reactions occur; the sole purpose of a postcolumn system is to introduce sufficient electrolytes (which would cause
problems if incorporated in the eluant) to increase background conductivity as necessary for electrochemical detection [647]. In the
discussions below, attention will be focused primarily on the first
type of postcolumn reaction mentioned above.

Reaction Requirements and Reactor Designs

PCR detection schemes and reactor designs have been reviewed by a
number of authors [26,648-651]. Some desirable characteristics of
such a system are:

1. The reaction proceeds to a high degree of completion (if not
   quantitatively, then to a reproducible extent) within a short
   period of time, preferably under 1 min.
2. The reagent volumetric flow rate is small relative to eluate
   volumetric flow rate, to minimize dilution and band dispersion.
3. In the absence of the analyte, the reagent itself produces a
   very low detector background, whereas the reagent-analyte
   reaction product produces a very high detector response.
4. The PCR reagent is reasonably stable.

Not all, but quite a few PCR detection systems come close to ideally
fulfilling these requirements. In less-than-ideal cases, difficulties are
usually encountered from incompatibilities of the ideal chromatographic
eluant system and the best PCR medium, and gradient elution may
prove to be especially difficult. Also, some postcolumn reactions are
much slower than desirable, and this significantly influences the reactor design. For reaction times less than 30 sec, no additional reactor other than the mixer is necessary; a length of small-bore tubing
sufficient to produce the necessary delay is adequate. The mixer design is critical, however, and will be addressed in the next section.
For reaction times between 30 and 120 sec, a packed-bed reactor is
generally the design of choice [26,652]. This is essentially a small
column packed with small-diameter, inert beads.

Long-Residence-Time PCRs

For reaction times that require more than 120 sec, segmented reactors
are desirable. Air-segmented reactors, following the design of Technicon autoanalyzers, have been used [653]. Of greater interest is
the more recent introduction of liquid-segmented reactors [654], which
have been shown, both theoretically and experimentally, to produce

less extracolumn dispersion than air-segmented or nonsegmented reactors for large reaction times. Instead of air, an immiscible liquid is introduced for segmenting. However, desegmentation is more troublesome than air segmentation and may itself be the largest source of dispersion. Instead of physical desegmentation, electronic processing of the detector output to isolate the signal of interest should not be too difficult; flow cells that can optically discriminate (on the basis of differences in the minimum angle of incidence necessary for total reflection) different glass(quartz)/fluid interfaces have already been designed [655-657]. An intriguing prospect of liquid segmentation is to actually extract the PCR product of interest into the segmenting phase, thus concentrating it simultaneously. Hydrophobic porous membranes can also be used for desegmenting a segmented stream before it enters the detector. Membrane-based flow-through phase separators have been described [658]. Very small dead-volume versions of such separators have since been designed as well [659].

Design of PCR detection systems that require large reaction times may also benefit from the multichannel, parallel storage concept utilized in flow injection analysis (described in detail by Ruzicka and Hansen [36]).

Short-Residence-Time PCRs

For small reaction times, which are generally found in PCR systems in which ionic reactions are involved, the necessary length of a small-diameter tube (no larger than 0.5 mm i.d.) is preferentially used as the delay tube. The dispersion expected from such a nonsegmented reactor as a function of tube internal diameter has been graphically presented by Deelder et al. [26]. A single bead-string reactor [38, 39] for the delay tube is of potential utility for achieving good mixing without introducing large amounts of dispersion, for postcolumn reactions of short-to-intermediate duration. Coiling the reactor into a small-diameter helix should both reduce axial dispersion and increase radial mixing [26,32]; the effect may be significant, even for a very short-residence-time PCR system. It has recently been shown that serpentine or waved tubes actually perform significantly better than coiled tubes [659]. This has been verified for actual PCR detection systems [660]. At least one commercial detector employs serpentine connecting tubes [101].

Mixing Noise in PCR Systems: Mixer Designs

For short-residence-time PCR systems, most often the primary source of detector noise is incomplete or inhomogeneous mixing of the reagent and the column effluent. Mixing is hardly a trivial topic; volumes have been written on it [661]. Most mixers used to date are static types, modifications of a basic three-port tee. The "run-of-the-mill" tee is both too large in dead volume and not particularly good in producing efficient mixing. Since the first design of a low-dead-volume mixing tee (the term *tee* is being used here in the sense of a three-port union; it does not necessarily imply a T configuration) by Stahl et al. [662], efforts to design better and lower-dead-volume tees continue [663]. For any given flow rate, reducing the bore of the connecting passages proportionately increases the velocity of the flowing fluids, leading to greater turbulence and more efficient mixing at the confluence point. Kucera and Umagat [664] used a tee with a 30-nl dead volume, laser-drilled to form 150-μm connecting passages, for microbore applications. Mixing noise or inhomogeneity associated with a well-designed tee is usually traced back to pulsations in the fluid streams. Typically, the main chromatographic pumping system already contains a pulse dampener, and the column itself also dampens the amplitude of pump pulsations. Use of a reciprocating pump to deliver the postcolumn reagent is generally unacceptable. First, because of the smaller flow rate (compared to the eluate stream) and high reagent concentration in the reagent stream, the detector is much more susceptible to pulsations in this pumping system. Second, because back pressure downstream of the mixing point is minimal, introduction of conventional pulse dampeners is of little avail, because most of these do not work well against low back pressures. Although pulseless gear-driven syringe type pumps are ideal for the job, inexpensive gas-pressurized reagent delivery systems are often adequate for the task, because of the low back pressure. Helium is preferred over air as the pressurizing gas because of its low solubility (and consequent lack of problems due to bubble formation in the detector) and also because some PCR reagents (e.g., alkaline PAR) are quite susceptible to oxidative degradation. Kucera and Umagat's results, shown in Figure 40, illustrate the importance of eliminating flow pulsations. Methanol and 0.1% benzene in methanol were, respectively, pumped at 35 μl/min through opposing ports of the special mixing tee described above, and the UV absorbance at 254 nm was monitored following a capillary reactor. The bottom trace, D, shows high mixing noise, originating from pump pulsations. The noise is greatly reduced with the introduction of pulse dampeners (trace C) and reduced even further as 2-μm stainless steel frits are introduced at the outlet of each port (trace B). Finally, in trace A, the noise has been essentially eliminated by insertion of a piece of fused-silica capillary in the internal volume of the tee. The biggest noise reduction is due to a reduction in the pump pulsations. Further improvements come from introducing flow obstacles in the stream paths, which likely generate eddies and improve mixing efficiency. In this context, it is worthwhile to note the author's favorable experience with one type of commercially available mixing tee, ostensibly

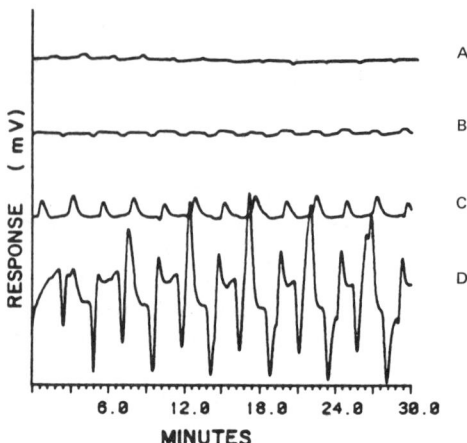

Figure 40. Mixing noise as a function of different mixing designs. (D) 30-nl dead-volume tee; each flow rate, 35 µl/min. (C) Conditions as in D, except pulseless flow. (B) Conditions as in C, except that 2-µm stainless steel frits have been introduced at each port. (A) Conditions as in C, except that a 20-µm-diameter fused-silica capillary piece was introduced in the connecting channel. (From Ref. 664.)

designed for nonchromatographic applications [665]. These tees are designed for rapid mixing of flow streams for investigations in rapid kinetics, where uniform mixing must be achieved in subsecond time scales. This low-volume tee employs a series of closely spaced graded grids (used for electron microscopy) in the confluence arm of the tee for achieving uniform mixing. With respect to noise measurements under conditions similar to those employed by Kucera and Umagat [664], except at higher flow rates, this device exhibits very substantially less mixing noise compared to a conventional "zero-dead-volume tee" when the confluent outlet ports are directly connected to the detector.

Frei et al. have made a series of investigations on the physical configuration of the tee union, i.e., non-T tee designs. Mixing inhomogenieties become particularly severe if there are significant differences in the densities and viscosities of the eluate and the reagent, due to the "layering effect." Compared to the conventional T-design, band dispersion was significantly reduced by adapting a configuration wherein eluate inlet and confluent outlet were on diametrically opposite ports (180° to each other) and the reagent entered at an angle of 150° relative to the eluate inlet (i.e., 30° relative to the confluent outlet). Further design improvements involved an arrow configuration, where the shaft of the arrow was the confluent port and exactly bisected a 60° "V" of the arrow tip, which formed the eluate and the reagent inlets [666]. In later work, a 90° "V", with or without a small packed-bed reactor following the confluent port, was studied in order to induce even better mixing [647,667]. In yet another study [668], the 60°-V arrow design was compared with the conventional 180° "T" and a 120° "Y." For both the T and the Y design, the vertical arm was used as the confluent port. Few differences in performance were found. The Y design performed slightly better in terms of induced band dispersion, but only a single piece of each type were studied, thus limiting firm conclusions. Cassidy and Elchuk [402] have reported an ingenious and simple "zero-dead-volume" tee design. The body of the mixer was bored out of a Swagelok T-union, and capillary stainless-steel tubing (1.59 mm o.d., 0.15 mm i.d.) was used to transfer solutions in and out of the mixer. The two inlet tubes were cut at a 45° angle and then butted against one another; the inlet solutions flow between and around the inlet tubes and then into the exit tubing (Figure 41). Scott and Kucera's tee design, published in the same year, is somewhat similar [669]. Inlet tube ends were cut obliquely at a 45° angle and arranged to meet at the center of the T, thus forming a V-shaped gap, as shown in Figure 42. The ends of the exit tube were shaped to a point by two oblique cuts, also at 45°, which were arranged to fit into the V-shaped gap as shown. The dead volumn of the overall assembly was less than 2 μl. The latest tee from Cassidy et al. [715] incorporates one or more stainless steel screens and reportedly produces better mixing than the previous design [402]. A tangential entry mixing chamber with inlet and outlet ports located at the bottom and top, respectively, has been patented [670].

All mixers described thus far are of the static type because of the obvious difficulty in designing very small dead-volume dynamic mixers. In a recent study [671], a number of static and dynamic mixing systems were designed and evaluated for application as a gradient mixing chambers in microbore LC. Mixing noise from dynamic mixing systems was found to be lower than that from static mixers. The dead

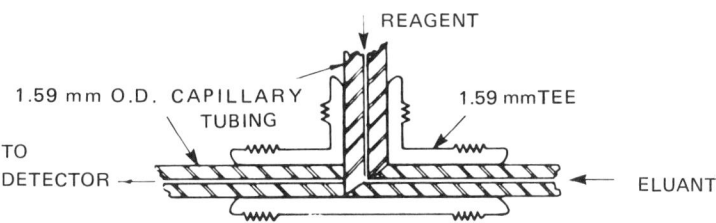

Figure 41. Cross section of tee for postcolumn application. All tubing is 150 μm i.d. (From Ref. 402.)

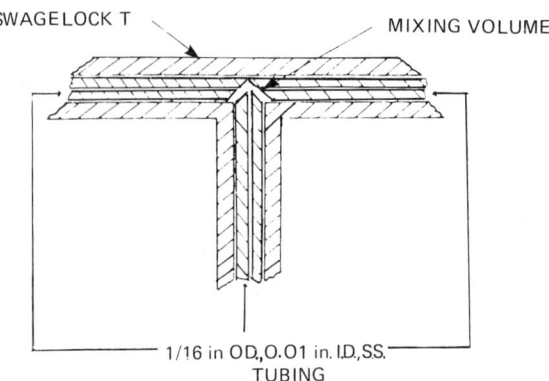

Figure 42. Cross section of tee for postcolumn application. (From Ref. 669.)

volume of such mixers is reasonably low and may be of potential utility in PCR systems. The use of an ultrasonic field to induce mixing may also be useful; such a device has been patented [627]. It should not be regarded, however, that static mixers lack sophistication. The "Visco-Jet Micro Mixer" (Lee Company) boasts of no less than 36 clockwise and counterclockwise spin chambers in a volume of 10 µl.

Membrane Reactors

The latest entry in the field of postcolumn reagent introduction systems is membrane based and simultaneously displays the virtue of excellent radial mixing. These devices can be divided into two broad types, passive and active. The first type has been patented [673, 674], and reported in the literature by Davis and Peterson [675]. The passive reactor consists of membrane tube(s) immersed in the reagent solution and allows the reagent to diffuse into the flow stream through the membrane. In an active membrane reactor, the reagent is forced through the porous membrane by application of external pressure.

Passive Reactors--The passive design employs a (semi)permeable hollow membrane tube (or a number of tubes in parallel) immersed in a relatively concentrated solution of the postcolumn reagent. Concentration differential of the reagent between the two sides of the membrane provides the motive force for the permeative transport of the reagent into the inner fluid stream. Because the tubes are of narrow bore, typically less than 300-µm i.d. and because reagent enters radially, uniformly along the length of the reactor, good radial mixing is achieved. Three basic types of membranes were studied by Davis and Peterson: sulfonated polyethylene cation exchanger, α-methylstyrene-poly(methylsiloxane) copolymer, and microporous cellulose. Several practical examples included (a) the raising of pH of the column effluent for the visible detection of nitrophenols and the fluorescence detection of phenols, (b) addition of fluorescamine or ninhydrin for the detection of amines and amino acids, and (c) addition of iodide for the detection of peroxides. The required length and number of membrane tubes depend on the particular application, specifically on factors such as requisite permeation flux (determined in turn by the chromatographic flow rate), nature of the membrane, its diameter and wall thickness, and especially the intrinsic permeability of the membrane to the reagent. For the applications stated above, typically 10- to 200-cm lengths were required. When long lengths are necessary, multiple fiber configurations are employed to avoid pressure-induced rupture of the membrane tubes.

Sample Loss in Passive Reactors--The outward permeation of some of the sample from the flowing stream into the external reagent is clearly a matter of concern. Even though Davis [673] correctly opines that 90% loss of sample would not necessarily preclude highly sensitive analysis from being performed if the sensitivity gain by the postcolumn reaction is a thousand fold, it would be even better to achieve the same sensitivity gain without losing any sample whatsoever. In a number of situations, this is indeed possible. The major class of such examples involves oppositely charged ionic analytes and reagents. For example, an anionic chromogenic reagent may be introduced through an anion exchange membrane to react with a cationic analyte of interest; the cation is effectively Donnan forbidden from permeating out of the membrane. (Note, however, that if the complex formed has a overall negative charge, there is no barrier for the product to cross the membrane.) A second class of examples involves a size-selective (dialytic) membrane which can prevent a large solute molecule from coming out while allowing small reagent molecules to permeate in. Admittedly, this is not likely to be of common occurrence in ionic chromatography. In yet another type of example, the solute of interest and the reagent are similarly charged. The Donnan barrier may be overcome for a formally Donnan-forbidden reagent species by using a large concentration of the reagent. Because the ionic concentration of the analyte species (which is also Donnan forbidden by the membrane) inside the membrane is typically quite low, Donnan forces are sufficient to prevent its outward transport. To cite an example, formaldehyde stabilizes sulfite from oxidative degradation by forming hydroxymethanesulfonate. A popular reaction to determine sulfite is the Schiff reaction involving the reaction of pararosaniline, formaldehyde, and sulfite [676]. However, hydroxymethanesulfonate must be decomposed back to formaldehyde and sulfite, which can be done by making the solution alkaline, before it can react with pararosaniline. This is accomplished by passing the hydroxymethanesulfonate sample through a 5-cm length of a Nafion 811x(s) tube (see Table 3) immersed in 50% NaOH. Sufficient NaOH penetrates to raise the sample pH to 12 (flow rate, 0.5 ml/min) while no loss of sulfite is detectable [677].

In considering loss of sample by permeation outward through the membrane, it should be noted that the sample must first be transported *to* the membrane before it can be transported *through* the membrane. If one is dealing with a passive membrane-based PCR system in which sample loss through the membrane is not preventable by any of the conditions mentioned in the paragraph above, then the system should be designed so that very little of the sample is actually transported to the wall. Of tubular geometries, the hollow linear tube has the poorest mass transport efficiency to the wall; it would not be possible to do any worse. (Certainly it would be a folly to use helical, annular helical, or packed bed designs!) So the length of the membrane tube remains the sole factor. The length should be the possible minimum, which in turn demands that maximum possible reagent concentration be used.

<u>Operating Characteristics of a Passive Reactor</u>--Mass transport of the reagent *to* the membrane is generally not of particular concern. The transport of the reagent through the membrane wall is usually sufficiently slow, relative to reagent concentration, for diffusive replenishment of the reagent to occur at the membrane surface. In a case where membrane transport is unusually fast (this should be regarded as favorable, because necessary reactor length is proportionately shortened), the simple expedient of stirring the reagent solution in which the membrane is immersed is sufficiently remedial. The more common reason for which stirring is needed is due to outward osmotic transport of the solvent. In a typical case, external ionic strength is high, internal ionic strength is low, and the membrane is highly permeable to the solvent. Consequently, significant amounts of solvent (water) are transported outward through the membrane, thus diluting the reagent in the immediate vicinity of the membrane. (Of course, if there is no sample loss, this leads to the enviable situation of the simultaneous possession and consumption of the cake, because the postcolumn reagent addition is achieved with simultaneous volume reduction!)

If a large amount of reagent is used in the reservoir, dilution of the reagent due to osmosis as well as its contamination with the sample (if finite sample loss does occur) can be tolerated for substantially long operational periods, with periodic reagent replacement being necessary. Alternatively, like membrane suppressor configurations (e.g., Figure 1), the reagent may simply be pumped or gravity fed through an external jacked housing the membrane assembly. In many cases, it may be possible to use a saturated reagent solution containing excess solid reagent.

Active Membrane Reactors--An active membrane PCR system is distinguished from a passive system in that external pressure is used to force the reagent into the fluid stream flowing through a porous membrane tube. This type of PCR system has been commercially developed for Dionex chromatographic instrumentation [233,678] and independently in the author's laboratory. Although superior noise characteristics, compared to conventional tee-based reagent addition systems, have been reported by Dionex scientists, details on the nature of the membrane, etc. are not available. The following discussion therefore is limited to reactors investigated in the author's laboratory.

The reagent can be introduced through the membrane in a closed system by pumping with a syringe pump [679]. However, gas-pressurized reagent introduction is simple, economical, and highly flexible. The membrane reactor is immersed in the reagent solution in a closed bottle and pressurized by regulated pressure from a suitable gas cylinder. Again, helium is the pressurizing gas of choice. With purely aqueous eluants, bubble formation in the detector is rarely a problem. Consequently, only a small amount of back pressure at the detector outlet (10-15 psi) is adequate, and the membrane reactor can be conveniently operated slightly above this back pressure (typically under 30 psi). The disadvantage of a pneumatic reagent introduction system, the relatively long equilibration times, can be avoided by not releasing the pneumatic pressure when the rest of the system is shut down. Flow can be prevented by putting suitable plugs at the membrane reactor inlet/outlet or by simply connecting them together. Fluorocarbon pressurizing (up to 10 lb/in.^{2}) reservoirs are commercially available (Lazar Research). Three types of microporous membranes have thus far been investigated: (a) cellulose (Spectrapor, Spectrum Medical Industries), (b) polypropylene (Celgard, see Ref. 50), and (c) PTFE (Gore-Tex TB 001, see Section II.I). The pore sizes increase in the order cited. The cellulose fibers were found by-and-large to be too delicate to work with as individual strands. Further details of this type of membrane reactor are therefore omitted. Because in active reactor designs, the reagent input can be easily varied by controlling the pressure, only a single length is employed to reduce dispersion. For a typical reactor this length is quite small, between 1 and 5 cm. Note that although reagent volumetric input rate can be varied, it is useful to keep it relatively small by using a high reagent concentration to avoid large dilutions of the eluate. Obviously, there is no possibility of sample loss with an active membrane reactor.

Operational Characteristics of an Active Membrane Reactor--We have observed that if a number of species are being introduced through the membrane from the same reagent solution, the input flux of the different species in the permeate may not be in exactly the same ratio as they are in the reagent solution. This is especially true at low volumetric input rates and where there is a large difference in the size of the different species involved. For example, in an enzyme-mediated fluorometric assay of peroxide, p-hydroxyphenylacetate is employed as a fluorogenic substrate along with peroxidase (MW, 40,000) as the enzyme catalyst. Both substances are introduced from the same solution through a PTFE membrane reactor into the flow stream [680]. Experiments have clearly established that the peroxidase/p-hydroxyphenylacetate ratio in the permeate is significantly lower than that in the reagent solution. Although this is indeed expected in a passive reactor, it suggests that diffusive transport cannot be totally ignored in active reactors either. Although in the above example the size difference between the two species involved is rather extreme, the same phenomenon, to a smaller extent, is observed for many other solutions, especially when one of the species involved is the fast-diffusing hydrogen ion.

Active membrane reactors are easily fabricated in low volume configurations. A comparative study [716] shows no decrease in plate height even with 3 μm column packing. The mixing efficiency is the same as that with a screen-tee reactor [715]; in both systems baseline noise is largely due to pump pulsations.

Gaseous Reagents in a PCR

Porous membrane reactors can be used in the passive mode for diffusing in gaseous reagents. This is particularly useful for raising or lowering pH of the column effluent via suspension of the membrane above concentrated NH_3 or HCl, respectively. The gaseous reagent permeates rapidly through the membrane. The polypropylene membrane reactor is particularly good for this purpose. Pressure required to force an aqueous solution through the small pores of this hydrophobic membrane is quite large. Thus, unless the detector represents a very significant back pressure, no liquid leakage occurs. Many nonaqueous solvents, however, wet the membrane, resulting in liquid leakage, and hence cannot be used in this manner.

Hybrid Membrane Reactors

For the introduction of two mutually incompatible reagents (i.e., where a mixed solution is unstable), a dual membrane annular reactor has been developed in the author's laboratory, somewhat similar in design to the dual membrane suppressor (Figure 16). The preferred configuration, if possible for the particular combination of reagents, is porous/ion exchanger. The inner membrane is the ion exchange membrane, sheathed by the outer porous membrane. All inlet/outlet tubes are PTFE. Ionic reagent 1 flows continuously through the central channel. The entire assembly is immersed in reagent 2, which is introduced pneumatically through the porous membrane. The principal fluid stream of interest flows in the annular channel. A porous/porous combination is also possible. In this case, one end of the inner membrane is sealed off. The other end, connected to a PTFE inlet tube, is led out of the reservoir in which the assembly is immersed (and in which reagent 1 is pneumatically introduced through the outer membrane) and dips into reagent 2 contained in a second reservoir, independently pneumatically pressurized.

Applications of PCR Detection Schemes

Examples of postcolumn reactions used in ionic chromatography are many and varied. By and large, the principal emphasis thus far has been in metal ion chromatography. Since the original work of Fritz and Story [399] with arsenazo I, arsenazo III, and PAR as postcolumn reagents, PCR detection of metal ions (particularly with PAR) has come into wide use. Better sensitivities for alkaline earth metals are obtained with arsenazo I. Pohlandt and Fritz [499] have preferred the use of arsenazo III for detecting a variety of other metals as well. Details of the reactions of a large number of metal ions with PAR-Zn(EDTA) have been published by Jezorek and Freiser [681]. PAR reacts usefully with over 40 metal ions. The review of Anderson and Nickless [682] on the analytical use of PAR and related reagents, although somewhat dated, remains the best. Cassidy and Elchuk have made extensive use of PAR for trace analysis [155,388,402,631,683], and many others have used it as well. In more recent years, Cassidy et al. [199] have found it advantageous to use arsenazo III as the postcolumn reagent, instead of PAR, for the detection of lanthanides and Th and U. Hayakawa et al. [684] have also preferred arsenazo III for rare earth detection in conjunction with a citrate eluant. According to Wang et al. [85], PAR and arsenazo I are both suitable for transition metal determination in conjunction with lactate, methyllactate, and tartrate eluants. Among common metal ions, PAR does not react with aluminum, though tiron is useful for detecting this ion.

The rationale for the popularity of PAR as a reagent for postcolumn PCR detection is not eminently clear to this author. EDTA-type ligands form complexes with a larger number of metal ions, and these often have greater formation constants. Chromogenic ligands bearing dual iminodiacetate functionalities include xylenol orange (XO) and methylthymol blue (MTB). Both of these form complexes with a greater number of metal ions than does PAR, and these complexes are frequently intensely colored. These complexes are typically anionic and are therefore susceptible to cationic micellar absorptivity enhancement [604]. Equally importantly, a large number of MTB complexes absorb in the red (they are blue in color), opening the future for stable and inexpensive He-Ne laser sources for highly sensitive detection both conventionally, and by the thermal lens effect [685].

Alkaline MTB and XO are not as susceptible to oxidative degradation as alkaline PAR. Except for a report by Ishii et al. [686] in which 15 rare earths were separated on a 0.5-mm i.d. microcolumn and detected by a PAR scheme involving XO and a cationic micelle, this author is not aware of other work utilizing XO or MTB in PCR detection.

Postcolumn reaction with dithizone has also been used for metal ion detection [687]. An acid-base indicator has been introduced in the PCR for the detection of eluting acids [133,662].

Electrochemical detection schemes involving postcolumn reactions by ligand displacement have been extensively employed by Takata et al. [256-260,262] and were discussed in detail earlier (Section V.B). These authors have also developed an electrocatalytic detection method for Co^{2+} in a PCR scheme [263] and detected silicate, phosphate, and germanate electrochemically after a postcolumn reaction with molybdate [261]. Indirect electrochemical methods involving the introduction of a metal ion [112] or quinone/hydroquinone [120,121] through a PCR have also been discussed earlier (Section III.B).

The molybdate reaction is also useful in visible absorptiometric detection of the anions mentioned above, especially after reduction to molybdenum blue by reaction with a reducing agent in a second PCR. Both air-segmented [688,689] and nonsegmented [690,691] PCR systems have been used. The reaction of ferric ion with thiocyanate is well known, and urinary thiocyanate can be determined directly with a $Fe(ClO_4)_3/HClO_4$ eluant [692]. The related application involving the introduction of $Fe(ClO_4)_3/HClO_4$ in a PCR makes possible the detection of a large number of anions when nitrate or acetate is used as eluant [693]. Phosphates, phosphonates, EDTA, NTA, etc.--typical components of commercial detergents--are also detectable optically with the use of ferric ion as the PCR reagent [694].

Fluorescence detection of metal ions via fluorescamine addition after their elution as p-aminophenyl - EDTA complexes has been reported [348]. Reports in the literature on fluorometric PCR detection schemes for metal ions are scant. Although there are a number of natively fluorescent reagents and although the fluorescence is quenched by several metal ions (fluorescein complexone, calcein, being a notable example [604]), the preferred situation would be a nonfluorescent reagent that forms fluorescent complexes with a large number of metal ions. Calcein will react with alkaline earth metals to form fluorescent complexes at high pH, where calcein itself is nonfluorescent. Several rare earths form fluorescent complexes with certain β-diketones [604]. Kojic acid forms an intensely fluorescent complex with ferric iron [695].

In early analytical chemistry, the feeling was that "salvation [of analytical chemists] lies in the development of truly selective methods of analysis, and [that their] final resting place will be heaven, in which [they have] a shelf containing 92 reagents, one for each element, where No. 13 is the infallible specific for aluminum, No. 26 is the sure shot for iron, No. 39 the unfailing relief for yttrium, and so on to uranium..." [696]. Consequently, reagents that nonspecifically react with a large number of species (e.g., PAR, MTB, XO, etc.) were not regarded as particularly valuable. Today we can depend on the chromatographic system to provide the speciation, and often prefer a universal, rather than a specific, reagent. Among fluorogenic complexing agents, reagents of the oxine (8-hydroxyquinoline) family represent such an example. Although oxine itself forms fluorescent complexes with a large number of metal ions, most of these are insoluble. It is possible to solubilize these in micellar media, e.g., Triton X-100 [231]. Thioxine reacts similarly to oxine [604]. In any case, sulfoxine (oxine-5-sulfonic acid) reacts analogously and forms water-soluble fluorescent complexes with over 40 metal ions [604,697].

The reaction of nonfluorescent Ce^{4+} with reducing anions to form fluorescent Ce^{3+} in a PCR detection scheme has already been cited [644].

Chemiluminescence from the oxidation of luminol by H_2O_2 is catalyzed by several transition metal ions and forms the basis of a chemiluminescence PCR detection scheme that is extraordinarily sensitive; a detection limit of 0.6 pg/ml for Co^{2+} has been estimated [698]. The linear range of the method, however, is quite limited; it extends to only 10 ng/ml for Co^{2+}. It is the only method known to this author that is of limited utility due to its extraordinary sensitivity and the consequent need for extreme purity of all reagents and solvents.

Induced chemiluminescence is also an excellent way to detect eluting fluorophores [699]. Peroxyoxalate is unstable and decomposes to CO_2. It can be generated in situ by reacting bis(2,4,6-trichlorophenyl)oxalate with hydrogen peroxide. In the presence of a fluorophore, the excited peroxyoxalate intermediate transfers its energy to the fluorophore, which then emits light at its characteristic wavelength.

Precipitation detection has been shown to be useful for chromatographic detection of biomolecules [700], and there is potential for such a PCR detection scheme for inorganic anions. A mixture of Ba^{2+} and Ag^+ will precipitate a large number of inorganic anions, whereas alkaline phosphate will precipitate virtually all cations except the alkali metals. Alkaline ferrocyanide will similarly precipitate all but the alkali and alkaline earth metals. Trace detection is likely to be hampered by large induction times for precipitation in very dilute solutions, although the sensitivity of nephelometric detection itself can be greatly improved with inexpensive laser sources. If a very small concentration of nucleating sites, e.g., colloidal silica, can be reproducibly introduced by a porous membrane PCR, induction times could potentially be shortened sufficiently to permit trace analysis.

C. Gradient Elution in Conductimetric Ion Chromatography

The term *gradient elution* implies an increase in eluant strength during the course of a chromatographic separation. The benefits of gradient elution in liquid chromatography (which is exactly parallel to temperature programing in gas chromatography) are well known and described in any textbook covering the general topic and do not need to be repeated here. Gradient elution poses no special problems if the detector is insensitive to changes in the eluant background. Negligible baseline drift, for example, is observed for UV detection above 210 nm with a methanol:water gradient, a common practice in reverse phase LC. Conductivity detection is by far the most widely used in IC. Gradient elution in IC, whether the system is ion exchanger or IIR based, requires a change in the ionic composition of the eluant. Typically, this change is reflected in the effluent reaching the detector, which therefore responds to the change in the form of irregularities in the baseline, usually an increase in conductance. Tackling such a problem is significantly more difficult for a nonsuppressed single-column system than for the suppressed system, because the changes in the eluant composition are *directly* reflected in the effluent. Consequently, the limited number of studies that have appeared thus far on gradient elution conductimetric IC all deal with suppressed systems. The first published study in this area, by Sunden et al. [701], concerns a gradient separation of sulfur oxyanions on ion exchange columns via bicarbonate/carbonate eluant systems. The initial premise was to use a gradient between equimolar concentrations of $NaHCO_3$ and Na_2CO_3. At first sight, it may seem that the detector should see the same concentration of H_2CO_3 after the eluant has been suppressed. Two separate factors complicate the real situation. First, initially the ion exchange sites are occupied by the singly charged HCO_3^-. During the course of the gradient run, the HCO_3^- is displaced by doubly charged CO_3^{2-}, one CO_3^{2-} taking the place of two HCO_3^- ions. Consequently, there is a temporary increase in H_2CO_3 concentration reaching the detector, which responds by producing a "hump" in the baseline.

The second problem is more subtle and is akin to the problems encountered in methanol:water gradient elution reverse-phase LC with UV detection, if the water contains traces of UV-absorbing impurities. All reagents contain varying amounts of impurities. As the bicarbonate solution is pumped through the column, an equilibrium is eventually established between the concentration of the impurity ion(s) in the mobile phase and that adsorbed on the stationary phase. As the eluant is switched to carbonate, containing, say, the same level of impurities, a new equilibrium is established. Because carbonate is a stronger eluant than bicarbonate, the new equilibrium requires a lower stationary-phase concentration of the impurity ion(s). Consequently, the amount of impurities represented by the difference in the adsorbed amounts for the two eluants appears over the background composition as a broad hump. The severity of this phenomenon is typically less than that originating from the first cause, addressed in the previous paragraph. The actual extent will depend on a number of factors, including

1. The concentration of the impurity ion(s).
2. The nature of the impurity ion. The more strongly it is retained, the greater the adsorbed amount and the worse off the analyst.
3. The pK of the acid corresponding to the impurity anion. The change in conductance after suppression increases with decreasing pK.
4. The total ion exchange capacity of the column (or total equivalents of IIR adsorbed, for IIR-based systems).
5. The extent of change in the eluant strength during the gradient run.
6. The absolute concentrations of eluants being used. This problem has not been generally addressed and is likely to limit the capabilities of gradient elution conductimetric IC for true trace analysis.

Sunden et al. [701] did, however, recognize the problem arising when doubly charged CO_3^{2-} replaces HCO_3^-. In any case, the changes in elution pattern, when the eluant changes from 5 mM $NaHCO_3$ to 5 mM Na_2CO_3, were judged inadequate for the intended separation of SO_3^{2-}, SO_4^{2-}, and $S_2O_3^{2-}$ (which elute in that order) in under 15 min. The authors therefore abandoned the quest for a nonshifting baseline and applied a much more drastic step-gradient from 4.8 mM $NaHCO_3$/4.7 mM Na_2CO_3 to 7.2 mM $NaHCO_3$/9.1 mM Na_2CO_3, commencing at 3 min from sample injection. Although the baseline shift was noticeable, the intended separation was achieved with good resolution within the desired time. The extent of baseline shift was more than 10% of the 25-ppm sulfate peak, indicating that trace analysis will not generally be feasible.

As a result of this study, it became apparent that if the CO_2 in the suppressed effluent can be removed, converting the effluent to essentially pure water, baseline shifts with varying concentrations of H_2CO_3 reaching the detector can be eliminated. Membrane-based postsuppressors were therefore developed by Sunden et al. [70] to remove the CO_2, and very substantial improvement in baseline stability was attained for gradient analysis of routine anion mixtures. The starting eluant was 0.25 mM $NaHCO_3$/0.24 mM Na_2CO_3 and ended with an eluant composition of 18 mM $NaHCO_3$/3 mM Na_2CO_3; this represents a 40-fold change in the H_2CO_3 concentration before the postsuppressor. The overall change in background conductance reaching the detector after the postsuppressor was only 4 µS/cm, whereas without the postsuppressor the change was an order of magnitude greater. A better postsuppressor design has since been reported by Siemer and Johnson [71]. The postsuppressor design and its benefits, quite aside from gradient elution, have been discussed earlier (Section II.I). The modeling of analyte retention behavior due to varying amounts of Na_2CO_3/$NaHCO_3$ in the eluant has been carried out by Jenke and Pagenkopf [711] and should be useful for future experiments with such a gradient system.

Tarter has studied the prospect of gradient elution with a large number of analyte ions without any postsuppressor [122]. Retention times were reported under gradient elution conditions for a large number of analyte anions including propionate, fluoride, acetate, formate, sulfide, hypochlorite, chloride, nitrite, chlorate, bromide, nitrate, phosphate, sulfite, sulfate, iodide, thiocyanate, dichromate, and thiosulfate, which elute in the order cited. Typical gradient runs utilized 3 mM $NaHCO_3$ as starting eluant and 4 mM Na_2CO_3 or 2 mM Na_2CO_3 / 2 mM NaOH as final eluants. This study revealed that an electrochemical detector, placed after the suppressor column, successfully monitored the elution of anions that are not electroactive, indirectly via changes in effluent pH; this has been mentioned earlier (Section III.B). What is additionally interesting is that the baseline shift of the electrochemical detector under gradient elution conditions was smaller than that exhibited by the conductivity detector. Also, from a signal/noise standpoint, the electrochemical detector apparently performed better than the conductivity detector. Additionally, Tarter found that a smaller change in the eluant strength early in the gradient program is much more effective than large changes later in the run. This is quite explicable, because changes late in the gradient program have only a marginal effect on the elution of analytes that have already traversed most of the column length.

Dasgupta [45] has studied the utility of gradient elution in IIR-based systems with a PSDVB stationary phase. A starting eluant of 0.87 mM NPr_4OH/1.3 mM NaOH was changed to 10 mM NaOH, using a linear gradient starting at 6 min from injection, over a period of 4 min. Fluoride, chloride, bromide, nitrate, chlorate, sulfate, iodide, thiocyanate, and perchlorate were separated under isocratic conditions with the starting eluant, the separation requiring nearly 70 min. The same separation was accomplished in approximately half the time under the above gradient conditions. An annular helical suppressor [32] was used for postcolumn cation exchange. The shift in the background conductance during the gradient run was substantial and was likely due to the difficulty of preparing and storing hydroxide eluants free from CO_2 and indigenous impurities.

Approaches to Ionic Chromatography 337

In terms of instrumentation used to generate the gradient program, Dasgupta [45] used a conventional two-pump system with dynamic mixing at the high-pressure end, whereas Sunden et al. [70] as well as Tarter [122] used low-pressure mixing techniques before the eluant pump. In the study of Sunden et al., a stepper motor-driven peristaltic pump, controlled by a frequency-sweep generator, was used to introduce the strong eluant at increasing rates into a mixing tee at the eluant pump inlet. The balance of the flow requirement for the eluant pump was supplied by the weak eluant. Tarter [122] used a small glass mixing chamber at the eluant pump inlet, filled with variable amounts of the weak eluant. One or two separate pumps then introduced one or two additional eluants into this chamber at constant rates, generating a continuous gradient.

Gradient conductimetric IC is still in its infancy. The future doubtless holds exciting prospects.

VII. CONCLUSIONS

In a field as rapidly developing as ionic chromatography, it is foolhardy to spend too much time with one's crystal ball. The significant time lag between the author's last involvement with this chapter and its actual appearance into print may well make some predictions retrospective. Still, it is hard to resist the temptation.

The mainstay of IC is still the conductimetric detector. Different ions show different dependence in a frequency versus conductance plot; this is referred to as the *Debye-Falkenhagen effect* [702]. The potential exists for employing swept-frequency sources to determine this dependence on-the-fly, and for identifying the eluting ion on a basis other than its retention time. Because of the high frequencies involved, electrodeless cells may be practical; careful design will, however, be necessary to minimize cell capacitance.

The emergence of flow injection analysis in its rightful place as an inexpensive rapid and precise technique to perform repetitive assays for individual species suggests that IC applications will primarily blossom in areas where analysis, rather than assay of an individual species, is warranted. Thanks to the new membrane reactors, IC will compete on a realistic basis with atomic spectrometric techniques; applications of chromogenic reagents that can be directly incorporated in the eluant will also increase. Diode array detectors will permit accurate quantitation, even when chromatographic separation is incomplete, through the use of appropriate combinations of chromogenic reagents. New fluorogenic reagents will be developed both for incorporation in the eluant and for utilization in a PCR.

The potential of getting to a background of pure water through postsuppressor-induced removal of CO_2 not only will benefit gradient-elution techniques in anionic chromatography, but also will open up new vistas in sensitivity. Once a nonionic background is achieved, eluting ions can be quantitatively exchanged for a fluorescent ion in a membrane-based ion exchanger. In this context, zwitterionic eluants [703], including amino acids [4], are potentially useful. For both cation and anion analysis, ionic ammonium cyanate as eluant, followed by postcolumn thermal conversion to essentially nonionic urea (the Wohler reaction, see Ref. 704) may also be potentially advantageous.

The electrochemical detector has already been shown to be superior to conductance detection [122,123] for anion detection in suppressed IC. Its utility in the potential-differentiated (or even potential-swept) mode for transition metal analysis is untapped. The introduction of suitable, colored or fluorescent acid/base indicators after the suppressor in suppressed anionic IC may lead to excellent sensitivities, especially if a pure-water background can be achieved.

The revolution in ionic analysis through chromatography has just begun. Its eventual potential and utility will be limited only by the reader's creative imagination and not by this author's cloudy crystal ball.

ACKNOWLEDGMENTS

The author thanks his wife Sarah Jean for bearing with him the many months spent in writing this chapter. This manuscript greatly benefited from expert reviews by R. M. Cassidy, J. C. Davis, D. T. Gjerde, J. P. Lodge, Jr., C. A. Pohl, H. Small, and T. S. Stevens who should rightfully be credited for much that is good in it. The shortcomings are the author's own. Even in the wake of computerized literature searching, comprehensive coverage of the literature has not been possible. The citations are reasonably complete till the end of 1984.

The ion chromatographic work of the author reported herein was primarily supported by the U. S. Department of Energy, Office of the Basic Energy Sciences, through Grant No. DE-FG05-84ER13281; however, any opinions, findings, conclusions, or recommendations expressed herein are those of the author and do not necessarily reflect the views of the DOE.

REFERENCES AND NOTES

1. H. Small, T. S. Stevens, and W. C. Bauman, *Anal. Chem.* 47, 1801-1809 (1975).
2. A. Walsh, *Spectrochim. Acta* 7, 108-117 (1955).

3. H. Small and T. S. Stevens, U.S. Patent 3,915,642, October 28, 1975. This technique is covered by similar patents in many other countries.
4. H. Small and J. Solc, in *The Theory and Practice of Ion Exchange*, Society of Chemical Industry, London, 1976, pp. 32-1 through 32-100.
5. R. E. Majors, H. G. Barth, and C. H. Lochmuller, *Anal. Chem.* 56, 300R-349R (1984).
6. I. Molnar, H. Knauer, and D. Wilk, *J. Chromatogr.* 201, 223-240 (1980).
7. P. R. Haddad and A. L. Heckenberg, *J. Chromatogr.* 300, 357-394 (1984).
8. G. Schmuckler, *J. Chromatogr.* 313, 47-57 (1984).
9. W. F. Koch, *Anal. Chem.* 51, 1571-1573 (1979).
10. A. Rembaum, S. Y. Shiao-Ping, and E. Klein, U.S. Patent 4,045,352, August 30, 1977.
11. T. S. Stevens, J. C. Davis, and H. Small, *Anal. Chem.* 53, 1488-1492 (1981).
12. P. G. Gormley and M. Kennedy, *Proc. R. Irish Acad. Sci. [A]* 52, 163-169 (1949).
13. L. R. Snyder and J. J. Kirkland, *Introduction to Modern Liquid Chromatography*, 2nd ed., Wiley, New York, 1982, p. 87.
14. Dionex Corporation, Anion Fiber Suppressor 2 (AFS-2) and Cation Fiber Suppressor 1 (CFS-1).
15. S. A. Bouyoucos, *J. Chromatogr.* 242, 170-176 (1982).
16. W. Stumm and J. J. Morgan, *Aquatic Chemistry*, 2nd ed., Wiley, New York, 1981, p. 212.
17. H. L. Yeager, *ACS Symp. Ser.* No. 180, 41-64 (1982).
18. E. I. du Pont de Nemours Co., Nafion Perfluorosulfonic Acid Products, Product Information Bulletin E-05569, 1976.
19. J. G. Eberhart and T. R. Sweet, *J. Chem. Educ.* 37, 422-425 (1966).
20. H. L. Yeager and A. Steck, *Anal. Chem.* 51, 862-865 (1979).
21. W. R. Dean, *Philos. Mag.* 4, 207-223 (1927).
22. W. R. Dean, *Philos. Mag.* 5, 673-695 (1928).
23. L. C. Truesdell, Jr., and R. J. Adler, *AIChE J.* 16, 1010-1015 (1970).
24. L. R. Austin and J. D. Seader, *AIChE J.* 19, 85-93 (1973).
25. L. M. Janssen, *Chem. Eng. Sci.* 31, 215-218 (1976).
26. R. S. Deelder, M. G. F. Kroll, A. J. B. Beeren, and J. H. M. Van den Berg, *J. Chromatogr.* 149, 669-682 (1978).
27. R. N. Trivedi and K. Vasudeva, *Chem. Eng. Sci.* 29, 2291-2295 (1974).
28. R. N. Trivedi and K. Vasudeva, *Chem. Eng. Sci.* 30, 317-325 (1975).

29. I. S. Krull, C. M. Selavka, R. J. Nelson, L.-R. Chen, and K. Bratin, Abstract No. 639, 36th Pittsburgh Conference on Analytical Chemistry and Applied Spectroscopy, New Orleans, LA, February, 1985.
30. G. T. Westbrook, AIChE Symp. Ser. 68(124), 283-293 (1971).
31. P. K. Dasgupta, Anal. Chem. 56, 97-103 (1984).
32. P. K. Dasgupta, Anal. Chem. 56, 103-105 (1984).
33. T. S. Stevens and M. A. Langhorst, Anal. Chem. 54, 950-953 (1982).
34. T. S. Stevens, G. L. Jewett, and R. A. Bredeweg, Anal. Chem. 54, 1206-1208 (1982).
35. J. A. Cox and G. R. Litwinski, Anal. Chem. 55, 1640-1642 (1983).
36. J. Ruzicka and E. H. H. Hansen, Flow Injection Analysis, Wiley, New York, 1981.
37. J. M. Reijn, W. E. Van der Linden, and H. Poppe, Anal. Chim. Acta 123, 229-237 (1981).
38. J. M. Reijn, H. Poppe, and W. E. Van der Linden, Anal. Chim. Acta 145, 59-70 (1983).
39. J. F. K. Huber, K. M. Jonker, and H. Poppe, Anal. Chem. 52, 2-9 (1980).
40. T. S. Stevens, G. L. Jewett, R. A. Bredeweg, L. B. Westover, and H. Small, European patent application no. 82201155.7, publication no. 0 075 371, A1, September 17, 1982.
41. D. T. Gjerde, D. E. Burge, and T. Jupille, Wescan Instruments, Inc., U. S. Patent applied for.
42. Waters Associates, Milford, MA.
43. P. K. Dasgupta, R. Q. Bligh, and M. Mercurio, Anal. Chem. 57, 484-489 (1985).
44. C. A. Pohl, U. S. Patent 4,265,634, May 5, 1981.
45. P. K. Dasgupta, Anal. Chem. 56, 769-772 (1984).
46. J. Stillian, LC Liq. Chromatogr. HPLC Mag. 3, 802-812 (1985).
47. Y. Hanaoka, M. Takeshi, S. Muramoto, M. Tamizo, and A. Nanba, J. Chromatogr. 239, 537-548 (1982).
48. S. Rokushika, Z. Y. Qiu, and H. Hatano, J. Chromatogr. 260, 81-87 (1983).
49. V. D'Agostino, J. Lee, and G. Orban, Grafted membranes, in Zinc – Silver oxide Batteries (A. Fleischer and J. Lander, eds.), Wiley, New York, 1971.
50. Celanese Corporation, Celgard Division, Charlotte, NC. Celgard X-10 microporous polypropylene hollow fiber, Product Data Sheet, May, 1983.
51. E. I. du Pont de Nemours Company, Experimental perfluorinated tubing data sheet. Polymer Products Division, Wilmington, DE 19898. Nafion tubing is available from Perma Pure Products, Inc., Toms River, NJ, and special dimensions are possibly available on a custom basis.

52. RAI Research Corporation, Hauppage, NY.
53. J. Lee, V. D'Agostino, J. Santisi, and C. Perini, paper presented at the Electrochemical Society Meeting, Pittsburgh, PA, 1978.
54. T. D. Gierke, G. E. Munn, and F. C. Wilson, ACS Symp. Ser. 180, 195-216 (1982).
55. P. K. Dasgupta, R. Q. Bligh, J. Lee, and V. D'Agostino, Anal. Chem. 57, 253-257 (1985).
56. Zeus Industrial Products, Raritan, NJ.
57. W. F. Sheehan, Physical Chemistry, Prentice Hall, New York, 1963, pp. 328-330.
58. S. Rokushika, Z. L. Sun, and H. Hatano, J. Chromatogr. 253, 87-94 (1982).
59. F. Helfferich, Ion Exchange, McGraw-Hill, New York, 1962, p. 143.
60. Stepan Chemical Company, Surfactants Department, Northfield, IL. The trade name is Bio-soft 100; the supplied reagent is about 80% in concentration and contains a small amount of sulfuric acid as impurity.
61. The disodium salt is available from Aldrich Chemical Company, Milwaukee, WI. Repeated ion exchange through a cation exchanger in the H^+ form is necessary to make the pure acid in sodium-free form.
62. The trisodium salt is available from Pfaltz and Bauer, Inc., Stamford, CT. Preparation of pure acid is as in [61], but it is extremely difficult to prepare this material completely sodium free.
63. Available from Polysciences Inc., Warrington, PA. The average MW of this material is 70,000. The reagent is supplied in aqueous solution, and the small amount of sulfuric acid present in it as an impurity may be removed by exhaustive dialysis against high-purity water. See [55] for the details of this and other methods of purifying this reagent.
64. F. Y. Lo, B. M. Escott, J. H. Fendler, E. T. Adams, Jr., R. D. Larsen, and P. W. Smith, J. Phys. Chem. 79, 2609-2621 (1975).
65. C. A. Pohl, Dionex Corporation, personal communication, May, 1985.
66. Yokagawa Electric Works Ltd., Japanese Patents 82 69,251 and 82 69,252, April 27, 1982.
67. Hitachi Ltd., Japanese Patent 84 133,59, July 31, 1984.
68. T. Murayama, S. Muramoto, and Y. Hanaoka, Yokagawa Electric Works Ltd., French Patent 24,929,83, April 30, 1982; U. S. Patent 4,403,039, September 6, 1983.
69. K. H. Jansen, K. H. Fischer, and B. Wolf, Biotronik Wissenschaftliche Geraete G.m.b.H., European Patent 69,285, January 12, 1983; U. S. Patent 4,459,357, July 10, 1984.

70. T. Sunden, A. Cedergren, and D. D. Siemer, *Anal. Chem.* 56, 1085-1089 (1984).
71. D. D. Siemer and V. J. Johnson, *Anal. Chem.* 56, 1033-1034 (1984).
72. J. T. Edsall, in CO_2: Chemical, Biochemical, and Physiological Effects (R. E. Foster et al., eds.), NASA SP-188, Washington DC, 1969.
73. R. A. Robinson and R. H. Stokes, *Electrolyte Solutions*, Butterworths, London, 1968.
74. American Chemical Society Committee on Environmental Improvement, *Anal. Chem.* 52, 2242-2249 (1980).
75. J. S. Fritz, D. T. Gjerde, and R. M. Becker, *Anal. Chem.* 52, 1519-1522 (1980).
76. J. S. Fritz, D. T. Gjerde, and C. Pohlandt, *Ion Chromatography*, Alfred Huthig, Heidelberg, 1982.
77. T. Okada and T. Kuwamoto, *Anal. Chem.* 55, 1001-1004 (1983).
78. T. Okada and T. Kuwamoto, *Anal. Chem.* 57, 829-833 (1985).
79. D. T. Gjerde, G. Schmuckler, and J. S. Fritz, *J. Chromatogr.* 187, 35-45 (1980).
80. Bio-Rad Laboratories, Richmond, CA. New Anion Separator Column for Suppressed or Nonsuppressed IC, Technical Note 1, 1983.
81. T. Okada and T. Kuwamoto, *J. Chromatogr.* 284, 149-156 (1984).
82. S. Matsushita, *J. Chromatogr.* 312, 327-336 (1984).
83. R. K. Pinschmidt, in *Ion Chromatographic Analysis of Environmental Pollutants*, Vol. 2 (J. D. Mulik and E. Sawiciki, eds.), Ann Arbor Science, Ann Arbor, 1979, pp. 41-50.
84. E. L. Johnson and K. K. Haak, in *Liquid Chromatography in Environmental Analysis* (J. F. Lawrence, ed.), Humana, Clifton, NJ, 1984, pp. 263-300.
85. W.-N. Wang, Y.-J. Chen, and M.-T. Wu, *Analyst* 109, 281-286 (1984).
86. C. A. Chang and K. L. Fong, *J. Chromatogr.* 312, 99-107 (1984).
87. J. M. Keller, *Anal. Chem.* 53, 344-345 (1981).
88. Dionex Corporation, Sunnyvale, CA; ESA Inc., Bedford, MA.
89. H. Sato, *Bunseki Kagaku 31*, T23-T28 (1982).
90. Toyo Soda Company, Tokyo, Japan; Bio-Rad Laboratories, Richmond, CA; Waters Associates, Milford, MA.
91. V. Svoboda and J. Marsal, *J. Chromatogr.* 148, 111-116 (1978).
92. J. F. Alder and A. Thoer, *J. Chromatogr.* 178, 15-26 (1979).
93. J. F. Alder, P. K. P. Drew, and P. R. Fielden, *J. Chromatogr.* 212, 167-177 (1981).

94. J. F. Alder, P. K. P. Drew, and P. R. Fielden, *Anal. Chem.* 55, 256-262 (1983).
95. E. Pungor, F. Pal, and K. Toth, *Anal. Chem.* 55, 1728-1731 (1983).
96. J. F. Alder, P. R. Fielden, and A. J. Clark, *Anal. Chem.* 56, 985-988 (1984).
97. A. E. Buchholz, C. I. Verplough, and J. L. Smith, *J. Chromatogr. Sci.* 20, 499-501 (1982).
98. J. A. Glatz and J. E. Girard, *J. Chromatogr. Sci.* 20, 266-273 (1982).
99. D. R. Jenke and G. K. Pagenkopf, *Anal. Chem.* 54, 2603-2604 (1982).
100. Wescan Instruments, Santa Clara, CA.
101. Tri-Det., Perkin-Elmer Corp., Norwalk, CT. See J. R. Gant and P. A. Perone, *Am. Lab.* 17(3), 104-111 (1985).
102. H. Mackie, S. J. Speciale, L. J. Throop, and T. Yang, *J. Chromatogr.* 242, 177-180 (1982).
103. J. Slanina, F. P. Bakker, P. A. C. Jongejean, L. Van Lamoen, and J. J. Mols, *Anal. Chim. Acta* 130, 1-8 (1981).
104. K. Suzuki, H. Aruga, and T. Shirai, *Anal. Chem.* 55, 2011-2013 (1983).
105. M. C. Franks and D. A. Pullen, *Analyst* 99, 503-514 (1974).
106. R. S. Deelder, H. A. J. Linssen, J. G. Koen, and A. J. B. Beeren, *J. Chromatogr.* 203, 153-163 (1981).
107. F. A. Schultz and D. E. Mathis, *Anal. Chem.* 46, 2253-2255 (1974).
108. V. V. Bardin, Yu. M. Ivanov, and O. F. Shartukov, *Zh. Anal. Khim.* 33, 1732-1737 (1978).
109. T. Deguchi, T. Kuma, and H. Nagai, *J. Chromatogr.* 152, 349-355 (1978).
110. H. Hershcovitz, Ch. Yarnitzsky, and G. Schmuckler, *J. Chromatogr.* 252, 113-119 (1982).
111. G. D. Christian, *Analytical Chemistry*, 3rd ed., Wiley, New York, 1980, p. 287.
112. C. R. Loscombe, J. B. Cox, and J. A. W. Dalziel, *J. Chromatogr.* 166, 403-410 (1978).
113. P. R. Haddad and P. W. Alexander, *J. Chromatogr.* 294, 397-402 (1984).
114. P. R. Haddad, P. W. Alexander, and M. Trozanowicz, *J. Chromatogr.* 315, 261-270 (1984).
115. P. R. Haddad, P. W. Alexander, and M. Trojanowicz, *J. Chromatogr.* 321, 363-374 (1985).
116. T. Yoshinori, M. Haruo, T. Hiromichi, and M. Giichi, *Bunseki Kagaku* 15, 573-577 (1967).
117. R. D. Rocklin and E. L. Johnson, *Anal. Chem.* 55, 4-7 (1983).

118. A. M. Bond, I. D. Heritage, G. G. Wallace, and M. J. McCormick, *Anal. Chem.* 54, 582-585 (1982).
119. E. A. Ostrovidov, L. A. Kuleshova, S. I. Mitina, R. G. Vinogradova, A. M. Vorontsov, A. S. Kanev, and O. A. Rys'ev, *Zh. Anal. Khim.* 35, 1677-1681 (1980).
120. K. Tanaka, Y. Ishihara, and H. Sunahara, *Bunseki Kagaku* 24, 235-238 (1975).
121. J. E. Girard, *Anal. Chem.* 51, 836-839 (1979).
122. J. G. Tarter, *Anal. Chem.* 56, 1264-1268 (1984).
123. J. G. Tarter, *J. Liq. Chromatogr.* 7, 1559-1566 (1984).
124. S. Egashira, *J. Chromatogr.* 202, 37-43 (1980).
125. T. Kaimura and M. Tanaka, *Anal. Chim. Acta* 110, 117-122 (1979).
126. A. Takahashi, *Bunseki Kagaku* 29, 508-512 (1980).
127. J. R. Thayer and R. C. Huffaker, *Anal. Biochem.* 102, 110-119 (1980).
128. H. Terada, T. Ishihara, and Y. Sakabe, *Eisei Kagaku* 26, 136-139 (1980).
129. H. Small and T. E. Miller, Jr., *Anal. Chem.* 54, 462-469 (1982).
130. J. R. Larson and C. D. Pfeiffer, *Anal. Chem.* 55, 393-396 (1983).
131. H. J. Cortes and T. S. Stevens, *J. Chromatogr.* 295, 269-275 (1984).
132. J. Andrasko, *J. Chromatogr.* 314, 429-435 (1984).
133. A. Wada, M. Bonoshita, Y. Tanaka, and K. Hibi, *J. Chromatogr.* 291, 111-118 (1984).
134. D. R. Jenke, *Anal. Chem.* 56, 2468-2470 (1984).
135. S. A. Wilson and E. S. Yeung, *Anal. Chim. Acta* 151, 53-63 (1984).
136. R. A. Cochrane and D. E. Hillman, *J. Chromatogr.* 241, 392-394 (1982).
137. P. R. Haddad and A. L. Heckenberg, *J. Chromatogr.* 252, 177-184 (1982).
138. P. R. Haddad and C. E. Cowie, *J. Chromatogr.* 303, 321-330 (1984).
139. F. A. Buytenhaus, *J. Chromatogr.* 218, 57-66 (1981).
140. J. P. Ivey, *J. Chromatogr.* 267, 218-221 (1983).
141. G. P. Ayers and R. W. Gillette, *J. Chromatogr.* 284, 510-514 (1984).
142. P. E. Jackson, P. R. Haddad, and S. Dilli, *J. Chromatogr.* 295, 471-478 (1984).
143. L. Eek and N. Ferrer, *J. Chromatogr.* 322, 491-497 (1985).
144. R. J. Williams, *Anal. Chem.* 55, 851-854 (1983).
145. S. W. Downey and G. M. Hieftje, *Anal. Chim. Acta* 153, 1-13 (1983).
146. J. H. Knox and G. R. Laird, *J. Chromatogr.* 122, 17-34 (1976).
147. J. H. Knox and J. Jurand, *J. Chromatogr.* 125, 89-101 (1976).

148. *Paired-Ion Chromatography* is a registered trade name of the technique, used by Waters Associates, Milford, MA. Waters Bulletin F61, May, 1976.
149. B. Fransson, K. G. Wahlund, I. M. Johansson, and G. Schill, *J. Chromatogr. 125*, 327-344 (1976).
150. J. C. Kraak, K. M. Jonker, and J. F. K. Huber, *J. Chromatogr. 142*, 671-688 (1976).
151. C. P. Terweij-Groen, S. Heemstra, and J. C. Kraak, *J. Chromatogr. 161*, 69-82 (1978).
152. E. Tomlinson, T. M. Jefferies, and C. M. Riley, *J. Chromatogr. 159*, 315-358 (1978).
153. C. Horvath, W. Melander, I. Molnar, and P. Molnar, *Anal. Chem. 49*, 2295-2305 (1977).
154. R. P. W. Scott and P. Kucera, *J. Chromatogr. 125*, 51-63 (1979).
155. R. M. Cassidy and S. Elchuk, *Anal. Chem. 54*, 1558-1563 (1982).
156. B. A. Bidlingmeyer, S. N. Deming, W. P. Price, B. Sachok, and M. Petrusek, *J. Chromatogr. 186*, 419-434 (1979).
157. T. D. Rotsch and D. J. Pietrzyk, *J. Chromatogr. Sci. 19*, 88-95 (1981).
158. B. A. Bidlingmeyer, *LC Liq. Chromatogr. HPLC Mag. 1*, 344-349 (1983).
159. R. G. Pearson, *J. Am. Chem. Soc. 85*, 3533-3539 (1963).
160. R. H. A. Sorel and A. Hulshoff, *Adv. Chromatogr. 21*, 87-129 (1983).
161. M. A. Mercurio and P. K. Dasgupta, Paper No. 15, 40th Southwest regional meeting of the American Chemical Society, Lubbock, TX, December 5-7, 1984.
162. W. E. Barber and P. W. Carr, *J. Chromatogr. 260*, 89-96 (1983).
163. L. Hackzell and G. Schill, *Chromatographia 15*, 437-444 (1982).
164. B. A. Bidlingmeyer and F. V. Warren, *Anal. Chem. 54*, 2351-2356 (1982).
165. F. F. Cantwell and S. Puon, *Anal. Chem. 51*, 623-632 (1979).
166. The capacity factor, often denoted by the symbol k, is defined as $(t_R - t_M)/t_M$, where t_R and t_M are the retention times (or volumes) of the sample and an unretained solute, respectively. For a more detailed discussion, see any standard text on chromatography, e.g., Ref. 176.
167. Z. Iskandarani and D. J. Pietrzyk, *Anal. Chem. 54*, 1065-1071 (1982).
168. Z. Iskandarani and D. J. Pietrzyk, *Anal. Chem. 54*, 2427-2431 (1982).
169. R. S. Deelder and J. H. M. Van den Berg, *J. Chromatogr. 218*, 327-339 (1981).
170. J. H. Knox and R. A. Hartwick, *J. Chromatogr. 204*, 3-21 (1981).

171. C. M. Riley, L. A. Sternson, and A. J. Repta, *J. Chromatogr. 219*, 235-244 (1981).
172. R. C. Kong, B. Sachok, and S. N. Deming, *J. Chromatogr. 199*, 307-316 (1980).
173. S. N. Deming and R. C. Kong, *J. Chromatogr. 217*, 421-434 (1981).
174. J. J. Stranahan and S. N. Deming, *Anal. Chem. 54*, 2251-2256 (1982).
175. R. M. Cassidy and S. Elc̲̅ ̲̅, *J. Chromatogr. Sci. 21*, 454-459 (1983).
176. L. R. Snyder an̲̅ ̲̅ ̲̅d, *An Introduction to Modern Liquid Chrom̲̅ ̲̅*, Wiley, New York, 1979, pp. 323-348.
177. B. A ̲̅ ̲̅. *Sci. 18*, 525-539 (1980).
178. B ̲̅ ̲̅, *J. Chromatogr. 122*, 30̲̅
179. A. ̲̅ ̲̅ *J. Chromatogr. 134*, 459-4̲̅
180. J. L. ̲̅ ̲̅ ̲̅. M. Hendrikx, and R. S. Deelder ̲̅ ̲̅ ̲/, 1-16 (1978).
181. E. Tomli̲̅ ̲̅ ̲̅ey, and T. M. Jefferies, *J. Chromatogr. 173*, 89-10̲̅
182. R. N. Reev̲̅ ̲̅. *Chromatogr. 177*, 393-397 (1979).
183. A. Bartha and Gy. Vigh, *J. Chromatogr. 265*, 171-182 (1983).
184. A. Bartha, H. A. H. Billiet, L. De Galan, and Gy. Vigh, *J. Chromatogr. 291*, 91-102 (1984).
185. A. Bartha, Gy. Vigh, H. A. H. Billiet, and L. De Galan, *J. Chromatogr. 303*, 29-38 (1984).
186. N. E. Skelly, *Anal. Chem. 54*, 712-715 (1982).
187. F. G. P. Mullins and G. F. Kirkbright, *Analyst 109*, 1217-1221 (1984).
188. D. W. Armstrong and F. Nome, *Anal. Chem. 53*, 1662-1666 (1981).
189. J. P. de Kleijn, *Analyst 107*, 223-224 (1982).
190. R. Vespalec, J. Neca, and M. Vrchlabsky, *J. Chromatogr. 286*, 171-183 (1984).
191. P. K. Dasgupta, Unpublished studies, Texas Tech University, 1982.
192. M. Denkert, L. Hackzell, G. Schill, and E. Sjogren, *J. Chromatogr. 218*, 31-43 (1981).
193. L. Hackzell, M. Denkert, and G. Schill, *Acta Pharm. Suec. 18*, 271-282 (1981).
194. W. E. Barber and P. W. Carr, *J. Chromatogr. 301*, 25-38 (1984).
195. W. E. Barber and P. W. Carr, *J. Chromatogr. 316*, 211-225 (1984).
196. P. Helboe, *J. Chromatogr. 261*, 117-122 (1984).

197. M. Dreux, M. Lafosse, and M. Pequignot, *Chromatographia 15*, 653-656 (1982).
198. R. M. Cassidy and S. Elchuk, *J. Chromatogr. 262*, 311-315 (1983).
199. C. H. Knight, R. M. Cassidy, B. M. Recoskie, and L. W. Green, *Anal. Chem. 56*, 474-478 (1984).
200. B. Kagedal, M. Kallberg, J. Martensson, and J. Sorbo, *J. Chromatogr. 274*, 95-102 (1983).
201. B. B. Wheals, *J. Chromatogr. 262*, 61-76 (1983).
202. G. Schmuckler, B. Roessner, and G. Schwedt, *J. Chromatogr. 302*, 15-20 (1984).
203. G. Chiu, *Anal. Chem. 52*, 1157-1158 (1980).
204. S. Kok, K. A. Buckle, and M. Wooton, *J. Chromatogr. 260*, 189-192 (1983).
205. J. W. Lethbridge and R. N. Reeve, *J. Chromatogr. 314*, 387-395 (1984).
206. C. M. Riley, L. A. Sternson, and A. J. Repta, *J. Chromatogr. 219*, 235-244 (1981).
207. K. C. Marsh, L. A. Sternson, and A. J. Repta, *Anal. Chem. 56*, 491-497 (1984).
208. S. Motomizu, I. Sawatani, M. Oshima, and K. Toel, *Anal. Chem. 55*, 1629-1631 (1983).
209. C. H. Gast, J. C. Kraak, H. Poppe, and F. J. M. Maessen, *J. Chromatogr. 185*, 549-561 (1979).
210. T. L. Chester, C. A. Smith, and S. Culshaw, *J. Chromatogr. 287*, 447-451 (1984).
211. R. G. Julin, H. W. Vanderborn, and J. J. Kirkland, *J. Chromatogr. 112*, 443-453 (1975).
212. D. Bushee, I. S. Krull, R. N. Savage, and S. B. Smith, Jr., *J. Liq. Chromatogr. 5*, 463-478 (1982).
213. I. S. Krull, D. Bushee, R. N. Savage, and S. B. Smith, Jr., *Anal. Lett. 15A*, 267-281 (1982).
214. D. Bushee, D. Young, I. S. Krull, R. N. Savage, and S. B. Smith, Jr., *J. Liq. Chromatogr. 5*, 693-706 (1982).
215. H. J. Cortes, *J. Chromatogr. 234*, 517-520 (1982).
216. U. Leuenberger, R. Gauch, K. Rieder, and E. Baumgartner, *J. Chromatogr. 202*, 461-468 (1980).
217. J. J. Stranahan, S. N. Deming, and B. Sachok, *J. Chromatogr. 202*, 233-237 (1980).
218. V. V. Berry, *J. Chromatogr. 321*, 33-43 (1985).
219. A. Berthod, M. Kolosky, J.-L. Rocca, and O. Vittori, *Analusis 7*, 395-400 (1979).
220. B. W. Hoffmann and G. Schwedt, *J. High Res. Chromatogr. Chrom. Commun. 5*, 439-440 (1982).
221. A. M. Bond and G. G. Wallace, *Anal. Chem. 53*, 1209-1213 (1981).

222. A. M. Bond and G. G. Wallace, *Anal. Chem.* **54**, 1706-1712 (1982).
223. A. M. Bond and G. G. Wallace, *Anal. Chem.* **55**, 718-723 (1983).
224. A. M. Bond and G. G. Wallace, *Anal. Chem.* **56**, 2085-2090 (1984).
225. R. M. Smith and L. E. Yankey, *Analyst* **107**, 744-748 (1982).
226. J. W. O'Laughlin, *Anal. Chem.* **54**, 178-181 (1982).
227. J. W. O'Laughlin and R. S. Hanson, *Anal. Chem.* **52**, 2263-2268 (1980).
228. S. J. Valenty and P. E. Behnken, *Anal. Chem.* **50**, 834-837 (1978).
229. A. Mangia and M. T. Lugari, *J. Liq. Chromatogr.* **6**, 1073-1080 (1983).
230. H. Hoshino, T. Yotsuyanagi, and K. Aomura, *Bunseki Kagaku* **27**, 315-316 (1978).
231. H. Hoshino and T. Yotsuyunagi, *Bunseki Kagaku* **29**, 807-808 (1980).
232. H. Yamada and T. Hattori, *J. Chromatogr.* **320**, 403-407 (1985).
233. J. M. Riviello and C. A. Pohl, Paper presented at the 25th Rocky Mountain Conference of the American Chemical Society, Denver, CO., August 14-17, 1983.
234. D. A. Buckingham, C. R. Clark, and R. F. Tasker, *J. Liq. Chromatogr.* **4**, 689-700 (1981).
235. H. Lam-Thanh, S. Fermandijan, and P. Fromageot, *J. Liq. Chromatogr.* **4**, 681-688 (1981).
236. T. Tsuda, T. Nozu, and G. Nakagawa, *J. Chromatogr.* **242**, 331-336 (1985).
237. G. Schwedt, *Chromatographia* **12**, 613-619 (1979).
238. G. Schwedt, *Top. Current Chem.* **85**, 159-212 (1979).
239. G. Schwedt, *Chromatographic Methods in Inorganic Analysis*, Hutling, Heidelberg, 1981.
240. B. R. Willeford and H. Veening, *J. Chromatogr.* **251**, 61-88 (1982).
241. R. M. Cassidy, in *Trace Analysis*, Vol. 1 (J. F. Lawrence, ed.), Academic Press, New York, 1981.
242. I. S. Krull, in *Liquid Chromatography in Environmental Analysis* (J. F. Lawrence, ed.), Humana, Clifton, NJ, 1984, pp. 169-262.
243. G. Nickless, *J. Chromatogr.* **313**, 129-159 (1985).
244. H. G. Barth, W. E. Barber, C. H. Lochmuller, R. E. Majors, and F. E. Regnier, *Anal. Chem.* **58**, 211R-250R (1986).
245. F. R. Nordmeyer, L. D. Hansen, D. J. Eatough, D. K. Rollins, and J. D. Lamb, *Anal. Chem.* **52**, 852-856 (1980).

246. J. D. Lamb, L. D. Hansen, G. G. Patch, and F. R. Nordmeyer, *Anal. Chem.* 53, 749-750 (1981).
247. J. W. Wimberley, *Anal. Chem.* 53, 2137-2138 (1981).
248. R. Kunin, *Elements of Ion Exchange*, Reinhold, New York, 1960.
249. G. J. Sevenich and J. S. Fritz, *Anal. Chem.* 55, 12-16 (1983).
250. K. Brunt, in *Trace Analysis*, Vol. 1 (J. F. Lawrence, ed.), Academic Press, New York, 1981.
251. P. T. Kissinger, in *Liquid Chromatography Detectors* (T. M. Vickrey, ed.), Marcel Dekker, New York, 1983, pp. 125-164.
252. P. T. Kissinger, *Anal. Chem.* 447A-456A (1977).
253. R. J. Rucki, *Talanta* 27, 147-156 (1980).
254. K. Stulik, V. Pacakova, and M. Podlak, *J. Chromatogr.* 298, 225-230 (1984).
255. Y. Takata and G. Muto, *Bunseki Kagaku* 14, 543-547 (1965).
256. Y. Takata and G. Muto, *Anal. Chem.* 45, 1864-1868 (1973).
257. Y. Takata and Y. Arikawa, *Bunseki Kagaku* 22, 312-318 (1973).
258. Y. Takata and Y. Arikawa, *Bunseki Kagaku* 24, 762-767 (1975).
259. G. Muto and Y. Takata, *Bunseki Kagaku* 26, 407-411 (1977).
260. Y. Takata, H. Miyagi, K. Hirota, and Y. Arikawa, *Bunseki Kagaku* 26, 752-757 (1977).
261. Y. Takata and G. Muto, *Bunseki Kagaku* 28, 15-20 (1979).
262. Y. Takata and K. Fujita, *J. Chromatogr.* 108, 255-263 (1975).
263. Y. Takata, F. Mizuniwa, and C. Maekoya, *Anal. Chem.* 51, 2337-2339 (1979).
264. L. R. Taylor and D. C. Johnson, *Anal. Chem.* 46, 262-266 (1974).
265. Y. Takata and G. Muto, *Nippon Kaisui Gakkaishi* 33, 84-92 (1979).
266. R. M. Cassidy and S. Elchuk, *Int. J. Envir. Anal. Chem.* 10, 287-294 (1981).
267. J. Y. Lewis, J. P. Zodda, E. Deutsch, and W. R. Heineman, *Anal. Chem.* 55, 708-713 (1983).
268. I. S. Krull, S. Braverman, C. S. Selavkra, F. Hochberg, and L. A. Sternson, *J. Chromatogr. Sci.* 21, 166-173 (1983).
269. S. J. Bannister, L. A. Sternson, and A. J. Repta, *J. Chromatogr.* 273, 301-318 (1983).
270. W. A. MacCrehan, R. A. Durst, and J. M. Bellama, *Anal. Lett.* 10, 1175-1188 (1977).
271. W. A. MacCrehan and R. A. Durst, *Anal. Chem.* 50, 2108-2112 (1978).
272. W. A. MacCrehan, R. A. Durst, and J. M. Bellama, in *Trace Organic Analysis: A New Frontier in Analytical Chemistry*, NBS Special Publication No. 519, U. S. National Bureau of Standards, Washington DC, 1979, pp. 57-63.

273. W. A. MacCrehan, *Anal. Chem.* 53, 74-77 (1981).
274. K. Fuwa, M. Haraguchi, M. Morita, and J. C. Van Loon, *Bunko Kenkyu* 31, 289-305 (1982).
275. H. Sun, *Fenxi Huaxue* 10, 117-124 (1982).
276. D. R. Jones and S. E. Manahan, *Anal. Lett.* 6, 745-753 (1973).
277. D. R. Jones and S. E. Manahan, *Anal. Lett.* 8, 569-574 (1975).
278. D. R. Jones and S. E. Manahan, *Anal. Chem.* 48, 7-10 (1976).
279. D. R. Jones and S. E. Manahan, *Anal. Chem.* 48, 502-505 (1976).
280. W. Fanasaka, T. Hanai, and K. Fujimara, *J. Chromatogr. Sci.* 12, 517-529 (1974).
281. R. M. Cassidy, M. T. Hurteau, J. P. Mislan, and R. W. Ashley, *J. Chromatogr. Sci.* 14, 444-447 (1976).
282. C. Botre, F. Cacace, and R. Cozzani, *Anal. Lett.* 9, 825-830 (1976).
283. F. J. Fernandez, *Perkin-Elmer Chromatogr. Newslett.* 5(2), 17-21 (1977).
284. J. Treit, J. S. Nielsen, B. Kratochvil, and F. F. Cantwell, *Anal. Chem.* 55, 1650-1653 (1983).
285. J. A. Koropchak and G. N. Coleman, *Anal. Chem.* 52, 1252-1255 (1980).
286. D. Naranjit, Y. Thomassen, and J. C. Van Loon, *Anal. Chim. Acta* 110, 307-312 (1979).
287. A. A. Grabinski, *Anal. Chem.* 53, 966-968 (1981).
288. Y. Chang, L. A. Sternson, and A. J. Repta, *Anal. Lett.* B11, 449-459 (1978).
290. C. M. Riley, L. A. Sternson, and A. J. Repta, *J. Chromatogr.* 217, 405-420 (1981).
291. A. R. Tills and B. J. Alloway, *J. Soil Sci.* 34, 769-781 (1983).
292. D. T. Burns, F. Glockling, and M. Harriott, *Analyst* 106, 921-930 (1981).
293. G. R. Ricci, L. S. Shepard, G. Colovos, and N. E. Hester, *Anal. Chem.* 53, 610-613 (1981).
294. F. E. Brinckman, W. R. Blair, K. L. Jewett, and W. P. Iverson, *J. Chromatogr. Sci.* 15, 493-503 (1977).
295. N. Kahn and J. C. Van Loon, *J. Liq. Chromatogr.* 2, 23-36 (1979).
296. B. W. Renoe, C. E. Shideler, and J. Savory, *Clin. Chem.* 27, 1546-1550 (1981).
297. R. H. Fish and F. E. Brinckman, *Preprint, ACS Div. Fuel Chem.* 28(3), 177-180 (1983).
298. R. H. Fish, *ACS Symp. Ser.* 230, 477-492 (1983).
299. R. H. Fish, R. S. Tannous, W. Walker, C. S. Weiss, and F. E. Brinckman, *Chem. Commun.* 490-492 (1983).
300. F. E. Brinckman, C. S. Weiss, and R. H. Fish, in *Chemical and Geochemical Aspects of Fossil Energy Extraction* (T. F.

Yen et al., eds.), Ann Arbor Science, Ann Arbor, MI, 1983, pp. 197-224.
301. C. S. Weiss, E. J. Parks, and F. E. Brinckman, in *Arsenic: Industrial, Biomedical and Environmental Perspectives. Proceedings of the Arsenic Symposium* (W. H. Lederer and R. J. Fensterheim, eds.), Van Nostrand, New York, 1983, pp. 309-337.
302. E. J. Parks, F. E. Brinckman, and W. E. Blair, *J. Chromatogr. 185*, 563-572 (1979).
303. F. E. Brinckman, *J. Organometal. Chem. Chem. Libr. 12*, 343-376 (1981).
304. K. L. Jewett and F. E. Brinckman, *J. Chromatogr. Sci. 19*, 583-593 (1981).
305. W. R. Blair, G. J. Olson, F. E. Brinckman, and W. P. Iverson, *Microb. Ecol. 8*, 241-251 (1982).
306. E. J. Parks, R. B. Johannesen, and F. E. Brinckman, *J. Chromatogr. 255*, 439-454 (1983).
307. R. H. Fish and J. J. Komlenic, *Anal. Chem. 56*, 510-517 (1984).
308. K. L. Jewett and F. E. Brinckman, in *Liquid Chromatography Detectors* (T. M. Vickrey, ed.), Marcel Dekker, New York, 1983, pp. 205-241.
309. P. Titarelli and A. Mascherpa, *Anal. Chem. 53*, 1466-1469 (1981).
310. H. Koizumi, R. D. McLaughlin, and T. Hadeishi, *Anal. Chem. 51*, 387-392 (1979).
311. T. M. Vickrey, H. E. Howell, and M. T. Paradise, *Anal. Chem. 51*, 1880-1883 (1979).
312. D. Chakraborti, D. C. Hillman, K. J. Irgolic, and R. A. Zingaro, *J. Chromatogr. 249*, 81-92 (1982).
313. R. A. Stockton and K. J. Irgolic, *Int. J. Envir. Anal. Chem. 6*, 313-319 (1979).
314. F. E. Brinckman, K. L. Jewett, W. P. Iverson, K. J. Irgolic, K. C. Ehrhardt, and R. A. Stockton, *J. Chromatogr. 191*, 31-46 (1980).
315. E. A. Woolson and N. Aharonson, *J. Assoc. Off. Anal. Chem. 63*, 523-528 (1980).
316. R. Iadevia, N. Aharonson, and E. A. Woolson, *J. Assoc. Off. Anal. Chem. 63*, 742-746 (1980).
317. J. C. Van Loon, J. Lichwa, and B. Radziuk, *J. Chromatogr. 136*, 301-305 (1977).
318. J. C. Van Loon, *Anal. Chem. 51*, 1139A-1150A (1979).
319. J. C. Van Loon, *Am. Lab. 13*(5), 47-53 (1981).
320. D. D. Siemer, P. Koteel, D. T. Haworth, W. J. Taraszewski, and S. R. Lawson, *Anal. Chem. 51*, 575-579 (1979).
321. D. J. Freed, *Anal. Chem. 47*, 186-187 (1975).

322. I. S. Krull and S. Jordan, *Am. Lab. 12*(10), 21-33 (1980).
323. D. M. Fraley, D. Yates, and S. E. Manahan, *Anal. Chem. 51*, 2225-2229 (1979).
324. D. M. Fraley, D. A. Yates, S. E. Manahan, D. Stalling, and J. Petty, *Appl. Spectrosc. 35*, 525-531 (1981).
325. M. Morita, T. Uehiro, and K. Fuwa, *Anal. Chem. 52*, 349-351 (1980).
326. M. Morita, T. Uehiro, and K. Fuwa, *Anal. Chem. 53*, 1806-1808 (1981).
327. M. Morita and T. Uehiro, *Anal. Chem. 53*, 1997-2000 (1981).
328. L. T. Taylor, D. W. Hausler, and A. M. Squires, *Science 213*, 644-646 (1981).
329. D. W. Hausler and L. T. Taylor, *Anal. Chem. 53*, 1227-1231 (1981).
330. D. W. Hausler and L. T. Taylor, *Anal. Chem. 53*, 1223-1227 (1981).
331. K. Yoshida and H. Haraguchi, *Anal. Chem. 56*, 2580-2585 (1984).
332. K. Jinno and H. Tsuchida, *Anal. Lett. 15*, 427-437 (1982).
333. M. Ibrahim, W. Gilbert, and J. A. Caruso, *J. Chromatogr. Sci. 22*, 111-115 (1984).
334. J. P. McCarthy, J. A. Caruso, and F. L. Fricke, *J. Chromatogr. Sci. 21*, 389-393 (1983).
335. I. S. Krull, K. W. Panaro, and L. L. Gershman, *J. Chromatogr. Sci. 21*, 460-462 (1983).
336. P. C. Uden and I. D. Bigley, *Anal. Chim. Acta 94*, 29-34 (1977).
337. P. C. Uden, I. E. Bigley, and F. H. Walters, *Anal. Chim. Acta 100*, 555-561 (1978).
338. P. C. Uden, B. D. Quimby, R. M. Barnes, and P. C. Eliott, *Anal. Chim. Acta 101*, 99-109 (1978).
339. K. Tajima, M. Fujita, F. Kai, and M. Takamatsu, *J. Chromatogr. Sci. 22*, 244-248 (1984).
340. M. D. Seymour, J. P. Sickafoose, and J. S. Fritz, *Anal. Chem. 43*, 1734-1737 (1971).
341. J. S. Fritz and J. P. Sickafoose, *Talanta 19*, 1573-1579 (1972).
342. M. D. Seymour and J. S. Fritz, *Anal. Chem. 45*, 1394-1399 (1973).
343. M. D. Seymour and J. S. Fritz, *Anal. Chem. 45*, 1632-1636 (1973).
344. J. S. Fritz and L. Goodkin, *Anal. Chem. 46*, 959-962 (1974).
345. K. Kawazu and H. Kakiyama, *J. Chromatogr. 151*, 339-345 (1978).
346. L. Goodkin, M. D. Seymour, and J. S. Fritz, *Talanta 22*, 245-251 (1975).
347. M. Zenki, *Anal. Chem. 53*, 968-971 (1981).

348. J. R. Beckett and D. A. Nelson, *Anal. Chem. 53*, 909-910 (1981).
349. L. N. Mackey, P. A. Rodriguez, and F. B. Schroeder, *J. Chromatogr. 208*, 1-8 (1981).
350. M. J. Kessler, *Am. Lab. 14*(8), 52-63 (1982).
351. D. Ishii, A. Hirose, and Y. Iwasaki, *J. Chromatogr. 157*, 43-50 (1978).
352. D. Ishii, A. Hirose, and Y. Iwasaki, *J. Radioanal. Chem. 46*, 41-49 (1978).
353. D. Ishii, A. Hirose, and I. Horiuchi, *J. Radioanal. Chem. 45*, 7-14 (1978).
354. J. F. K. Huber and A. M. Van Urk-Schoen, *Anal. Chim. Acta 58*, 395-409 (1972).
355. W. Smulek and W. Lada, *Radiochem. Radioanal. Lett. 30*, 199-207 (1977).
356. W. Lada and W. Smulek, *Radiochem. Radioanal. Lett. 34*, 41-49 (1978).
357. W. Smulek and W. Lada, *J. Radioanal. Chem. 50*, 169-178 (1979).
358. E. P. Horwitz and C. A. A. Bloomquist, *J. Inorg. Nucl. Chem. 34*, 3851-3871 (1972).
359. E. P. Horwitz and C. A. A. Bloomquist, *J. Chromatogr. Sci. 12*, 11-22 (1974).
360. E. P. Horwitz and C. A. A. Bloomquist, *J. Chromatogr. Sci. 12*, 200-205 (1974).
361. E. P. Horwitz and C. A. A. Bloomquist, *J. Inorg. Nucl. Chem. 37*, 425-434 (1975).
362. E. P. Horwitz, W. H. Delphin, C. A. A. Bloomquist, and G. F. Vandergrift, *J. Chromatogr. 125*, 203-218 (1976).
363. S. M. Qaim, *Radiochem. Radioanal. Lett. 25*, 335-344 (1976).
364. E. P. Horwitz, C. A. A. Bloomquist, and W. H. Delphin, *J. Chromatogr. Sci. 15*, 41-46 (1977).
365. N. R. Larsen and W. B. Pedersen, *J. Radioanal. Chem. 45*, 135-140 (1978).
366. M. Schaedel, W. Bruchle, B. Haefner, J. U. Kratz, W. Schorstein, N. Trautmann, and G. Hermann, *Radiochim. Acta 25*, 111-117 (1978).
367. S. M. Qaim, H. Ollig, and G. Blessing, *Radiochim. Acta 26*, 59-62 (1979).
368. N. R. Larsen, *J. Radioanal. Chem. 52*, 85-91 (1979).
369. J. G. van Raaphorst and H. Haremaker, *J. Radioanal. Chem. 53*, 71-80 (1979).
370. A. Billon, *J. Radioanal. Chem. 51*, 297-305 (1979).
371. S. H-Y. Wong, *Adv. Chromatogr. 19*, 1-32 (1981).
372. T. C. Pinkerton, W. R. Heineman, and E. Deutch, *Anal. Chem. 52*, 1106-1110 (1980).

373. R. E. Needham and M. F. Delaney, *Anal. Chem.* 55, 148-150 (1983).
374. R. E. Needham and M. F. Delaney, *LC Liq. Chromatogr. HPLC Mag.* 2, 760-765 (1984).
375. Y. Yamamoto, M. Yamamoto, and S. Ebisui, *Anal. Lett.* 6, 451-460 (1973).
376. K. Saitoh, M. Saitoh, and N. Suzuki, *J. Chromatogr.* 92, 291-297 (1974).
377. K. Saitoh and N. Suzuki, *J. Chromatogr.* 109, 333-339 (1975).
378. N. Suzuki and K. Saitoh, *Bull. Chem. Soc. Jpn.* 50, 2907-2910 (1977).
379. N. Suzuki, K. Saitoh, and M. Shibukawa, *J. Chromatogr.* 138, 79-87 (1977).
380. H. Noda, K. Saitoh, and N. Suzuki, *Chromatographia* 14, 189-191 (1981).
381. H. Noda, K. Saitoh, and N. Suzuki, *J. Chromatogr.* 168, 250-254 (1979).
382. M. Saito, R. Kuroda, and M. Shibukawa, *Anal. Chem.* 55, 1025-1029 (1983).
383. M. K. L. Bicking, *Anal. Chem.* 56, 2671-2673 (1984).
384. N. Suzuki, J. Suzuki, and K. Saitoh, *J. Chromatogr.* 177, 166-169 (1979).
385. F. Gasparrini, D. Misti, G. Natile, and B. Galli, *J. Chromatogr.* 161, 356-359 (1978).
386. K. A. M. Creber and J. K. S. Wan, *J. Am. Chem. Soc.* 103, 2101-2102 (1981).
387. D. E. Henderson, R. Chaffee, and F. P. Novak, *J. Chromatogr. Sci.* 19, 79-83 (1981).
388. R. M. Cassidy and S. Elchuk, *J. Chromatogr. Sci.* 19, 503-507 (1981).
389. K. Dorfner, *Ion Exchangers, Properties and Applications*, Ann Arbor Science, Ann Arbor, MI, 1972.
390. K. A. Kraus, in *Trace Analysis* (J. H. Yoe and H. J. Koch, eds.), Wiley, New York, 1957.
391. C. D. Scott, in *Modern Practice of Liquid Chromatography* (J. J. Kirkland, ed.), Wiley, New York, 1971.
392. A. K. De, *Separation of Heavy Metals*, Pergamon, New York, 1961.
393. J. Inczedy, *Analytical Applications of Ion Exchangers*, Pergamon, New York, 1966.
394. W. Rieman, III and H. F. Walton, *Ion Exchange in Analytical Chemistry*, Pergamon, New York, 1970.
395. O. Samuelson, *Ion Exchange Separations in Analytical Chemistry*, Wiley, New York, 1963.
396. J. Schubert, *Ion Exchange: Theory and Practice* (F. C. Nachod, ed.), Academic Press, New York, 1949, pp. 167-222.

397. J. A. Marinsky and Y. Marcus, eds., *Ion Exchange and Solvent Extraction: A Series of Advances*, Marcel Dekker, New York.
398. H. F. Walton in *Chromatography — Fundamentals and Applications of Chromatographic and Electrophoretic Methods* (E. Heftmann, ed.), Elsevier, Amsterdam, 1983, pp. A225-A255.
399. J. S. Fritz and J. N. Story, *Anal. Chem.* 46, 825-829 (1974).
400. J. S. Fritz, *Pure Appl. Chem.* 49, 1547-1554 (1977).
401. R. M. Cassidy and S. Elchuk, *J. Liq. Chrom.* 4, 379-398 (1981).
402. R. M. Cassidy and S. Elchuk, *Anal. Chem.* 51, 1434-1438 (1979).
403. R. W. Slingsby and J. M. Rivello, *LC Liq. Chromatogr. HPLC Mag.* 1, 354-356 (1983).
404. R. D. Rocklin, *LC Liq. Chromatogr. HPLC Mag.* 1, 504-507 (1983).
405. R. D. Rocklin, *Anal. Chem.* 56, 1959-1962 (1984).
406. R. Eberhardt, H. Lehner, and K. Schlogl, *Monatsh. Chem.* 104, 1409-1420 (1973).
407. R. Eberhardt, G. Glotzmann, H. Lehner, and K. Schlogl, *Tetrahed. Lett.* 49/50, 4365-4368 (1974).
408. D. T. Haworth and T. Liu, *J. Chem. Educ.* 53, 730-731 (1976).
409. D. T. Haworth and T. Liu, *J. Chromatogr. Sci.* 14, 519-520 (1976).
410. D. G. Gresham, C. P. Lillya, P. C. Uden, and F. H. Walters, *J. Organometal Chem.* 142, 123-131 (1977).
411. A. Pryde, *J. Chromatogr.* 152, 123-129 (1978).
412. Z. Plzak, J. Plesek, and B. Stibr, *J. Chromatogr.* 168, 280-283 (1979).
413. H. Veening, J. M. Greenwood, W. H. Shanks, and B. R. Willeford, *Chem. Commun.* 1305 (1969).
414. J. M. Greenwood, H. Veening, and B. R. Willeford, *J. Organometal Chem.* 38, 345-348 (1972).
415. S. A. Gardner, R. J. Seyler, H. Veening, and B. R. Willeford, *J. Organometal Chem.* 60, 271-278 (1973).
416. T. J. Cardwell and T. Caridi, *J. Chromatogr.* 288, 357-364 (1984).
417. J. R. Fox, W. L. Gladfelter, and G. L. Geoffroy, *Inorg. Chem.* 19, 2574-2578 (1980).
418. W. J. Evans and M. F. Hawthorne, *J. Chromatogr.* 88, 187-189 (1974).
419. C. H. Gast and J. C. Kraak, *Int. J. Envir. Anal. Chem.* 6, 297-312 (1979).
420. M. N. Bocharev, G. N. Bortnikov, N. P. Makarenko, L. P. Mairova, A. V. Kiselev, and Y. I. Yashin, *J. Chromatogr.* 170, 53-63 (1979).

421. A. Casoli, A. Mangia, G. Predieri, and E. Sappa, *J. Chromatogr.* 303, 404-411 (1984).
422. P. Heizmann and K. Ballschmitter, *J. Chromatogr.* 137, 153-163 (1977).
423. M. Moriyasu and Y. Hashimoto, *Anal. Lett.* A11, 593-602 (1978).
424. J. W. O'Laughlin and T. P. O'Brien, *Anal. Lett.* A11, 829-844 (1978).
425. E. Gaetani, C. Francesco, and A. Mangia, *Ann. Chim. (Rome)* 69, 181-187 (1979).
426. O. Liska, G. Guiochon, and H. Colin, *J. Chromatogr.* 171, 145-151 (1979).
427. O. Liska, J. Lehotay, E. Brandsteterova, and G. Guiochon, *J. Chromatogr.* 171, 153-159 (1979).
428. J. Lehotay, O. Liska, E. Brandsteterova, and G. Guiochon, *J. Chromatogr.* 172, 379-383 (1979).
429. O. Liska, J. Lehotay, E. Brandsteterova, G. Guiochon, and H. Colin, *J. Chromatogr.* 172, 384-387 (1979).
430. M. Moriyasu and Y. Hashimoto, *Chem. Lett.* 117-120 (1980).
431. M. Moriyasu and Y. Hashimoto, *Bull. Chem. Soc. Jpn.* 53, 3590-3595 (1980).
432. E. B. Edward-Inatimi and J. A. W. Dalziel, *Anal. Proc.* 17, 40-42 (1980).
433. E. B. Edward-Inatimi, *J. Chromatogr.* 256, 253-266 (1983).
434. M. Lohmuller, P. Heizmann, and R. Ballschmitter, *J. Chromatogr.* 137, 165-170 (1977).
435. T. Deguchi, R. Takeshita, I. Sanemesa, and H. Nogai, *J. Chromatogr.* 173, 271-279 (1979).
436. S. J. Bannister, L. A. Sternson, and A. J. Repta, *J. Chromatogr.* 173, 333-342 (1979).
437. H. Eggers and H. A. Russell, *Chromatographia* 17, 486-490 (1983).
438. C. A. Tollinche and T. H. Risby, *J. Chromatogr. Sci.* 16, 448-454 (1978).
439. S. V. Galushko, I. P. Shishkina, and Yu. I. Usatenko, *Zh. Anal. Khim.* 37, 1833-1836 (1982).
440. J. F. K. Huber, J. C. Kraak, and H. Veening, *Anal. Chem.* 44, 1544-1549 (1972).
441. P. Heizmann and K. Ballschmitter, *Z. Anal. Chem.* 266, 206-207 (1973).
442. P. C. Uden and F. H. Walters, *Anal. Chim. Acta* 79, 175-183 (1975).
443. C. S. Hambali and P. R. Haddad, *Chromatographia* 13, 633-634 (1980).
444. K. H. Konig, E. H. Ehmeke, G. Schneeweis, and B. Steinbrech, *Z. Anal. Chem.* 297, 411-413 (1979).

445. M. L. Richter and K. G. Rienits, *FEBS Lett.* 116, 211-216 (1980).
446. K. H. Koenig, G. Schneeweis, and B. Steinbrech, *Z. Anal. Chem.* 316, 13-15 (1983).
447. S. S. Minor and J. W. Everett, Jr., *Inorg. Chem.* 15, 1526-1530 (1976).
448. R. E. Graf and C. P. Lillya, *J. Organometal Chem.* 47, 413-416 (1973).
449. R. E. Graf and C. P. Lillya, *J. Organometal Chem.* 122, 377-391 (1976).
450. C. T. Enos, G. L. Geoffroy, and T. H. Risby, *J. Chromatogr. Sci.* 15, 83-84 (1977).
451. H. T. McKone, *J. Chem. Educ.* 57, 380-381 (1980).
452. C. H. Gast, F. Nooitgedacht, and J. C. Kraak, *J. Organometal Chem.* 184, 221-229 (1980).
453. E. D. Sternberg and K. P. C. Vollhardt, *J. Am. Chem. Soc.* 102, 4839-4841 (1980).
454. J. M. Huggins, J. A. King, Jr., K. P. C. Vollhardt, and M. J. Winter, *J. Organometal Chem.* 217, 105-118 (1981).
455. E. R. F. Gesing and K. P. C. Vollhardt, *J. Organometal Chem.* 217, 105-118 (1981).
456. D. A. Buckingham, *J. Chromatogr.* 313, 93-127 (1984).
457. C. H. Gast, J. C. Kraak, L. H. Staal, and K. Vrieze, *J. Organometal Chem.* 208, 225-238 (1981).
458. D. T. Burns, F. Glockling, and M. Harriott, *J. Chromatogr.* 200, 305-308 (1980).
459. G. L. Geoffroy and R. A. Epstein, *Inorg. Chem.* 16, 2795-2799 (1977).
460. G. Schwedt, *Z. Anal. Chem.* 288, 50-55 (1977).
461. G. Schwedt, *Chromatographia* 11, 145-148 (1978).
462. G. Schwedt and A. Schwartz, *J. Chromatogr.* 160, 309-312 (1978).
463. G. Schwedt, *Chromatographia* 12, 289-293 (1979).
464. G. Schwedt, *Z. Anal. Chem.* 295, 382-387 (1979).
465. T. Tande, J. E. Petterson, and T. Torgrimsen, *Chromatographia* 13, 607-610 (1980).
466. G. Drasch, L. V. Meyer, and G. Gauert, *Z. Anal. Chem.* 311, 695-696 (1982).
467. P. C. Uden, D. M. Parees, and F. H. Walters, *Anal. Lett.* 8, 795-805 (1975).
468. E. Gaetani, C. F. Laureri, A. Mangia, and G. Parolari, *Anal. Chem.* 48, 1725-1727 (1976).
469. Y. T. Shih and P. W. Carr, *Anal. Chim. Acta* 142, 55-62 (1982).
470. D. A. Roston, *Anal. Chem.* 56, 241-244 (1984).

471. K. Ohashi, S. Iwai, and M. Hariguchi, *Bunseki Kagaku 31*, E285-E290 (1982).
472. R. C. Gierira and P. W. Carr, *J. Chromatogr. Sci. 20*, 461-465 (1982).
473. A. Mangia, G. Parolari, E. Gaetani, and C. F. Laureri, *Anal. Chim. Acta 92*, 111-116 (1977).
474. G. L. Wheeler and P. F. Lott, *Microchem. J. 19*, 390-405 (1974).
475. G. Weber and G. Schwedt, *Z. Anal. Chem. 314*, 114-118 (1983).
476. G. Weber and G. Schwedt, *Z. Anal. Chem. 316*, 594-599 (1983).
477. B. W. Hoffmann and G. Schwedt, *Z. Anal. Chem. 316*, 623-633 (1983).
478. G. Schwedt and R. Budde, *Chromatographia 15*, 527-529 (1982).
479. E. B. Jessen, K. Taugbol, and T. Griebokk, *J. Chromatogr. 168*, 139-142 (1979).
480. C. H. Gast and J. C. Kraak, *J. Liq. Chromatogr. 4*, 765-783 (1981).
481. S. Igarashi, M. Nakano, and T. Yotsuyanagi, *Bunseki Kagaku 32*, 67-68 (1983).
482. D. W. Armstrong, W. DeMond, and B. P. Czech, *Anal. Chem. 57*, 481-484 (1985).
483. O. Szabadka, K. W. Burton, and J. Inczedy, *J. Chromatogr. 201*, 67-72 (1980).
484. H. Loewenschuss and G. Schmuckler, *Talanta 11*, 1399-1408 (1964).
485. K. Kawabuchi, M. Kanke, T. Muraoka, and M. Yamaguchi, *Bunseki Kagaku 25*, 213-218 (1976).
486. P. Figura and B. McDuffle, *Anal. Chem. 49*, 1950-1953 (1977).
487. C. Lee, N. B. Kim, I. Lee, and K. S. Chung, *Talanta 24*, 241-245 (1977).
488. W. H. Ficklin, *Anal. Lett. 15*, 865-871 (1982).
489. K. Brajter and I. Miazek, *Talanta 28*, 759-764 (1981).
490. R. E. van Grieken, C. M. Bresseleers, and B. M. Vanderborght, *Anal. Chem. 49*, 1326-1331 (1977).
491. E. M. Moyers and J. S. Fritz, *Anal. Chem. 49*, 418-423 (1977).
492. M. G. Gimpel and K. K. Unger, *Chromatographia 17*, 200-204 (1983).
493. D. E. Leyden and G. H. Luttrell, *Anal. Chem. 47*, 1612-1617 (1975).
494. R. Raja, *Am. Lab. 14*(7), 47,35-37 (1982).
495. E. Marton-Schmidt, J. Inczedy, Z. Laki, and O. Szabadka, *J. Chromatogr. 201*, 73-77 (1980).
496. D. E. Leyden, G. H. Luttrell, A. E. Sloan, and N. J. deAngelio, *Anal. Chim. Acta 84*, 97-108 (1976).
497. J. N. King and J. S. Fritz, *J. Chromatogr. 153*, 507-516 (1978).
498. G. M. Orf and J. S. Fritz, *Anal. Chem. 50*, 1328-1330 (1978).

499. C. Pohlandt and J. S. Fritz, *J. Chromatogr. 176*, 189-197 (1979).
500. R. J. Phillips and J. S. Fritz, *Anal. Chim. Acta 121*, 225-232 (1980).
501. J. P. Ghosh and J. R. Das, *Talanta 28*, 274-276 (1981).
502. J. L. Lundgren and A. A. Schilt, *Anal. Chem. 49*, 974-980 (1977).
503. D. W. Lee and M. Hallman, *Anal. Chem. 48*, 2214-2218 (1976).
504. M. B. Colella, S. Siggia, and R. M. Barnes, *Anal. Chem. 52*, 967-972 (1980).
505. M. B. Collela, S. Siggia, and R. M. Barnes, *Anal. Chem. 52*, 2347-2350 (1980).
506. I. I. Autokol'skaya, G. V. Myaesoedova, L. L. Bol'shakova, M. G. Ezermitskaya, M. P. Volymets, A. V. Karyakin, and S. B. Savvin, *Zh. Anal. Khim. 31*, 742-745 (1976).
507. T. Sheshadri and A. Kettrup, *Z. Anal. Chem. 296*, 247-252 (1979).
508. J. R. Parrish and R. Stevenson, *Anal. Chim. Acta 70*, 189-198 (1974).
509. J. R. Parrish, *Lab. Pract. 24*, 399-400 (1975).
510 J. R. Parrish, *Anal. Chem. 49*, 1189-1192 (1977).
511. J. R. Parrish, *Anal. Chem. 54*, 1890-1892 (1982).
512 J. M. Hill, *J. Chromatogr. 76*, 455-458 (1973).
513. K. F. Sugawara, H. H. Weetall, and G. D. Schucker, *Anal. Chem. 46*, 489-492 (1974).
514. R. Jezorek and H. Freiser, *Anal. Chem. 51*, 366-372 (1979).
515. F. Vernon and K. M. Nyo, *J. Inorg. Nucl. Chem. 40*, 887-891 (1978).
516. M. M. Guedes de Mota, F. G. Roemer, and B. Griepink, *Z. Anal. Chem. 287*, 19-22 (1977).
517. A. J. Bauman, H. H. Weetall, and N. Weliky, *Anal. Chem. 39*, 932-935 (1967).
518. Z. Slovak, S. Slovakova, and M. Smrz, *Anal. Chim. Acta 75*, 127-138 (1975).
519. B. M. Vanderborght and R. E. van Grieken, *Anal. Chem. 49*, 311-316 (1977).
520. H. Watanabe, K. Goto, S. Taguchi, J. W. McLaren, S. S. Berman, and D. S. Russell, *Anal. Chem. 53*, 738-739 (1981).
521. R. E. Sturgeon, S. S. Berman, and S. N. Willie, *Talanta 29*, 167-171 (1982).
522. I. I. Autokol'skaya, G. V. Myaesoedova, L. I. Bol'shakova, and O. P. Shvoeva, *J. Chromatogr. 102*, 287-291 (1974).
523. E. M. Moyers and J. S. Fritz, *Anal. Chem. 48*, 1117-1120 (1976).

524. R. J. Phillips and J. S. Fritz, *Anal. Chem.* **50**, 1504-1508 (1978).
525. J. N. King and J. S. Fritz, *Anal. Chem.* **57**, 1016-1020 (1985).
526. T. Braun and A. B. Farag, *Anal. Chim. Acta* **69**, 85-96 (1974).
527. M. Forster and K. H. Lieser, *Z. Anal. Chem.* **309**, 177-180 (1981).
528. M. Forster and K. H. Lieser, *Z. Anal. Chem.* **309**, 352-354 (1981).
529. L. A. Saari and W. R. Seitz, *Anal. Chem.* **56**, 810-812 (1984).
530. J. L. Hern and J. H. Strohl, *Anal. Chem.* **50**, 1954-1959 (1978).
531. J. W. Lin and G. E. Janauer, *Anal. Chim. Acta* **79**, 219-227 (1975).
532. T. Braun and A. B. Farag, *Anal. Chim. Acta* **65**, 115-126 (1973).
533. I. P. Alimarin, T. A. Bolshova, N. I. Ershova, and M. B. Polinskaya, *Zh. Anal. Khim.* **24**, 26-30 (1969).
534. M. Grote, A. Schwalk, and A. Kettrup, *Z. Anal. Chem.* **313**, 297-303 (1982).
535. H. Tanaka, M. Chikuma, and M. Nakayama, in *Analytical Techniques in Environmental Chemistry* (J. Albaiges, ed.), Pergamon, 1982, pp. 381-388.
536. C. J. Pedersen, *J. Am. Chem. Soc.* **89**, 2495-2496 (1967).
537. C. J. Pedersen, *J. Am. Chem. Soc.* **89**, 7017-7036 (1967).
538. G. W. Gokel, Abstract No. 286, paper presented at the 40th Southwest regional meeting of the American Chemical Society, Lubbock, TX, Dec 5-7, 1984.
539. P. K. Dasgupta, Unpublished studies, Texas Tech University, 1982.
540. E. Blasius, K.-P. Janzen, W. Klein, V. B. Nguyen, T. Nguyen-Tien, R. Pfeiffer, G. Scholten, H. Simon, H. Stockemer, and A. Touissant, *J. Chromatogr.* **201**, 147-166 (1980).
541. E. Blasius, K.-P. Janzen, W. Adrian, W. Klein, H. Klotz, H. Luxenburger, E. Mernke, V. B. Nguyen, T. Nguyen-Tien, R. Rausch, J. Stockemer, and A. Touissant, *Talanta* **27**, 127-141 (1980).
542. M. Igawa, K. Saito, J. Tsukamoto, and M. Tanaka, *Anal. Chem.* **53**, 1942-1944 (1981).
543. M. Nakajima, K. Kimura, and T. Shono, *Anal. Chem.* **55**, 463-467 (1983).
544. E. Blasius, K.-P. Janzen, W. Adrian, G. Klautke, R. Lorscheider, P.-G. Maurer, V. B. Nguyen, T. Nguyen, G. Scholten, and J. Strockenen, *Z. Anal. Chem.* **284**, 337-360 (1977).
545. E. Blasius, K.-P. Janzen, H. Luxenburger, V. B. Nguyen, H. Klotz, and J. Stockemer, *J. Chromatogr.* **167**, 307-320 (1978).

546. K. Kimura, M. Nakajima, and T. Shono, *Anal. Lett.* 13, 741-750 (1980).
547. M. Shen, S. Zhao, and M. Wang, *He Huaxue Yu Fangshe Huaxue* 4, 52-55 (1982).
548. E. Shchori, J. Jagur-Grodzinski, Z. Luz, and M. Shporer, *J. Am. Chem. Soc.* 93, 7133-7138 (1971).
549. T. Funck, F. Eggers, and E. Grell, *Chimia* 26, 637-644 (1972).
550. E. Shchori, J. Jagur-Grodzinski, and M. Shporer, *J. Am. Chem. Soc.* 95, 3842-3846 (1973).
551. H. Degani, *Biophys. Chem.* 6, 345-349 (1977).
552. L. J. Rodriguez, G. W. Liesegang, R. D. White, M. M. Farrow, N. Purdie, and E. M. Eyring, *J. Phys. Chem.* 81, 2118-2128 (1977).
553. J. Boquant, A. Delville, J. Grandjean, and P. Laszlo, *J. Am. Chem. Soc.* 104, 686-691 (1982).
554. E. Blasius and K.-P. Janzen, *Top. Current Chem.* 98, 163-189 (1981).
555. E. Blasius and K.-P. Janzen, *Pure Appl. Chem.* 54, 2115-2128 (1982).
556. R. B. King and P. R. Heckley, *J. Am. Chem. Soc.* 96, 3118-3123 (1974).
557. K. G. Heumann and H.-P. Schiefer, *Angew Chem. Int. Ed. Eng.* 19, 406-407 (1980).
558. K. Yagi and M. Sanchez, *Makromol. Chem. Rapid Comm.* 2, 311-315 (1981).
559. L. A. Fernando, M. L. Miles and L. H. Bowen, *Anal. Chem.* 52, 1115-1119 (1980).
560. P. Grossmann and W. Simon, *J. Chromatogr.* 235, 351-363 (1982).
561. W. H. Delphin and E. P. Horwitz, *Anal. Chem.* 50, 843-847 (1978).
562. P. K. Dasgupta, Unpublished studies, Texas Tech University, 1982.
563. T. Nakagawa, A. Shibukawa, and T. Uno, *J. Chromatogr.* 254, 27-34 (1983).
564. T. Nakagawa, A. Shibukawa, and H. Murata, *J. Chromatogr.* 280, 31-42 (1983).
565. D. W. Armstrong, A. Alak, K. Bui, W. DeMond, T. Ward, T. E. Riehl, and W. L. Hinze, *J. Inclusion Phenomena* 2, 533-545 (1985).
566. D. W. Armstrong and T. Ward, Unpublished studies, Texas Tech University, 1985.
567. S. Matsushita, Y. Toda, N. Baba, and K. Hosako, *J. Chromatogr.* 259, 459-464 (1983).
568. See, for example, R. L. Stevenson and K. Harrison, *Am. Lab.* 13(5), 76-81 (1981).

569. P. A. Asmus, C. Low, and M. Novotny, *J. Chromatogr. 119*, 25-32 (1976).
570. P. A. Asmus, C. Low, and M. Novotny, *J. Chromatogr. 123*, 109-118 (1976).
571. D. H. Saunders, R. A. Barford, P. Magidman, L. T. Olszewiski, and H. L. Rothbart, *Anal. Chem. 46*, 834-840 (1974).
572. S. D. Fazio, J. B. Crowther, and R. A. Hartwick, *Chromatographia 18*, 216-220 (1984).
573. R. B. Moore, III, J. E. Wilkerson, and C. R. Martin, *Anal. Chem. 56*, 2572-2575 (1984).
574. H. Small, *J. Inorg. Nucl. Chem. 18*, 232-244 (1961).
575. P. Hajos and J. Inczedy, *J. Chromatogr. 201*, 253-257 (1980).
576. J. S. Fritz and J. N. Story, *J. Chromatogr. 90*, 267-274 (1974).
577. H. W. Gibson and F. C. Bailey, *Macromolecules 13*, 34-41 (1980).
578. F. Doerscher, J. Klein, F. Pohl, and H. Widdecke, *Makromol. Chem. 184*, 1585-1596 (1983).
579. M. Caude and R. Rosset, *J. Chromatogr. Sci. 15*, 405-412 (1977).
580. D. J. Mackey, *J. Chromatogr. 242*, 275-287 (1982).
581. S. G. Weber and W. G. Tramposch, *Anal. Chem. 55*, 1771-1775 (1983).
582. R. L. Smith and D. J. Pietrzyk, *Anal. Chem. 56*, 610-614 (1984).
583. S. Siergiez and N. Danielson, *J. Chromatogr. Sci. 21*, 362-366 (1983).
584. O. Mikes, P. Strop, and J. Coupek, *J. Chromatogr. 153*, 23-36 (1978).
585. O. Mikes, P. Strop, J. Zbrozek, and J. Coupek, *J. Chromatogr. 180*, 17-30 (1979).
586. O. Mikes, P. Strop, M. Smrz, and J. Coupek, *J. Chromatogr. 192*, 159-172 (1980).
587. O. Mikes, P. Strop, Z. Hostomska, M. Smrz, J. Coupek, A. Frydrychova, and M. Bares, *J. Chromatogr. 261*, 363-379 (1983).
588. O. Mikes, P. Strop, Z. Hostomska, M. Smrz, S. Slovakova, and J. Coupek, *J. Chromatogr. 301*, 93-105 (1984).
589. D. P. Lee, *LC Liq. Chromatogr. HPLC Mag. 2*, 828-832 (1984).
590. D. T. Gjerde and J. S. Fritz, *J. Chromatogr. 176*, 199-206 (1979).
591. R. E. Barron and J. S. Fritz, *J. Chromatogr. 284*, 13-25 (1984).
592. R. E. Barron and J. S. Fritz, *J. Chromatogr. 316*, 201-210 (1984).
593. R. E. Barron and J. S. Fritz, *React. Polym. Ion Exch. Sorbents 1*, 215-226 (1983).

594. R. C. Buechele and D. J. Reuter, *J. Chromatogr. 240*, 502 507 (1982).
595. H. Sato, *Bunseki Kagaku 31*, 97-99 (1982).
596. M. J. Van Os, J. Slanina, C. L. De Ligny, W. E. Hammers, and J. Agterdenbos, *Anal. Chim. Acta 144*, 73-82 (1982).
597. J. W. Wimberley, *Anal. Chem. 53*, 1709-1710 (1981).
598. E. Schulek, Z. Remport-Horvath, A. Laszity, and E. Koros, *Magy. Kem. Foly. 75*, 58-62 (1969).
599. E. Schulek, Z. Remport-Horvath, A. Laszity, and E. Koros, *Talanta 16*, 323-329 (1969).
600. S.-Y. Shiao, J. S. Johnson, Jr., G. Mohiuddin, W. Y. Hata, J. S. Tolan, and W. W. Doerr, *J. Chromatogr. 303*, 13-28 (1984).
601. J. Reuter and G. Schwedt, *Z. Anal. Chem. 311*, 112-115 (1982).
602. I. M. Kolthoff and P. J. Elving, eds., *A Treatise on Analytical Chemistry*, Part II: *Analytical Chemistry of the Elements*, Vols. 1-17, Wiley Interscience, New York, 1961-1980.
603. K. L. Cheng, K. Ueno, and T. Imamura, *Handbook of Organic Analytical Reagents*, CRC Press, Boca Raton, FL, 1982.
604. J. Strzelbicki and R. A. Bartsch, *Anal. Chem. 53*, 1894-1899 (1981).
605. J. Strzelbicki and R. A. Bartsch, *Anal. Chem. 53*, 2247-2250 (1981).
606. J. Strzelbicki and R. A. Bartsch, *Anal. Chem. 53*, 2251-2253 (1981).
607. W. A. Charewicz, G. S. Heo, and R. A. Bartsch, *Anal. Chem. 54*, 2094-2097 (1982).
608. W. A. Charewicz and R. A. Bartsch, *Anal. Chem. 54*, 2300-2306 (1982).
609. J. F. Koszuk, B. P. Czech, W. Walkowiak, D. A. Babb, and R. A. Bartsch, *Chem. Commun.* 1504-1505 (1984).
610. W. Walkowiak, L. E. Stewart, and R. A. Bartsch, Abstract No. 2, Paper presented at the 40th Southwest regional meeting of the American Chemical Society, Lubbock, TX, Dec 5-7, 1984.
611. H. Watanabe, in *Solution Behavior of Surfactants*, Vol. 2 (K. L. Mittal and E. J. Fendler, eds.), Plenum, New York, 1982.
612. J. Miura, H. Ishii, and H. Watanabe, *Bunseki Kagaku 25*, 808-809 (1976).
613. H. Ishii, J. Miura, and H. Watanabe, *Bunseki Kagaku 26*, 252-256 (1977).
614. H. Tanaka and H. Watanabe, *Bunseki Kagaku 27*, 189-191 (1978).
615. H. Watanabe, N. Yamaguchi, and H. Tanaka, *Bunseki Kagaku 28*, 367-370 (1979).
616. H. Watanabe, K. Tachikawa, and H. Ohmori, *Bunseki Kagaku 31*, 471-473 (1982).

617. H. Watanabe and N. Yamaguchi, *Bunseki Kagaku 33*, E211-E214 (1984).
618. H. Hoshino, T. Saitoh, H. Taketomi, T. Yotsuyanagi, H. Watanabe, and K. Tachikawa, *Anal. Chim. Acta 147*, 339-345 (1983).
619. R. P. W. Scott, in *Small Bore Liquid Chromatographic Columns: Their Properties and Uses* (R. P. W. Scott, ed.), Wiley, New York, 1984, pp. 194-198.
620. J. J. Kirkland, *Analyst 99*, 859-885 (1974).
621. P. R. Haddad and A. L. Heckenberg, *J. Chromatogr. 299*, 301-305 (1984).
622. T. Jupille, D. Burge, and D. Togami, *Chromatographia 16*, 312-316 (1982).
623. M. C. Harvey and S. D. Stearns, in *Liquid Chromatography in Environmental Analysis* (J. F. Lawrence, ed.), Humana, Clifton, NJ, 1984, pp. 301-340.
624. P. R. Haddad and A. L. Heckenberg, *J. Chromatogr. 318*, 279-288 (1985).
625. M. M. Plechaty, *LC Liq. Chromatogr. HPLC Mag. 2*, 684-688 (1984).
626. M. C. Harvey and S. D. Stearns, *Am. Lab. 14*(5), 68-74 (1982).
627. T. B. Hoover and G. D. Yager, *Anal. Chem. 56*, 221-225 (1984).
628. C.-C. Wan, S. Chiang, and A. Corsini, *Anal. Chem. 57*, 719-723 (1985).
629. D. D. Siemer, *Anal. Chem. 52*, 1874-1876 (1980).
630. K. M. Roberts, D. T. Gjerde, and J. S. Fritz, *Anal. Chem. 53*, 1691-1695 (1981).
631. R. M. Cassidy and S. Elchuk, *J. Chromatogr. Sci. 18*, 217-223 (1980).
632. F. R. Nordmeyer and L. D. Hansen, *Anal. Chem. 54*, 2605-2607 (1982).
633. J. E. DiNuzio and M. Jubara, *Anal. Chem. 55*, 1013-1016 (1983).
634. J. A. Cox and J. R. Litwinski, *Anal. Chem.* 2016 (1983).
635. J. A. Cox and J. E. DiNuzio, *Anal. Chem. 49*, 1272-1275 (1977).
636. A. H. M. T. Scholten and R. W. Frei, *J. Chromatogr. 176*, 349-357 (1979).
637. A. H. M. T. Scholten and R. W. Frei, *J. Chromatogr. 199*, 239-248 (1980).
638. Tracor Analytical Instruments, Austin, TX.
639. N. Nimura and K. Ishida, *J. Chromatogr. 221*, 249-255 (1980).
640. J. P. Ivey, *J. Chromatogr. 281*, 314-318 (1983).
641. I. S. Krull, K-H. Xie, S. Colgan, U. Neue, T. Izod, R. King, and B. Bidlingmeyer, *J. Liq. Chromatogr. 6*, 605-626 (1983).
642. J. F. Studebaker, *J. Chromatogr. 185*, 497-503 (1979).
643. J. F. Studebaker, S. A. Slocum, and E. L. Lewis, *Anal. Chem. 50*, 1500-1503 (1978).

644. S. H. Lee and L. R. Field, *Anal. Chem. 56*, 2647-2653 (1984).
645. C. E. Werkhoven-Goewie, W. M. A. Niessen, U. A. Th. Brinkmann, and R. W. Frei, *J. Chromatogr. 203*, 165-172 (1981).
646. E. P. Lankmayr, B. Maichin, G. Knapp, and F. Nachtmann, *J. Chromatogr. 224*, 239-248 (1981).
647. P. T. Kissinger, K. Bratin, G. C. Davis, and L. A. Pachla, *J. Chromatogr. Sci. 17*, 137-146 (1979).
648. R. W. Frei and A. H. M. T. Scholten, *J. Chromatogr. Sci. 17*, 152-160 (1979).
649. G. Schwedt, *Angew Chem. 91*, 192-198 (1979).
650. I. S. Krull and E. P. Lankmayr, *Am. Lab. 14*(5), 18-32 (1982).
651. S. J. van der Wal, *J. Liq. Chromatogr. 6*, 37-59 (1983).
652. C. J. Little, J. A. Whatley, and A. D. Dale, *J. Chromatogr. 171*, 63-72 (1979).
653. J. C. Gfeller, G. Frey, and R. W. Frei, *J. Chromatogr. 142*, 277-281 (1977).
654. A. H. M. T. Scholten, U. A. Th. Brinkmann, and R. W. Frei, *J. Chromatogr. 205*, 229-237 (1981).
655. W. J. Smythe, S. L. Bellinger, H. G. Diebler, and R. Dannewitz, U.S. Patent 3,026,615, July 30, 1974.
656. W. Vogt and S. Wilhelm, German Patent 2 537 177, 1975.
657. W. Vogt, S. L. Braun, S. Wilhelm, and H. Schwab, *Anal. Chem. 54*, 596-598 (1982).
658. K. Ogata, K. Taguchi, and T. Imanari, *Anal. Chem. 54*, 2127-2129 (1982).
659. E. D. Katz and R. P. W. Scott, *J. Chromatogr. 286*, 169-175 (1983).
660. G. J. Schmidt and R. P. W. Scott, cited in K. Ogan, in *Smallbore Liquid Chromatography Columns: Their Properties and Uses* (R. P. W. Scott, ed.), Wiley, New York, 1984, pp. 104-105.
661. V. W. Uhl and J. B. Gray, eds., *Mixing*, Vols. 1 and 2, Academic Press, New York, 1966, 1967.
662. K. W. Stahl, G. Schaefer, and W. Lamprecht, *J. Chromatogr. Sci. 10*, 95-102 (1972).
663. T. Takeuchi and D. Ishii, *J. Chromatogr. 239*, 633-641 (1982).
664. P. Kucera and H. Umagat, *J. Chromatogr. 255*, 563-579 (1983).
665. Ted Pella Inc., Tustin, CA 92680.
666. R. W. Frei, L. Michel, and W. Santi, *J. Chromatogr. 126*, 665-677 (1976).
667. R. W. Frei, L. Michel, and W. Santi, *J. Chromatogr. 142*, 261-270 (1977).
668. A. H. M. T. Scholten, U. A. Th. Brinkmann, and R. W. Frei, *J. Chromatogr. 218*, 3-13 (1981).
669. R. P. W. Scott and P. Kucera, *J. Chromatogr. 185*, 27-41 (1979).
670. H. Yonekawa, German Patent 3,227,884, February 10, 1983.

671. H. E. Schwartz, B. L. Karger, and P. Kucera, *Anal. Chem.* 55, 1752-1760 (1983).
672. G. Busch, E. Heidenreich, V. Kudwig, and P. Nitsche, French Patent 2,462,926, February 20, 1981.
673. J. C. Davis, U.S. Patent 4,448,691, May 15, 1984.
674. D. P. Peterson and J. C. Davis, U.S. Patent 4,451,374, May 29, 1984.
675. J. C. Davis and D. P. Peterson, *Anal. Chem.* 57, 768-771 (1985).
676. P. K. Dasgupta, K. DeCesare, and J. C. Ullrey, *Anal. Chem.* 52, 1912-1922 (1980).
677. P. K. Dasgupta and V. K. Gupta, *Environ. Sci. Technol.* 20, 524-526 (1986).
678. J. M. Riviello and C. A. Pohl, Paper No. 506, 35th Pittsburg Conference on Analytical Chemistry and Applied Spectroscopy, March, 1984.
679. P. K. Dasgupta, *J. Liq. Chromatogr.* 7, 2367-2382 (1984).
680. H. Hwang and P. K. Dasgupta, *Anal. Chem.* 58, 1521-1524 (1986).
681. J. R. Jezorek and H. Freiser, *Anal. Chem.* 51, 373-376 (1979).
682. R. G. Anderson and G. Nickless, *Analyst* 92, 207-238 (1967).
683. R. M. Cassidy and S. Elchuk, *J. Liq. Chromatogr.* 4, 379-398 (1981).
684. T. Hayakawa, M. Moriyasu, and Y. Hashimoto, *Bunseki Kagaku* 32, 130-133 (1983).
685. See, for example, Y. Yang, *Anal. Chem.* 56, 2336-2338 (1984).
686. A. Hirose, Y. Iwasaki, I. Iwata, K. Ueda, and D. Ishii, *J. High Resol. Chromatogr. Chromatogr. Commun.* 4, 530-531 (1981).
687. P. Jones, P. J. Hobbs, and L. Ebdon, *Anal. Chim. Acta* 149, 39-46 (1983).
688. H. Yamaguchi, T. Nakamura, Y. Hirai, and S. Ohashi, *J. Chromatogr.* 172, 131-140 (1979).
689. N. Yoza, K. Ito, and S. Ohashi, *J. Chromatogr.* 196, 471-480 (1980).
690. Y. Hirai, N. Yoza, and S. Ohashi, *Anal. Chim. Acta* 115, 269-277 (1980).
691. Y. Hirai, N. Yoza, and S. Ohashi, *J. Chromatogr.* 206, 501-509 (1981).
692. T. Toida, K. Ogata, S. Tanabe, and T. Imanari, *Bunseki Kagaku* 29, 764-768 (1980).
693. T. Imanari, S. Tanabe, T. Toida, and T. Kawanishi, *J. Chromatogr.* 250, 55-61 (1982).
694. A. W. Fitchett and A. Woodruff, *LC Liq. Chromatogr. HPLC Mag.* 1, 48-49 (1983).
695. B. Walker and P. K. Dasgupta, Unpublished studies, Texas Tech University, 1984.
696. G. E. F. Lundell, *Ind. Eng. Chem.* 5, 1-15 (1933).

697. D. A. Phillips, K. Soroka, R. S. Vithanage, B. Walker, and P. K. Dasgupta, submitted for publication in *Anal. Chem.* (1984).
698. M. P. Neary, R. Seitz, and D. M. Hercules, *Anal. Lett.* 7, 583-590 (1974).
699. K. W. Sigvardson, J. M. Kennish, and J. W. Birks, *Anal. Chem.* 56, 1096-1102 (1984).
700. J. W. Jorgenson, S. L. Smith, and M. Novotny, *J. Chromatogr.* 142, 233-240 (1977).
701. T. Sunden, M. Lindgren, A. Cedregren, and D. D. Siemer, *Anal. Chem.* 55, 2-4 (1983).
702. S. Glasstone, *An Introduction to Electrochemistry*, Van Nostrand, Princeton, 1962, pp. 101-103.
703. J. P. Ivey, *J. Chromatogr.* 287, 128-132 (1984).
704. W. J. Svirbely and S. Peterson, *J. Am. Chem. Soc.* 65, 166-170 (1943).
705. W. J. Blaedel, T. J. Haupert, and M. A. Evenson, *Anal. Chem.* 41, 583-590 (1969).
706. H. Engelhardt and U. D. Neue, *Chromatographia* 15, 403-408 (1982).
707. D. L. Duval, J. S. Fritz, and D. T. Gjerde, *Anal. Chem.* 54, 830-832 (1982).
708. Yu. A. Zolotov, O. A. Shpigun, and L. A. Bubchikova, *Z. Anal. Chem.* 316, 8-12 (1983).
709. W. Ishibashi, R. Kikuchi, and K. Yamamoto, *Bunseki Kagaku* 31, 207-211 (1982).
710. M. A. Mercurio-Cason, P. K. Dasgupta, D. W. Blakeley, and R. L. Johnson, *J. Memb. Sci.* 27, 31-40 (1986).
711. D. R. Jenke and G. K. Pagenkopf, *J. Chromatogr.* 269, 202-207 (1983).
712. H. Hoshino and T. Yotsuyganagi, *Anal. Chem.* 57, 625-628 (1985).
713. Z. Iskandarani and T. E. Miller, Jr., *Anal. Chem.* 57, 1591-1594 (1985).
714. S.-T. Mho and E. S. Yeung, *Anal. Chem.* 57, 2253-2256 (1985).
715. R. M. Cassidy, S. Elchuk, N. L. Elliott, L. W. Green, C. H. Knight, and B. M. Recoskie, *Anal. Chem.* 57, 1181-1185 (1985).
716. R. M. Cassidy, S. Elchuk, and P. K. Dasgupta, submitted for publication in *Anal. Chem.*

7
A Review of Ion Chromatography: A Bibliography

James G. Tarter
Department of Chemistry
North Texas State University
Denton, Texas

I. INTRODUCTION

This chapter is designed primarily to serve as a reference for ion chromatographic applications and procedures. It is divided into seven sections including (a) books and dissertations, (b) review articles, (c) applications of ion chromatography according to type of sample analyzed, (d) applications of ion chromatography according to anions analyzed, (e) applications of ion chromatography according to cations analyzed, (f) theoretical and instrumental considerations, and (g) a bibliography of ion chromatography.

The references listed in this chapter can be found in *Chemical Abstracts*. All of the references are of published work and can be located through the abstracts. Personal communications, work in progress, and works presented at professional meetings are not included. Works presented at professional meetings that led to subsequent publication in a reviewed journal are included. A large source of information that was not included is the volume of material prepared for distribution by the manufacturers of ion chromatographic equipment and supplies. Again, much of the initial work in this area has been published in journals and is referenced accordingly. The nature of scientific literature is such that it is often difficult to assign an article to only one group among the seven sections listed above. Consequently, many of the articles will be listed under several categories. A paper will be referenced under a particular heading when the paper appears to have a significant, but not necessarily sole, interest in that area according to the abstract.

Table 1. Primary Journals for Ion Chromatographic Literature

Journal	No. of Refs.
Top five	
Analytical Chemistry	102
Journal of Chromatography	47
Bunseki Kagaku	39
Analytica Chimica Acta	15
LC, Liquid Chromatography HPLC Magazine	14
Total of top five	217
Next four	
American Laboratory	8
Journal of Chromatographic Science	8
Analytical Letters	8
Journal of American Industrial Hygiene Association	7
Total of top nine	248

The arrangement of the references will hopefully provide an impressive demonstration of the applicability of ion chromatography in a wide area of chemical analysis settings. The reference list, by itself, should provide a strong literature source for the literature search which should precede all sound chemical investigations.

The ion chromatographic literature spans a wide selection of international journals. In fact, journal articles from over 152 different journals are listed at the end of this chapter. There is, however, a relatively limited selection of journals that have published multiple papers on ion chromatography. Table 1 provides a listing of the top five journals, ranked according to the number of journal articles printed, gleaned from the bibliography section of this chapter. These five journals, representing only 3.3% of the total number of journals, account for over 37.4% of the total number of articles listed here.

II. BOOKS AND DISSERTATIONS

A limited number of published books and dissertations deal primarily with ion chromatography. Two books [323,438] are the proceedings of conferences dealing with the ion chromatographic analysis of environmental pollutants. In the past few years, several books have been published which treat ion chromatography in a general sense and cover both theory and applications [120,466]. In addition, Ref. 57 is a general reference on ion chromatography published in Chinese, whereas Ref. 328 is a general reference in Japanese.

Numerous dissertations containing significant ion chromatographic theory and/or applications have been written. These dissertations include the theoretical aspects of plate considerations for ion chromatography [210], retention time considerations for anions [192], ion chromatographic resins [18,136], the use of atomic spectroscopy as a detection system for ion chromatography [91], and the separation and electrochemical detection of various divalent transition metals [184]. Two dissertations [101,513] are primarily geared to those interested in the application of ion chromatography to the analysis of geological and environmental samples.

III. REVIEW ARTICLES

A large number of review articles concerning various aspects of ion chromatography have been written in numerous languages. Some of these articles are of a general nature, dealing with a wide range of ion chromatographic principles and practices, whereas others review a more restricted area of ion chromatography.

A. General Reviews

General review articles written in English include the following: Refs. 1, 66, 67, 117, 224, 228-230, 280, 318, 322, 371, 379, 399, 400, 404-406, 411, 445, 459, 460, 465, 470, 529, 534. General reviews written in ten languages, other than English, include those written in Italian [22,84], Polish [29], German [143, 206, 207, 223, 446, 473, 545-548], French [208], Japanese [158, 337, 345, 346, 418, 419, 522], Chinese [314,383], Russian [451], Norwegian [487], Finnish [530], and Spanish [50].

B. Specific Reviews

English

A very large body of review material is available on many specific areas of ion chromatographic theory and applications. The areas of applications include water analysis [59, 76, 95, 107, 359, 388, 389, 393, 395, 443, 525], environmental analysis [225, 436, 437, 455, 550], clinical applications [173], biochemical applications [402], food and beverage analysis [540], industrial uses [274, 298, 299, 301], and explosives [397]. In addition, there are review articles dealing with anion analysis [151,372], cation analysis [407,458], ion chromatography exclusion [300], and single-column ion chromatography [226]. Also available are reviews of various instrumental components such as columns [23,308,479,551,568] and detectors [247,424,474,489].

Non-English

Non-English-language selective reviews are also fairly common. The range of coverage includes environmental analysis in Japanese [326, 344,511], Chinese [221], and Italian [15]; anion reviews in Spanish [8] and Italian [292]; single-column ion chromatography reviews in German [39,40], Bulgarian [426], and Japanese [354]; reviews of ion chromatographic resins in Chinese [384,385]; and the analysis of power plant water in German [396].

IV. APPLICATIONS OF ION CHROMATOGRAPHY: SAMPLE TYPE

Ion chromatography has been successfully applied to a wide variety of analytical situations with a vast range of sample types. Many of the research articles cited in this section will appear under several headings because of the way the classifications scheme was designed and because many papers deal with more than one sample type.

A. Environmental Applications

Water

Cationic species in water are analyzed by IC less frequently than are anionic species in water because of the ready availability of other instrumentation for the trace analysis of the cations, especially the metals, by atomic absorption and atomic emission spectroscopy. However, several research articles are concerned with IC analysis of inorganic cations in water [3,166], and one paper discusses the analysis of methylamines in water [544]. The transition metals have occasionally been analyzed, but nearly always as anion or oxo complexes [105,576]. The most common cations analyzed by ion chromatography are the ammonium ion, the alkali metal cations, and the alkaline earth cations. The ammonium ion has been determined in various natural waters by several researchers [28,270,509,524,543]. The analysis of guanidine has also been reported [44].

Alkali and alkaline earth cations have been determined in various natural waters [28,270,339,524,543] as well as in industrial or treated water [104,125,365].

The analysis of anions by ion chromatography is widely gaining recognition and acceptance because of the speed of analysis and the number of anions that can be analyzed in one sample injection. Numerous research articles concern the analysis of anions in water [4, 61,134,166,168,182,227,236,253,275,352,356,357,495]. A small number of papers have dealt with the analysis of organics [12,485] or halogenated organics [579] in water. More specific papers have dealt with the analysis of anions in natural water [41,85,105,106,128,196, 213,237,238,261,270,286,311,335,347,382,444,457,463,498,524,543, 578], in drinking water [69,70,578], and in industrial-use water [104, 125,131,145,232,265,279,305,365,387-389,408,485,576]. One study has investigated the redox chemistry of sulfur in aqueous solution [312].

Air

As in water analysis, the analysis of cations in air is less common than the analysis of anions in air. Examples of cation analysis in the atmosphere include the analysis of ammonium ion and lower amines [10, 27,35,242,243,288,319,327,425] and the analysis of alkali metal cations [10,271].

The analysis of anions in atmospheric samples can basically be divided into two major groups, according to (a) the nature of the sample and (b) the ion or ions determined. In group (a), sample types include particulates [10,71,124,127,317,361,526], precipitates [65, 165,316], aerosols [156,271,287,288,320,321,510], high-volume air filters [46,262,362,499,557], and Palmes Tubes for NO_2 [296]. In group (b), the most commonly analyzed ions in atmospheric analysis are the oxides of sulfur and nitrogen, because of the deleterious effect of these species on environmental quality. The analysis of the various oxides of sulfur [10,11,34,65,71,116,124,127,156,165,262,271, 287,288,310,320,321,324,327,362,464,478,499,510,518,526,531,557], the analysis of the various oxides of nitrogen [10,65,71,124,127,156, 165,262,271,288,310,316,320,321,327,362,526,531,557], and the analysis of other inorganic anions [10,34,65,71,86,87,124,156,165,246,251, 262,288,316,317,325,327,340,361,362,499,549,557] are by far the most common. There are, however, a reasonable number of papers dealing with organic anion analysis in the atmosphere, primarily of formaldehyde and acetaldehyde [27,99,163,244,245,276,278,572] but also benzoic acid [126].

Soils

Compared to air and water analysis, the ion chromatographic analysis of the ionic constituents in soils is still a relatively uncommon practice. Only a few examples of organic amines [129] and inorganic cations [3,129,339] are available. A somewhat larger listing of various anions can be found [4,36,55,81,159,168,283,430,494, 495,497].

B. Industrial Applications

Ion chromatography has found a wide range of uses in the industrial workplace. One area of such application is in the analysis of flue gases and scrubbers. The most common species for analysis in flue gases and scrubbers are the sulfur oxides [79,123,154,177,179,476, 477] and the nitrogen oxides [121-123,476]. In addition, several of these papers also consider other inorganic anions [123,154,476] in flue gases and scrubbers. Another area that has seen widespread use of ion chromatography is the paper industry. Ions analyzed in pulping liquors and paper processing solutions include sulfur oxides [109,364,477], other inorganic anions [110-112,520], cations [110, 112], and organics [109]. The analysis of water and wastewater in an industrial setting has been undertaken and reported in many instances. Both cations [178,285,307] and anions [178,285,306,370, 387,394,408,485,498,502] have been analyzed.

In addition to the various applications listed above, there are several other areas where ion chromatography has proven useful, namely in caustics [403]; metal cutting fluids [569]; steel surfaces [9]; aqueous extracts of printed wiring [538,539]; metallurgical process solutions [378]; anions in pickling baths [92]; and anions in plating solutions [294] including gold, copper, and nickel solutions [147,148]; and tin electroplating fluids [254]. In the nuclear industry, ion chromatography has been used in fuel reprocessing situations [259,260, 452], in the determination of alkaline metals in uranium dioxide [3], as well as in the determination of other anions, including borate, in radioactive samples [37,174].

Ion chromatography has been applied to the analysis of various types of solid and liquid fuels. The analysis of anions in such liquid fuels as jet fuel and diesel has been extensive [17,47,250,277,303, 304,428,518,532], whereas the analysis of cations [17,580] and organics [25,26] in these liquid fuels is somewhat more limited. The solid fuels include such fuels as coal and oil shale. In these solid fuels, anions are the predominant species measured [263,279,291, 304,330,521].

One last area of industrial application involves the analysis of chemicals for the presence of specific elements. This type of procedure is quite common in a quality-control situation. The analysis of inorganic anions in organic chemicals, occasionally after sample decomposition, is quite common [43,60,157,180,198,199,231,309,315, 358,363,369,386,423,427,440,442,467,472,536,555,573], whereas the analysis of inorganic cations is rare [331]. The analysis of inorganic anions in inorganic materials [4] has been conducted for liquid bromine [392], for sodium hydroxide and sulfuric acid [176,469], for boron-containing chemicals [488,558], for boron in glass [468], and for electrolytes [64].

The usefulness of ion chromatography for the analysis of various photographic developing solutions has been demonstrated in several instances [233,234,341,342,343].

C. Biological and Biochemical Applications

The applications of ion chromatography to the biological and biochemical field are many and diverse. Examples of the wide range of applications include the analysis of inorganic ions in drugs [113,114, 161,540], acetate in pharmaceutics [256], and inorganics in cosmetics [113,114,334]. Inorganic ions in food [96,115,160,161,171,172, 195,204,250,302,349,350,429,517,542] and beer [205,248], as well as organic ions in food [183,302,412,575], have been analyzed.

Ion chromatography has also been applied to the analysis of biological samples for both organic [235,431,567] and inorganic [21,45, 161,167,266,428,503] anions. The various ions and fluids of the human body that have been subjected to ion chromatographic analysis include organic [401] and inorganic ions [77,281,338,339,391,450] in human serum, and oxalate in urine [282,293,360,409,410]. Inorganics have also been determined in urine [282], as have organic acids other than oxalic acid [38].

V. APPLICATIONS OF ION CHROMATOGRAPHY: IONS ANALYZED, ANIONS

The classification of articles according to the ions measured is a fairly simple task, but inevitably some articles are listed many times because of the large number of ions reported in some articles. A decision was therefore made to list an article under only those ions specifically mentioned in the abstract or title of the article. Articles using broad terms such as "the analysis of anions in..." are not included in this section.

A. Group VII Elements

Ions Containing Fluorine

By far the most common fluorine-containing ion is the fluoride ion itself. The analysis of the fluoride ion in environmental samples is quite common. The analysis of fluoride in water, including industrial-use water and treated water [69,106,145,196,232,236,261,265,275,305, 335,347,382,387,394,444,463,502,524,543], in air [17,34,46,65,124, 156,165,262,288,340], and in soil [159,283] are examples of environmental analyses. Fluoride is also quite commonly analyzed in geological samples because of the importance of fluorine in geological processes [103,241,255,261,382,430,513,560].

The analysis of the fluoride ion in industrial situations is quite varied, and includes the analysis of sulfur dioxide scrubbers [179], copper smelters [476], fly ash leachates [291], alloy pickling baths [92], plating solutions [254], and polymer combustion gases [472]. The amount of fluoride present in various fuels [47,303,330] has also been determined, as has the fluoride ion content of tantalum metal [190] and organic chemicals [231,536,579].

The fluoride ion has been determined in several materials of biological interest including color additives in food, drugs, and cosmetics [113,114]; domestic refuse [167]; and toothpaste [24].

There have been many studies concerning the effect on the fluoride ion of changing various instrumental parameters. These parameters include ion chromatographic resins [374,432], eluants [351], and detection systems [88,162,170,516]. In addition, the separation of the fluoride ion in relation to sample preparation and sample matrix has been described [197,264,267,375,380,535,570] including possible interferences [496]. The analysis of fluoride ion by ion chromatography has been compared to that obtained by other analytical techniques [83,203,272,295,329,340].

Fluorine-containing species other than the fluoride ion have also been investigated. These species include SO_2F_2 in air [34], fluorobenzoic acid in soil [36], fluoroboric acid [470], the hydrolysis of $CFCl_3$ [363], and $HFPO_4$ in toothpaste [24].

Ions Containing Chlorine

The chloride ion is one of the most frequently analyzed by ion chromatography. It has been analyzed in water [4,37,41,69,85,106,125, 128,135,145,168,196,213,227,232,236,253,261,265,270,275,305,335, 352,365,370,382,387,394,408,444,463,485,495,524,543], in air [10, 17,46,65,71,124,156,165,288,316,327,362,499,557], and in soil [4, 159,168,283,495]. Chloride has also been analyzed in various geological samples [102,103,261,382,431,513,561].

The analysis of chloride has been performed in industrial applications such as sulfur dioxide scrubbers and flue gases [123,179], copper smelters [476], coal fly ash [291], combustion of polymers [180, 472], analysis of pulp and paper processing fluids [111,112], utility wastes [178], sewage plant effluent [306], alloy pickling baths [92], electroplating fluids [254], and steel surfaces [9]. The chloride content of various fuels [47,303,330,428] has also been determined. The analysis of chemicals [4] including organic chemicals [60,157,231,358, 363,369,442,467,470,536,573,579], boric acid [488], tantalum metal [190], cadmium sulfide [252], liquid bromine [392], brines [403], sodium hydroxide [469], sulfuric acid [469,470], and electrolytes [64] is a common practice. Finally, chloride has been determined in printed wiring extracts [539], nuclear fuel reprocessing fluids [452], domestic refuse [167], and photographic developing solutions [233,234].

Biological areas where the ion chromatographic determination of chloride has been applied include color additives in foods, drugs, and cosmetics [113,114,161]; plants [21,45]; food [160,172,250,517]; and beer [205,248].

The instrumental operating conditions for chloride analysis, including resins [267,287,374,432,528], eluants [200,351,571], separations and interferences [49,197,209,264,380,535,570], and detection systems [5,13,88,162,164,168,170,214,332,434,456,503,516,566], have been discussed in the literature. General instrumental considerations [135,486], including concentration steps [552], equilibria processes [217-219], comparison with other techniques [83,133,203, 272,295,329], and comparison of various ion chromatographic columns [132], have been covered. One paper dealing with the linearity of calibration curves has also been published [527].

In addition to the chloride ion, various other chlorine-containing ions have been determined. Hypochlorite, chlorite, and chlorate ion concentrations have been analyzed in pulp and paper processing fluids [111,112]. The chlorate ion content of brines has also been studied [403]. The interaction of the chlorate ion with the resins [289] and the detection of the perchlorate ion [492] have also been discussed in the literature. The separation of the chlorate and perchlorate ions has also been reported [264].

Ions Containing Bromine

The bromide ion has been determined in water [69,85,106,196,236, 237,261,305,311,335,347,382,394], in air [46,165,316], in soil [36], and in photographic solutions [343]. In addition, the bromide ion has been measured in two geological samples [261,382].

Industrial applications for the ion chromatographic determination of the bromide ion are less common than for the fluoride and chloride ions. Such applications do include the combustion of polymers [180,472], the analysis of fuels [330,428], the analysis of nuclear fuel reprocessing fluids [452], the analysis of organic chemicals [60, 157,231,358,442,536,573,579], and the analysis of photographic developing solutions [233,234].

A limited amount of information is available concerning the determination of the bromide ion in samples of biological interest. These samples include color additives in food, drugs, and cosmetics [78, 113,114,250] and the analysis of bromide in human serum [77].

Studies of the processes involved in the ion chromatographic separation and detection of the bromide ion include the study of resins [267,284,289,390,432,528], columns [454], eluants [200,351], detectors [13,88,162,164,170,214,331,381,415,434,492], and equilibria considerations [181,218,219]. In addition, several papers deal with the separation of bromide in various sample matrices [78,197,264,380,535, 570]. General instrumental considerations [486] and a comparison of the ion chromatographic determination of bromide with that obtained via other analytical techniques [83,329] have also been published. The effect of concentration upon calibration curves has been investigated [220].

The bromate ion is the other bromine-containing ion that has been determined ion chromatographically. Bromate has been determined in bread [349,350] and other food products [78,564]. Resin interactions [284,289,390] and various separation procedures [78,264,541] have also been published.

Ions Containing Iodine

Iodine is the least frequently analyzed of the four common halogens. However, the iodide ion has been determined in air [246] and in soil [36]. Industrial applications of iodide analysis include fuels [330]; chemicals [157,358,442,536]; and food, drug, and cosmetic additives [78,114].

Instrumental considerations in the analysis of iodide that have been covered concern resins [284,289,374], eluants [200], and detectors [170,269,284,415,492]. Also, general instrumental considerations are discussed [486]. Separation [78,264,454,535] and sample preparation procedures [94] are also given.

The iodate ion content of food has also been measured via chromatography [78]; various resins for the separation of iodate have also been described [289,454] as have separation procedures [264].

B. Group VI Elements

Ions Containing Sulfur

Sulfur is an element having a wide range of possible valence states and ions. In fact, two of the sulfur-containing ions can exist in the same sample matrix, making determination of the sulfur content more difficult, especially if one wishes to differentiate between sulfur species. In addition, during sample preparation the valence of the sulfur may change, and thus the ion being measured may not necessarily be the ion that was initially present in the original sample. Consequently, many analyses are performed for "sulfur" generally, without specifying the exact nature or state of the original sulfur. Such general sulfur analyses have been performed on geological samples [103,513], fuel [47,263,277,303,304,330,428,532], organic chemicals [442,467,573], domestic refuse [167], cosmetics [334], and plants [21,45]. In addition, a comparison with the ASTM method has been performed [329], and the redox chemistry of sulfur has also been investigated [312].

The sulfate ion is the sulfur-containing ion that is most frequently analyzed by ion chromatography. Examples of the environmental analysis of sulfate include the analysis of water [4,37,41,69,85,106, 125,128,135,145,168,196,213,227,232,236,238,253,265,270,275,279, 286,305,347,352,365,370,382,387,394,408,444,463,485,495,524,543], of air [10,11,17,46,65,71,75,116,124,127,156,165,262,271,287,288, 310,320,321,324,327,362,464,478,499,510,518,526,531,557], and of soil [4,55,81,159,168,494,495]. The sulfate ion has also been measured in volcanic ash [430].

The analysis of sulfate in industrial situations is diverse and frequent: These applications include sulfur dioxide scrubbers and flue gases [79,123,154,177,179,477], fly ash samples [291], copper smelters [476], alloy pickling baths [92], pulp and paper processing fluids [109,111,112,364], utility wastes [178]; sewage plant effluent [306], printed wire extracts [539], and photographic developing solutions [233,234,343]. Analyses of sulfate in organic chemicals [60,64], in inorganic chemicals [4,176,369], in cadmium sulfide [252], in caustics [285], and in electrolytes [64] have all been reported.

Biological interest in the sulfate ion [266] has centered on color additives in food, drugs, and cosmetics [113,114]; food [160,161, 172,302]; beer [205,248]; drugs [161,554]; urine [282]; and human serum [77].

Instrumental parameters affecting the separation and detection of sulfate have been thoroughly investigated. These investigations include general instrumental considerations [135,486], resins [267,289, 336,374,390,432,528], eluants [571], equilibria processes [181,217-219], and detection systems [13,88,168,214,269,332,368,434,456,503, 516,566]. In addition, separation procedures [197,209,264,380,491], interferences [49], sample concentration [552], calibration curves [220,527], and comparisons with other analytical techniques [83,133, 272,295,348] have all been reported.

The second most common sulfur-containing ion, as far as ion chromatographic analysis is concerned, is the sulfite ion. The sulfite ion has been measured in water [279,347,387,485,543], in air [17,310, 478], and in samples of geological interest [255]. In addition, the sulfite ion has been measured in flue gases [123,154,177,179,476,477], polymer combustion gases [472], pulping liquors [109,111,112], utility wastes [178], brines [403], and photographic developing solutions [233,234,343]. Instrumental considerations include the stability of the sulfite ion [273], detection of the sulfite ion [188], and separation procedures [188,491].

The thiosulfate ion has been determined in water [85,279,521], in pulping liquors [109,112], in flue gases [177], and in photographic developing solutions [234,341,343]. Various equilibria processes [218, 219], resins [390], separation procedures [188,491], and detection systems [188,214] have all been covered. Polarographic detection of various polythionates has been reported [500].

The sulfide ion has been determined in pulping liquors [109,112]. Detection systems for the sulfide ion have been described [30,415, 537,556], as have resins [267] for the separation of sulfide.

In addition to the sulfur-containing species mentioned above, ion chromatography has been used to measure ammonium sulfamate in air [27] and SO_2F_2 in air [34]. A detector for organosulfur compounds has also been described [555].

Ions Containing Selenium

Selenite-selenate species are of interest and can be determined ion chromatographically. A procedure for the measurement of selenite-selenate species [577] and methods of detection [54,417] have been published. Selenite-selenate ions have been measured in water [182, 417,439,576,578] in several instances.

Ions Containing Tellerium

A procedure has been described that provides for the separation and detection of the tellurite and tellurate ions [56].

C. Group V Elements

Ions Containing Nitrogen

Along with chloride and sulfate, the nitrate ion is one of those more frequently analyzed by ion chromatography. The nitrate ion has been measured in many environmental samples including water [4,37,41,69, 70,85,106,128,135,145,168,196,227,232,236,253,265,270,275,286,305, 347,382,387,394,444,463,495,497,524,543], air [10,17,46,65,71,122, 124,127,156,165,262,271,288,310,316,320,321,327,362,526,557], and soil [4,55,81,159,168,494,495]. Several analyses have also been made on samples of geological interest [382,430].

Industrial applications of ion chromatographic nitrate analysis include sulfur dioxide scrubbers and flue gases [121,123,179], fly ash [291], copper smelters [476], combustion of polymers [472], alloy pickling baths [92], and sewage plant effluent [306]. The nitrate ion has also been measured in various fuels [47,263,303,330], chemicals [4,231,369], and electrolytes [64].

Biological applications of nitrate analysis include food, drug, and cosmetic color additives [113,114,161]; food [78,96,160,172,195,204, 302,429,517]; beer [205]; and human serum [77].

A rather extensive body of knowledge deals with the instrumental conditions and chemical equilibria that result in the separation and detection of the nitrate ion. Included are papers on resins [267,284, 289,374,390,432,528], on eluants [200,351,505,571], on detectors [13, 61,88,164,214,284,332,368,434,456,492,501,503,516,566], and on general equilibria processes [213,217-219]. Other instrumental work concerns general instrumental considerations [135,486], calibration curves [527], comparison with other techniques [85,98,133,272,295, 329,348], and comparison of ion chromatographic columns [132]. Articles dealing with sample concentration [552] and general preparation procedures [78,197,209,380,454,505] have also been published.

The other common nitrogen-containing ion is the nitrite ion. It has been determined in water [37,69,85,106,236,253,275,305,347, 387,543], in air [17,46,122,296,310,316,531], and in soil [159], as well as in samples of geological interest [430].

Examples of the industrial applications to which the ion chromatographic determination of nitrite has been applied include flue gases [121,123]; combustion of polymers [472], fuels [47,303,330], metal-cutting fluids [569], steel surfaces [9], and the analysis of sewage plant effluent [306]. Nitrite determinations have been performed on food [78,172,195,204,429,517], as well as on cola additives [113,114].

Various instrumental parameters, such as resins [284,289,432,528], eluants [200,351,505], and detectors [88,164,188,269,284,492,501, 503,566], have been investigated. General instrument considerations [486], comparisons with other techniques [98], comparison of ion chromatographic columns [132], and possible complications have been discussed [249]. Various separation procedures have also been described in the literature [78,188,380,454,505].

Several other nitrogen-containing compounds have also been measured by ion chromatography. These additional species include nitro compounds [97], and azide ion in human serum [281] and in air bag inflator effluent [549].

Ions Containing Phosphorus

By far the most common phosphorus-containing ion is the phosphate ion. The range of phosphate determinations is broad, but there are fewer publications in each category than for the chloride, nitrate, and sulfate ions. A fairly extensive selection of papers dealing with the analysis of water is available [69,85,106,232,236,265,275,305,306, 347,387,444,485,495,543], but only a limited number of papers deal with air [46,165] and soil [55,159,495] analysis. The analysis of P_2O_5, as phosphate, in rocks has been accomplished [255].

Industrial applications include polymer combustion gases [472], fuels [47], organic chemicals [60,231,573], electrolytes [64], and integrated circuit components [257]. Biological applications include color additives [113]; food, drug, and cosmetic additives [114,302, 517]; food [172]; beer [205,248]; human serum [77]; and toothpaste [24].

Various instrumental conditions have been investigated; these include resins [267,390,432], detection systems [13,88,164,332,503, 507,516,566], equilibria processes [181], and comparisons with other analytical techniques [83,272]. In addition to the instrumental factors [486], sampling and matrix effects dealing with concentration [552] and separation techniques [192,264,380,506,512] have been reported.

Both inorganic and organic phosphorus ions other than the phosphate ion have been investigated: The separation and analysis of various polyphosphate species [7,108,506], the analysis of phosphonic acids [440], the detection of organophosphorus compounds [5], the analysis of various dibutyl phosphates [240,259,260], the analysis of organophosphoric and phosphorothioic acids [33], and the analysis of sugar phosphates [302] have been performed. Monofluorophosphate has been determined in toothpaste [24].

Ions Containing Arsenic

The toxic nature of various arsenic compounds often makes the analysis of arsenic of prime importance in analytical chemistry. The arsenite and arsenate ions are the most commonly analyzed forms of arsenic in ion chromatography. Arsenite and arsenate have been analyzed in water [182,576], soil [497], and flue dust and smelters [154,476]. More theoretical considerations pertaining to these ions include equilibria processes [181,577], detection systems [398,556], and comparisons with wet chemical techniques [475]. One publication has dealt with the detection of organic arsenic species [398].

D. Group IV and Group III Elements

Group IV Elements

Only a limited amount of data in the literature concerns anions derived from the group IV elements. The carbonate ion, or bicarbonate ion depending upon the pH of the solution, is the most common of the group IV species. The volume of literature on carbonate analysis is limited by the fact that one of the more common eluants in suppressed ion chromatography is a carbonate-bicarbonate-based eluant. Examples of carbonate analysis do exist, however, and include the analysis of water [4,85], soil [4], brines [403], fly ash [291], pulp processing fluids [111], steel surfaces [9], and chemicals [4]. Instrumental and theoretical investigations include equilibrium processes [181], detection systems [368,503,566], and peak identification [191]. Separation of the carbonate peak is also discussed [375].

A very limited amount of material is available concerning the ion chromatographic analysis of silicates. The reported applications include paper processing fluids [111], borate products [558], and water [356,357]. A research area on silicates involves resin technology [267].

Bibliographic Review of Ion Chromatography

Group III Elements

The analysis of the borate ion, from the group III elements, includes samples from nuclear plant water [37,174], glass [257,468], and photographic solutions [343]. Detection of the boron species has been discussed [368], along with research techniques [175], separation procedures [375,377,559], and resins [267]. Fluoroboric acid has also been analyzed by ion chromatography [470].

E. Miscellaneous Inorganic Anions

A number of polyatomic anions that do not readily fit into the previous categories have been the subject of ion chromatographic research.

Various transition metals that exist as oxoanions, i.e., tungstate, molybdate, and chromate ions, can be detected by ion chromatography. The tungstate ion has been measured in water [105,576] and in carbonaceous material [520]. The molybdate ion has also been measured in water [105,576], as has the chromate ion [275,576]. One paper describes a technique for the ion chromatographic analysis of tungstate, molybdate, and chromate ions [577]. Chromate has been determined in lignosulfonates [366]. The interaction of the vanadate and molybdate ions with resins has been investigated [390].

Other inorganic species that have been the subject of ion chromatographic investigations include the cyanide and thiocyanate ions. Cyanide has been measured in water [275], air [86,87], metallurgical process solutions [378], and dust [251]. The thiocyanate ion has been measured in soil [36] and oil-shale retort water [279]. Methods of detection for cyanide [30,368,415,537] and for thiocyanate [170, 188,284,492] can be found in the literature. Finally, the separation of the cyanide ion in various sample matrices [94,375] and the interaction of the thiocyanate ion [284,289] and cyanide ion [267] with resins have been discussed. The redox behavior of the thiocyanate ion has been reported [312].

One last inorganic anion has been of interest in ion chromatography. The equilibrium processes occurring with the hydroxide ion have been reported [181].

F. Organic Anions

A large number of organic anions representing a large variety of organic functional groups have been subjected to ion chromatographic analysis [32]. In general, there are only a few papers dealing with any one subject or type of organic molecule.

Formate and Acetate Ions

Two of the more common organic species to be measured by ion chromatography are the formate and acetate ions. The formate ion is frequently produced from the oxidation of formaldehyde, and the acetate ion is often produced from the oxidation of the acetaldehyde molecules. The formate ion has been measured in air [25,26,99,163,244, 245,572], pulping liquors [109], and brines [403]. The acetate ion has been measured in air [26,276,572], pulping liquors [109], photographic developing solutions [234,343], brines [403], organic chemicals [199,315], and pharmaceuticals [256]. Instrumental considerations for the acetate ion [486], as well as separation procedures for the acetate ion, can be found in the literature [264,380].

Other Organic Acids

A large number of additional organic acids have been determined by ion chromatography. The types of analyses include organic acids in food and alcoholic beverages [183,412,575] and in plants [235]; pyruvate ion in meat [302] and in urine [401]; glycolate in water [485] and in ethylene and propylene glycols [423]; lactate ion in pulp liquors [109,111], in meat [302], in urine [401], in air [572], and in ethylene and propylene glycols [423]; oxalate in pulp liquors [109, 111] and in urine [282,293,360,409,410]; adipic acid in flue gases [79]; benzoic acid in air [126]; and pyruvic and methanesulfonic acid [146]. Also, vanillymandelic acid has been measured in urine [401]. Theoretical and procedural considerations for organic acid anions include resins [289], columns [420], detectors [198], and separations [264,422]. The separations of specific ions such as citrate [375,377], lactate [375,377], and polyamino carboxylic acids [56, 108] have been studied. Finally, a study of the propionate ion [315, 572] and a study of organic acid metabolites [567] have been undertaken.

Other Organic Anions

Phosphorus-containing organic compounds that have been analyzed by ion chromatography include dibutylphosphates [240,259,260], phosphonic acids [440], organophosphoric acids [33], sugar phosphates [302], and organic phosphorus compounds [5]. Detection of sulfonates in air [71] and of organosulfur compounds have been reported [555]. Nitro compounds [97] and the hydrolysis of halogenated methanes [363] have also been the object of scientific investigations.

Other types of organic compounds that have received attention include phenols [12,97], sugars [97,386], carbohydrates [96,412,416], ε-caprolactams [315], alcohols [412,416], saccharides [416], adenosine 5'-monophosphate (AMP) [302], chloroacetyl chloride [278], and nucleobases and nucleotides [421]. The interaction of analgesics with resins has been reported [267].

VI. APPLICATIONS OF ION CHROMATOGRAPHY: IONS ANALYZED, CATIONS

The analysis of cations by ion chromatography has lagged behind the analysis of anions because of the ready availability and accessability of atomic spectroscopic techniques. This area of ion chromatography, however, is now beginning to strive toward its full potential.

A. Group I Elements

The group I elements can be considered to include lithium, sodium, potassium, rubidium, and cesium, with the majority of the ion chromatographic research aimed at sodium and potassium. Analysis of the group I cations as a group includes general applications [458], analysis of water [3], analysis of soil [3], and analysis of uranium dioxide [3]. Theoretical considerations include the use of resins [185,333,432,435] and the use of detectors [493,504]. Several procedures for the separation of the group I cations have also been published [42,119].

Lithium

Instrumental considerations for the analysis of lithium have been reported [574].

Sodium

A fairly extensive selection of papers are concerned with the analysis of sodium. These papers include the analysis of water [28,104, 261,270,305,307,365,524,543], air [10,17,271,288]; sulfur dioxide scrubbers [179], pulp solutions [112], and utility wastes [178]. The sodium ion has also been measured in antibiotics [554], food and drugs [96,160,161], and organic chemicals [331]. Instrumental considerations [574], detection systems [162], and comparisons with other techniques [272] have all been reviewed. A separation procedure [137] has also been published.

Potassium

The literature for the analysis of potassium is similar to that for the analysis of sodium, and thus includes the analysis of water [28,125, 261,270,305,307,524,543], air [10], sulfur dioxide scrubbers [179], pulp solutions [112], food [96,160,161,349], antibiotics [554], and organic chemicals [331]. Papers concerning instrumental considerations [574], detection systems [162], comparisons with other analytical techniques [272], and separation procedures [137] are available in the literature.

B. Group II Elements

The ions of the group II elements can be considered to include magnesium, calcium, strontium, and barium. One paper dealing with the various applications of the ion chromatographic analysis of the group II cations has been published [458]. Discussion of resins [185,333, 432,435], detection systems [130,149,504], and separation procedures [119,258,565] are also available.

Magnesium

Reports of the analysis of magnesium in water [339,524], in soil [339], in utility wastes [178], in sulfur dioxide scrubbers [179], in blood [339,391], in serum [338], and in food and drugs [160,161] have been published. Additional information is available concerning eluants [571] and separation procedures [137].

Calcium

The analysis of calcium has been reported in such diverse sample matrices as water [339,524], soil [339], sulfur dioxide scrubbers [179], utility wastes [178], blood [339,391], serum [338], and food and drugs [160,161]. Eluants [571] and separation procedures [137] are similar to those used for magnesium.

Strontium

The analysis of strontium has been reported in water, soil, and blood [339].

C. Ammonium Ion and Amines

Sodium, potassium, and ammonium are the most common cations analyzed by ion chromatography. Reports of ammonium ion analyses include the analysis of water [3,28,31,261,270,305,307,509,524,543], air [10,17,27,35,46,242,243,288,319,327,425,580], soil [3], uranium dioxide [3], foams [470], and food [160]. Theoretical and procedural reports include instrumental considerations [574], resins [432, 435], detection systems [493,501,504], comparisons with other analytical techniques [272], and separation procedures [42,119,137].

The analysis of various amines has been performed in a wide variety of sample matrices. The analysis of amines, in general, has been considered [243] along with a discussion of applications [458]. Methylamines have been measured in water [31], air [35], and shellfish [431]. The analysis of lower amines [242] and of long-chain alkyl amines [544] has been reported. Additional amine species that have been measured include ethanolamine [223], ethylenediamine [43], cyclohexylamine [129], and guanidine [44].

D. Transition Elements

Work on the ion chromatographic analysis of the transition elements has just begun in the recent past. The detection of the heavy metals and rare earths [130] and the separation of trivalent lanthanides [447] have been reported. The parts-per-billion analysis of Au, Pd, and Pt complexes has also been reported [414].

Additional species that have been investigated include various first-, second-, and third-row transition elements. The analysis of manganese, iron, cobalt, nickel, zinc, and cadmium has been reported in soil, water, and blood [339]. Zinc and iron in food [96] have been measured, as have nickel, copper and gold plating solutions [148]. Sample preparation procedures for manganese, cobalt, nickel, copper, and cadmium [82] have been described, and the separation and detection of various divalent cations of manganese, iron, cobalt, nickel, copper, zinc, and cadmium [184,258] are available in the literature. A report has been published concerning the analysis of gold, palladium, platinum, bismuth, and mercury [137]. Resin interactions with divalent copper, mercury, nickel, lead, and cadmium ions have been reported [390]. The detection of the rare earths has also been discussed [537].

VII. INSTRUMENTAL, PROCEDURAL, AND THEORETICAL CONSIDERATIONS

A fairly exhaustive body of printed material is concerned with theoretical and instrumental considerations of ion chromatography. This information can be divided into several distinct groupings based upon the information involved: e.g., ion chromatography systems, ion chromatography components, comparison of ion chromatography with other techniques, theoretical considerations, and techniques involving sampling and analysis.

A. Ion Chromatography Systems

A reasonably diverse body of information has been published dealing with various ion chromatography systems, system descriptions, and system modifications. One source of such information which is always accessible includes the various manufacturers of ion chromatography instruments and components. These types of literature reports are not, however, included here because of the obvious bias found in manufacturers' literature.

System Descriptions

Basic system descriptions and adjustments to these basic systems that increase the applicability and versatility of the instruments have been

reported in relatively large numbers [14, 222, 223, 441, 486, 514, 574].
Use of conventional high-performance liquid chromatography equipment is treated in the literature as well [135]. The modifications to an ion chromatograph for use with radioactive samples have been described [68], as have various descriptions of ion chromatography exclusion (ICE) [375, 377, 541, 559].

System Modifications

Various types of modifications of, or alterations to, ion chromatographs have been made to increase the utility of the instrument with respect to a particular application problem. One such area of interest has been on-line operation and automation of the ion chromatograph [115, 165, 297, 313, 381, 456, 457, 518, 578]. Another area of interest has been the thermostat control of the instrument in order to minimize baseline instability when one is working at very low concentrations [215, 456]. Gradient elution is of immense value in high-performance liquid chromatography, and recent work has begun to show the applicability of gradient elution in ion chromatographic analysis [422, 491, 515].

B. Ion Chromatography Components

Resins and Columns

A general article dealing with use, care, and storage of ion chromatographic columns has been published [100]. An extensive body of literature has been published on the use, properties, and preparation of the resin materials employed in ion chromatographic columns [18-20, 52, 53, 62, 93, 132, 138, 140, 152, 153, 185, 187, 264, 267, 268, 289, 290, 333, 336, 341, 374, 376, 390, 420, 421, 432, 435, 454, 471, 482, 484, 528, 564].

Eluants

The key to any successful ion chromatographic analysis involves the successful selection of both column resin material and eluants. There are many research publications dealing with various uses and properties of eluants [118, 139-141, 142, 186, 193, 197, 200, 202, 219, 258, 289, 351, 353, 373, 433, 471, 505, 565, 571].

Suppressor Systems

The two-column method of ion chromatography involves the chemical suppression, or background conductivity reduction, of the eluant prior to the measurement of the conductance of the analyte ions. This area has been the object of much attention and improvement in the years since the inception of ion chromatography in the middle 1970s [72-74, 201, 453, 480, 481, 490].

Detectors

One of the more active areas of research involves the application of novel detection systems to analytical problem solving via ion chromatography. The standard detector used in ion chromatography is the conductivity detector. Various electrochemical detectors, other than the conductivity detector, have become quite common and important in ion chromatographic analysis because of the larger number of ions that can be analyzed. Numerous articles discuss the use of various electrochemical detectors for increased utility of the ion chromatographic technique [30, 97, 130, 149, 170, 184, 188, 239, 251, 368, 381, 413, 415, 434, 492, 493, 500, 501, 504, 507, 516, 519, 535, 537, 566]. Another rapidly growing detection mode involves the use of optical detectors such as ultraviolet and visible detectors. These types of detectors can be used to monitor (a) absorbing species directly, or (b) non-absorbing species through the use of an absorbing eluant [6, 13, 16, 58, 61, 88, 108, 161, 162, 164, 166, 189, 211, 214, 248, 264, 284, 332, 416, 422, 461, 501, 503, 537, 553, 555-557]. The use of a fluorescence detector has also been described [269]. A detection system that has received some attention, and is fundamentally related to the optical methods is atomic spectroscopy. Several research articles have discussed the use of atomic spectroscopy as a detection system in ion chromatography [54, 90, 366, 398, 417, 474]. The refractive index detector, which has found much use in high-performance liquid chromatography, has also been used in ion chromatography [48, 150, 214, 248, 553]. In addition, permittivity has been used as a means of detection in ion chromatographic analysis [5].

A growing trend in ion chromatographic analysis involves the use of multiple detectors to increase the efficiency of the procedure by providing the means for detecting and quantitating more species in each sample injection. Papers specifically describing the use of multiple detectors and the applications to which these specific combinations apply are numerous [5, 6, 168, 214, 248, 422, 456, 501, 553].

C. Comparison of Ion Chromatography with Other Techniques

When any new analytical technique is introduced, and as it grows in use and applicability, it will be compared to the methods currently in use for performing the same type of analyses. This has happened with ion chromatography in many different instances [83, 98, 133, 203, 205, 272, 295, 329, 340, 348, 352, 463, 475]. One other research paper, which uses ion chromatography as a check on analytical conditions [155], has been published but this paper has received some critical comments with respect to non-ion-chromatographic aspects [448].

D. Theoretical Considerations

A nontrivial volume of research has been published on theoretical aspects of ion chromatography and their effects on the application of ion chromatography to real-world samples. Several papers deal extensively with the linearity of the calibration curves and standardization procedures [144, 211, 220, 527, 562, 563]. One article explores the use of computer simulation [433] to better understand the phenomena involved. A relatively extensive list of publications cover equilibria processes and peak changes as a function of chromatographic conditions [2, 51, 89, 169, 181, 191, 194, 209, 212, 216-219, 249, 273, 355, 449, 533].

E. Techniques Involving Sampling and Analysis

Sample Preparation

One of the advantages of ion chromatography is that sample preparation is either minimized, or eliminated altogether, for many types of samples. In other analytical procedures, the sample preparation step has been significantly modified to make use of the ion-analyzing abilities of the ion chromatograph. General sample treatment considerations for microanalysis, including preconcentration of the sample, have been covered in several articles [250, 380, 552]. Some specific aspects of sample preparation covered in the literature include analysis of water [367], air [146], anions [257, 452, 462, 506], transition metals [82, 414, 447], biological samples [21, 45, 171, 248, 409, 431, 442], and geological materials [102, 103, 241, 255, 560, 561].

Analysis of Samples

A certain body of information is more concerned with analytical technique than with the application of the technique for problem solving. These papers include discussions of the analysis of inorganic species, predominantly anions [49, 63, 78, 80, 94, 491, 496, 501, 512, 570]; the analysis of alkali, alkaline earth, and amines [31, 42, 80, 119]; the analysis of transition metals [137, 447, 577]; and the analysis of organic compounds [32, 33, 56, 63, 416, 422, 508, 523]. Finally, the analysis of salt plus acid and salt plus base mixtures has been described [483].

REFERENCES

1. Tobias R. Acciani and Ray F. Maddalone, Technical Report, EPA/ 600/7-79/151 (1979). (Available from NTIS.)
2. S. Afrashtehfar and F. F. Cantwell, *Anal. Chem.* 54, 2422-2427 (1982).
3. Mukhtar Ahmad and Attiq-ur-Rehman Khan, *Nucleus (Karachi)* 18(4), 29-35 (1981).

4. Mukhtar Ahmad and Attiq-ur-Rehman Khan, *Nucleus (Karachi)* 19(3-4), 35-38 (1982).
5. J. F. Alder, P. R. Fielden, and A. J. Clark, *Anal. Chem.* 56, 985-988 (1984).
6. T. K. Al-Jorani and S. J. Lyle, *Anal. Proc. (London)* 20, 111-114 (1983).
7. C. Amano, K. Mizuki, and M. Suzuki, *Kanzei Chuo Bunseki-shoho* 24, 31-38 (1983).
8. Olga B. Amiano, *Bol. Inf. - DNQ* 8(38), 11-16 (1981).
9. B. E. Andrew, *HRC CC: J. High Resol. Chromatogr. Chromatogr. Commun.* 7(10), 580-581 (1984).
10. K. Anlauf, L. A. Barrie, H. A. Wiebe, and P. Fellin, *Canadian Res.* 13, 49-53 (1980).
11. B. R. Appel and W. J. Wehrmeister, in *Ion Chromatogr. Anal. Environ. Pollut.* (Eugene Sawicki and James D. Mulik, eds.), Vol. 2, Ann Arbor Sci., Ann Arbor, MI, 1979, pp. 223-233.
12. D. N. Armentrout, J. D. McLean, and M. W. Long, *Anal. Chem.* 51(7), 1039-1045 (1979).
13. G. P. Ayers and R. W. Gillett, *J. Chromatogr.* 284(2), 510-514 (1984).
14. Nobuyuki Baba, Keiichi Hosako, Susumu Matsushita, and Yoshimitsu Tada. *Toyo Soda Kenkyu Hokoku* 27(1), 21-36 (1983).
15. L. T. Baldassarri, E. Ferrari, and A. Rolla, *Chim. Ind. (Milan)* 66(3), 160-166 (1984).
16. W. E. Barber and P. W. Carr, *J. Chromatogr.* 260, 89-96 (1983).
17. N. P. Barkley, G. L. Contner, and M. Malanchuk, in *Ion Chromatogr. Anal. Environ. Pollut.* (Eugene Sawicki and James D. Mulik, eds.), Vol. 2, Ann Arbor Sci., Ann Arbor, MI, 1979, pp. 115-128.
18. Eugene Robert Barron, *Diss. Abstr. [B]* 44(12,Pt1), 3761 (1984).
19. R. E. Barron and J. S. Fritz, *React. Polym.* 1, 215-226 (1983).
20. R. E. Barron and J. S. Fritz, *J. Chromatogr.* 284, 13-25 (1984).
21. G. Bartonek and H. Werner, *GIT Fachz. Lab.* 27(12), 1075-1076 (1983).
22. N. Basili, Technical Report, CNEN-RT/CHI(82)3 (1982).
23. J. R. Benson and Dexter J. Woo, *J. Chromatogr. Sci.* 22, 386-399 (1984).
24. J. R. Benson, D. J. Woo, and N. Kitagawa, *LC, Liq. Chromatogr. HPLC Mag.* 2(5), 398-400 (1984).
25. Itamar Bodek and Kenneth T. Menzies, *ACS Symp. Ser.* 149, 599-613 (1981).
26. I. Bodek and K. T. Menzies, Technical Report, EPA-600/9-81-018 (1981).
27. Itamar Bodek and Richard H. Smith, *Am. Ind. Hyg. Assoc. J.* 41(8), 603-607 (1980).

28. D. C. Bogen and S. J. Nagourney, in *Ion Chromatogr. Anal. Environ. Pollut.* (Eugene Sawicki and James D. Mulik, eds.), Vol. 2, Ann Arbor Sci., Ann Arbor, MI, 1979, pp. 319-328.
29. Romuald Bogoczek and Grazyna Miemus, *Przem. Chem.* 59(9), 471-474 (1980).
30. A. M. Bond, I. D. Heritage, G. G. Wallace, and M. J. McCormick, *Anal. Chem.* 54(3), 582-585 (1982).
31. Spiros A. Bouyoucos, *Anal. Chem.* 49(3), 401-403 (1977).
32. Spiros A. Bouyoucos, *J. Chromatogr.* 242(1), 170-176 (1982).
33. Spiros A. Bouyoucos and David N. Armentrout, *J. Chromatogr.* 189(1), 61-71 (1980).
34. Spiros A. Bouyoucos, Richard G. Melcher, and James R. Vaccaro, *Am. Ind. Hyg. Assoc. J.* 44(1), 57-61 (1983).
35. Spiros A. Bouyoucos and Richard G. Melcher, *Am. Ind. Hyg. Assoc. J.* 44(2), 119-122 (1983).
36. Robert S. Bowman, *J. Chromatogr.* 285(3), 467-477 (1984).
37. F. Brandt and R. Trost, *VGB Kraftwerkstech.* 64(1), 74-77 (1984).
38. D. N. Buchanan and J. G. Thoene, *Anal. Biochem.* 124(1), 108-116 (1982).
39. P. Bucher, *SLZ, Schweiz. Lab.-Z.* 40(3), 91-93 (1983).
40. P. Bucher, *SLZ, Schweiz. Lab.-Z.* 40(5), 157-159 (1983).
41. Arlene E. Buchholz, Curtis I. Verplough, and James L. Smith, *J. Chromatogr. Sci.* 20(11), 499-501 (1982).
42. R. C. Buechele and D. J. Reutter, *J. Chromatogr.* 240(2), 502-507 (1982).
43. Richard C. Beuchele and Dennis J. Reutter, *Anal. Chem.* 54(12), 2113-2114 (1982).
44. E. P. Burrows, E. E. Brueggeman, and S. T. Hoke, *J. Chromatogr.* 294, 494-498 (1984).
45. L. M. Busman, R. P. Dick, and M. A. Tabatabai, *Soil Sci. Soc. Am. J.* 47(6), 1167-1170 (1983).
46. F. E. Butler, R. H. Jungers, L. F. Porter, A. E. Riley, and F. J. Toth, in *Ion Chromatogr. Anal. Environ. Pollut.* (Eugene Sawicki, James D. Mulik, and Eva Wittgenstein, eds.), Ann Arbor Sci., Ann Arbor, MI, 1978, pp. 65-76.
47. F. E. Butler, F. J. Toth, D. J. Driscoll, J. N. Hein, and R. H. Jungers, in *Ion Chromatogr. Anal. Environ. Pollut.* (Eugene Sawicki and James D. Mulik, eds.), Vol. 2, Ann Arbor Sci., Ann Arbor, MI, 1979, pp. 185-192.
48. F. A. Buytenhuys, *J. Chromatogr.* 218, 57-64 (1981).
49. Mary Ann O. Bynum, S. Y. Tyree, Jr., and William E. Weiser, *Anal. Chem.* 53(12), 1935-1936 (1981).
50. Jorge Calmet Fontane, *Tech. Lab.* 8(108), 936-939 (1983).
51. F. F. Cantwell and S. Puon, *Anal. Chem.* 51(6), 623-632 (1979).
52. R. M. Cassidy and Steven Elchuk, *Anal. Chem.* 54(9), 1558-1563 (1982).

53. R. M. Cassidy and S. Elchuk, *J. Chromatogr.* 262, 311-315 (1983).
54. Dipankar Chakraborti, Daniel C. J. Hillman, Kurt J. Irgolic, and Ralph A. Zingaro, *J. Chromatogr.* 249(1), 81-92 (1982).
55. Letian Chen and Shifen Mu, *Fenxi Huaxue* 11(2), 88-92 (1983).
56. S. G. Chen, K. L. Cheng, and Corazon R. Vogt, *Mikrochim. Acta* 1(5-6), 473-481 (1983).
57. Liangqi Chu, ed., *Application of Ion-Exchange Chromatography in Ore Analysis*, Dizhi Publ. House, Beijing, China, 1983.
58. R. A. Cochrane and D. E. Hillman, *J. Chromatogr.* 241(2), 392-394 (1982).
59. C. E. Coggan, *Anal. Proc. (London)* 19(12), 567-568 (1982).
60. J. F. Colaruotolo, in *Ion Chromatogr. Anal. Environ. Pollut.* (Eugene Sawicki, James D. Mulik, and Eva Wittgenstein, eds.), Ann Arbor Sci., Ann Arbor, MI, 1978, pp. 149-167.
61. M. Cooke, *HRC CC, J. High Resol. Chromatogr. Chromatogr. Commun.* 6, 383-385 (1983).
62. M. Cooke, *HRC CC, J. High Resol. Chromatogr. Chromatogr. Commun.* 7(9), 515-519 (1984).
63. H. J. Cortes, *J. Chromatogr.* 234, 517-520 (1982).
64. J. A. Cox and N. Tanaka, *Anal. Chem.* 57, 383-385 (1985).
65. Joan Crowther and Jenifer McBride, *Analyst (London)* 106(1263), 702-709 (1981).
66. Dennis Cuff, *CHEMSA* 7(10), 235-237 (1981).
67. Dennis Cuff, *Analytika (Johannesburg)*, Dec. 11-13,15 (1981).
68. L. L. Curfman and S. J. Johnson, Technical Report, RHO-SA-110 (1979). (Available from NTIS.)
69. T. Darimont, G. Schulze, and M. Sonneborn, *Fresenius' Z. Anal. Chem.* 314(4), 383-385 (1983).
70. T. Darimont, G. Schulze, and M. Sonneborn, *Fresenius' Z. Anal. Chem.* 317(3-4), 400-406 (1984).
71. Purnendu K. Dasgupta, *Atmos. Environ.* 16(5), 1265-1268 (1982).
72. Purnendu K. Dasgupta, *Anal. Chem.* 56(1), 103-105 (1984).
73. Purnendu K. Dasgupta, *Anal. Chem.* 56(4), 769-772 (1984).
74. P. K. Dasgupta, R. Q. Bligh, J. Lee, and V. D'Agostino, *Anal. Chem.* 57, 253-257 (1985).
75. P. K. Dasgupta, K. DeCesare, and J. C. Ullrey, *Anal. Chem.* 52, 1912-1922 (1980).
76. Hans Deister and Ernst Armin Runge, *Arch. Eisenhuettenwes.* 54(10), 405-410 (1983).
77. Peter De Jong and Marjolijn Burggraaf, *Clin. Chim. Acta* 132(1), 63-71 (1983).
78. J. P. deKleijn, *Analyst (London)* 107, 223-225 (1982).
79. J. H. Dempsey, P. Cruse, and K. Yates, in *Ion Chromatogr. Anal. Environ. Pollut.* (Eugene Sawicki and James D. Mulik, eds.), Vol. 2, Ann Arbor Sci., Ann Arbor, MI, 1979, pp. 89-97.

80. M. Denkert, L. Hackzell, G. Schill, and E. Sjogren, *J. Chromatogr.* *218*, 31-43 (1981).
81. W. A. Dick and M. A. Tabatabai, *Soil Sci. Soc. Am. J.* *43*(5), 899-904 (1979).
82. James E. DiNunzio and Michael Jubara, *Anal. Chem.* *55*(7), 1013-1016 (1983).
83. S. R. L. Dionex (Rome, Italy), *Chim. Oggi* (4), 25-27, 31 (1984).
84. J. S. Dits, *Gewaesserschutz, Wasser, Abwasser* 67, 287-310 (1983).
85. S. Dogan and W. Haerdi, *Chimia* 35, 339-342 (1981).
86. T. W. Dolzine, G. G. Esposito, and R. J. Gaffney, *ASTM Spec. Tech. Publ.* *786*, 142-152 (1982).
87. T. W. Dolzine, G. G. Esposito, and D. S. Rinehart, *Anal. Chem.* *54*(3), 470-473 (1982).
88. G. Domazetis, *Chromatographia* *18*(7), 383-386 (1984).
89. M. Doury-Berthod, D. Stammose, and C. Poitrenaud, *React. Polym., Ion Exch., Sorbents* *2*(1-2), 37-50 (1984).
90. S. W. Downey and G. M. Hieftje, *Anal. Chim. Acta* 153, 1-13 (1983).
91. Stephen Ward Downey, *Diss. Abstr. Int. [B]* *44*(5), 1447 (1983). (Available from Univ. Microfilms Int.)
92. Thomas R. Dulski, *Anal. Chem.* *51*(9), 1439-1443 (1979).
93. D. L. DuVal and J. S. Fritz, *J. Chromatogr.* *295*(1), 89-101 (1984).
94. D. L. DuVal, J. S. Fritz, and D. T. Gjerde, *Anal. Chem.* *54*(4), 830-832 (1982).
95. A. Eaton and J. Oppenheimer, *Proc.-AWWA Annu. Conf.*, 1079-1091 (1983).
96. Pamela Edwards, *Food Technol. (Chicago)* *37*(6), 53-56 (1983).
97. Pamela Edwards and Karen K. Haak, *Am. Lab. (Fairfield, Conn)* *15*(4), 78, 80, 83-87 (1983).
98. U.S. E.P.A., *Fed. Reg.* *48*(237) 55072-55074 (1983).
99. E. Estes, P. Grohse, W. S. Gutknecht, and R. K. M. Jayanty, Technical Report, EPA-600/4-83-031 (1983). (Available from NTIS.)
100. Denise R. Eubanks and John R. Stillian, *LC, Liq. Chromatogr. HPLC Mag.* *2*(2), 74, 76 (1984).
101. Keenan Lane Evans, *Diss. Abstr. Int. [B]* *42*(2), 617 (1981). (Available from Univ. Microfilms Int.)
102. Keenan L. Evans and Carleton B. Moore, *Anal. Chem.* *52*(12), 1908-1912 (1980).
103. Keenan L. Evans, James G. Tarter, and Carleton B. Moore, *Anal. Chem.* *53*(6), 925-928 (1981).
104. Wolfgang Fichte and Gerhard Mohr, *Maschinenschaden* *55*(2), 81-83 (1982).

105. Walter H. Ficklin, *Anal. Lett.* 15(A10), 865-871 (1982).
106. Marvin J. Fishman and Grace Pyen, Technical Report, USGS/WRI-79-101, USGS/WRD/WRI-79/083 (1979). (Available from NTIS.)
107. Arthur W. Fitchett, *Proc.-AWWA Water Qual. Technol. Conf.*, 149-160 (1982).
108. A. W. Fitchett and A. Woodruff, *LC, Liq. Chromatogr. HPLC Mag.* 1(1), 48-49 (1983).
109. G. O. Franklin, *Pulping Conf. (Proc.)* 255-261 (1981).
110. Greg Franklin, *Pulp Pap.* 56(2), 91-95 (1982).
111. G. O. Franklin, *Tappi* 65(5), 107-111 (1982).
112. G. O. Franklin and A. W. Fitchett, *Pulp Pap. Can.* 83(10), 40-44 (1982).
113. D. D. Fratz, in *Ion Chromatogr. Anal. Environ. Pollut.* (Eugene Sawicki, James D. Mulik, and Eva Wittgenstein, eds.), Ann Arbor Sci., Ann Arbor, MI, 1978, pp. 169-184.
114. D. D. Fratz, in *Ion Chromatogr. Anal. Environ. Pollut.* (Eugene Sawicki and James D. Mulik, eds.), Vol. 2, Ann Arbor Sci., Ann Arbor, MI, 1979, pp. 371-379.
115. Douglas D. Fratz, *J. Assoc. Off. Anal. Chem.* 63(4), 882-888 (1980).
116. Connie D. Frazier, in *Ion Chromatogr. Anal. Environ. Pollut.* (Eugene Sawicki and James D. Mulik, eds.), Vol. 2, Ann Arbor Sci., Ann Arbor, MI, 1979, pp. 211-221.
117. J. S. Fritz, *LC, Liq. Chromatogr. HPLC Mag.* 2, 446-452 (1984).
118. J. S. Fritz, D. L. DuVal, and R. E. Barron, *Anal. Chem.* 56, 1177-1182 (1984).
119. J. S. Fritz, D. T. Gjerde, and R. M. Becker, *Anal. Chem.* 52, 1519-1522 (1980).
120. J. S. Fritz, D. T. Gjerde, and C. Pohlandt, *Ion Chromatography*, Huethig, Heidelberg, FRG, 1982.
121. Toshiaki Fuji, *Bunseki Kagaku* 31(4), 214-216 (1982).
122. Toshiaki Fuji, *Kogai to Taisaku* 18(9), 901-904 (1982).
123. Toshiaki Fuji, *Bunseki Kagaku* 31(11), 677-679 (1982).
124. Norio Fukuzuki, *Niigata-ken Kogai Kenkyusho Kenkyu Hokoku* 8, 10-13 (1983).
125. M. A. Fulmer, J. Penkrot, and R. J. Nadalin, in *Ion Chromatogr. Anal. Environ. Pollut.* (James D. Mulik and Eugene Sawicki, eds.), Vol. 2, Ann Arbor Sci., Ann Arbor, MI, 1979, pp. 381-400.
126. K. Fung and D. Grosjean, *Anal. Lett.* 17(A6), 475-482 (1984).
127. K. K. Fung, S. L. Heisler, A. Price, B. V. Nuesca, and P. K. Mueller, in *Ion Chromatogr. Anal. Environ. Pollut.* (Eugene Sawicki and James D. Mulik, eds.), Vol. 2, Ann Arbor Sci., Ann Arbor, MI, 1979, pp. 203-209.

128. P. J. Galvin and J. A. Cline, *Atmos. Environ.* 12, 1163-1167 (1978).
129. Roland Gilbert, Reynald Rioux, and Souheil E. Saheb, *Anal. Chem.* 56(1), 106-109 (1984).
130. James E. Girard, *Anal. Chem.* 51(7), 836-839 (1979).
131. James E. Girard, Technical Report, DC/WRRC-37, W83-02292, OWRT-A-011-DC(1) (1982). (Available from NTIS.)
132. J. E. Girard and D. Y. Badio, *Anal. Chem.* 56, 2992-2994 (1984).
133. J. E. Girard and J. A. Glatz, *Int. Lab.* 11(8), 62,64,66-68 (1981).
134. J. E. Girard and J. A. Glatz, *ASTM Spec. Tech. Publ. 742*, 105-115 (1981).
135. J. E. Girard and J. A. Glatz, *Am. Lab. (Fairfield, Conn.)* 13(10), 26,28,30,33-35 (1981).
136. Douglas Thomas Gjerde, *Diss. Abstr. Int. [B]*, 41(6), 2166 (1980). (Available from Univ. Microfilms Int.)
137. D. T. Gjerde, Technical Report, IS-T-916 (1980). (Available from NTIS.)
138. D. T. Gjerde and J. S. Fritz, *J. Chromatogr.* 176, 199-206 (1979).
139. Douglas T. Gjerde and James S. Fritz, *Anal. Chem.* 53(14), 2324-2327 (1981).
140. D. T. Gjerde, J. S. Fritz, and G. Schmuckler, *J. Chromatogr.* 186, 509-519 (1979).
141. D. T. Gjerde, G. Schmuckler, and J. S. Fritz, *J. Chromatogr.* 187(1), 35-45 (1980).
142. Joseph A. Glatz and James E. Girard, *J. Chromatogr. Sci.* 20(6), 266-273 (1982).
143. Manfred Goebl, *GIT Fachz. Lab.* 27(4), 261-262,265 (1983).
144. Manfred Goebl, *GIT Fachz. Lab.* 27(5), 373-375 (1983).
145. L. W. Green and J. R. Woods, *Anal. Chem.* 53(14), 2187-2189 (1981).
146. D. Grosjean and J. D. Nies, *Anal. Lett.* 17(A2), 89-96 (1984).
147. Karen Haak, *Plat. Surf. Finish.* 70(9), 34-38 (1983).
148. Karen K. Haak and Gregory O. Franklin, *Annu. Tech. Conf. Proc. - Am. Electroplat. Soc. 70th*, 1-6 (1983).
149. P. R. Haddad, P. W. Alexander, and M. Trojanowicz, *J. Chromatogr.* 294, 397-402 (1984).
150. P. R. Haddad and A. L. Heckenberg, *J. Chromatogr.* 252, 177-184 (1982).
151. P. Haddad and A. Heckenberg, *Chem. Aust.* 50, 275-278 (1983).
152. Peter Hajos and Janos Inczedy, *J. Chromatogr.* 201, 253-257 (1980).
153. Yuzuru Hanaoka, Takeshi Murayama, Setsuo Muramoto, Tamizo Matsuura, and Akinori Nanba, *J. Chromatogr.* 239, 537-548 (1982).

154. L. D. Hansen, B. E. Richter, D. K. Rollins, J. D. Lamb, and D. J. Eatough, *Anal. Chem.* 51(6), 633-637 (1979).
155. L. D. Hansen, J. F. Ryder, N. F. Mangelson, M. W. Hill, K. J. Faucette, and D. J. Eatough, *Anal. Chem.* 52, 821-824 (1980).
156. Hiroshi Hara, Kenji Nagara, Kouichi Honda, and Atsuko Goto, *Taiki Osen Gakkaishi* 15(9), 380-383 (1980).
157. Tadashi Hara, Kaoru Fujinaga, and Fujio Okui, *Bull. Chem. Soc. Jpn.* 54(10), 2956-2959 (1981).
158. Tsutomu Hashimoto and Akiyoshi Miyanaga, *Yukagaku* 32(10), 574-578 (1983).
159. Ryusuke Hatano and Toshio Sakuma, *Nippon Dojo Hiryogaku Zasshi* 54(2), 161-163 (1983).
160. K. Hayakawa, R. Ebina, M. Matsumoto, and M. Miyazaki, *Bunseki Kagaku* 33(7), 390-392 (1984).
161. Kazuichi Hayakawa, Hiromi Hiraki, Byungki Choi, and Genichi Miyazaki, *Hokuriku Koshu Eisei Gakkaishi* 10(2), 24-25 (1983).
162. Kazuichi Hayakawa, Hiromi Hiraki, and Motoichi Miyazaki, *Bunseki Kagaku* 32(8), 504-505 (1983).
163. D. L. Haynes, in *Ion Chromatogr. Anal. Environ. Pollut.* (Eugene Sawicki and James D. Mulik, eds.), Vol. 2, Ann Arbor Sci., Ann Arbor, MI, 1979, pp. 157-169.
164. A. L. Heckenberg and P. R. Haddad, *J. Chromatogr.* 299(1), 301-305 (1984).
165. Arthur G. Hedley and Marvin J. Fishman, Technical Report, USGS/WRD/WRI-82/028, USGS/WRI-81-78 (1982). (Available from NTIS.)
166. O. Heisz, *GIT Fachz. Lab.* 27(7), 596-600 (1983).
167. P. Henschel, H. Keune, R. Kressner, T. Moehlmann, R. Schwabe, and M. Sonneborn, *Int. J. Environ. Anal. Chem.* 15(1), 19-24 (1983).
168. J. A. Hern, G. K. Rutherford, and G. W. VanLoon, *Talanta* 30(9), 677-682 (1983).
169. H. Hershcovitz, C. Yarnitzky, and G. Schmuckler, *J. Chromatogr.* 244(2), 217-224 (1982).
170. H. Hershcovitz, C. Yarnitzky, and G. Schmuckler, *J. Chromatogr.* 252, 113-119 (1982).
171. J. Hertz and U. Baltensperger, *LC, Liq. Chromatogr. HPLC Mag.* 2(8), 600-602 (1984).
172. J. Hertz and U. Baltensperger, *Fresnius' Z. Anal. Chem.* 318(2), 121-123 (1984).
173. G. M. Hieftje, *Clin. Chem. (Winston-Salem, N.C.)* 29, 1659-1664 (1983).
174. C. J. Hill and R. P. Lash, *Anal. Chem.* 52(1), 24-27 (1980).
175. C. J. Hill and R. P. Lash, *Canadian Res.* 13(1), 53-56 (1980).
176. R. A. Hill, *HRC CC, J. High Resol. Chromatogr. Chromatogr. Commun.* 6(5), 275-277 (1983).

177. L. J. Holcombe, B. F. Jones, E. E. Ellsworth, and F. B. Meserole, in *Ion Chromatogr. Anal. Environ. Pollut.* (Eugene Sawicki and James D. Mulik, eds.), Vol. 2, Ann Arbor Sci., Ann Arbor, MI, 1979, pp. 401-412.
178. L. J. Holcombe and F. B. Meserole, *Water Qual. Bull.* 6(2), 37-39, 55 (1981).
179. Larry J. Holcombe and John C. Terry, *Proc. Annu. Meet. - Air Pollut. Control Assoc.* 71(5), 78/71.6 (1978).
180. R. D. Holm and S. A. Barksdale, in *Ion Chromatogr. Anal. Environ. Pollut.* (Eugene Sawicki, James D. Mulik, and Eva Wittgenstein, eds.), Ann Arbor Sci., Ann Arbor, MI, 1978, pp. 99-110.
181. Thomas B. Hoover, *Sep. Sci. Technol.* 17(2), 295-305 (1982).
182. Thomas B. Hoover and George D. Yager, *Anal. Chem.* 56(2), 221-225 (1984).
183. Yousuke Hoshino, Hiroko Saitoh, and Kikuo Oikawa, *Bunseki Kagaku* 32(4), 273-276 (1983).
184. T. Hsi, *Diss. Abstr. [B]* 44/12, 3765 (1983).
185. Manabu Igawa, Ikuo Ito, and Masao Tanaka, *Bunseki Kagaku* 29(9), 580-584 (1980).
186. Manabu Igawa, Kiyochiko Saito, Masao Tanaka, and Takeo Yamabe, *Bunseki Kagaku* 32(5), E137-E141 (1983).
187. Manabu Igawa, Kimiko Saito, Jun Tsukamoto, and Masao Tanaka, *Anal. Chem.* 53(12), 1942-1944 (1981).
188. Toshio Imanari, Koreharu Ogata, Shinzo Tanabe, Toshihiko Toida, Takehiko Kawanishi, and Masaki Ichikawa, *Chem. Pharm. Bull.* 30(1), 374-375 (1982).
189. T. Imanari, S. Tanabe, T. Toida, and T. Kawanishi, *J. Chromatogr.* 250, 55-61 (1982).
190. Wataru Ishibashi, Ryoji Kikuchi, and Kozo Yamamoto, *Bunseki Kagaku* 30(9), 604-608 (1981).
191. Wataru Ishibashi, Ryoji Kikuchi, and Kozo Yamamoto, *Bunseki Kagaku* 31(4), 207-211 (1982).
192. Ziad Iskandarni, *Diss. Abstr. [B]* 43/04, 1088 (1982).
193. Z. Iskandarani and D. J. Pietrzyk, *Anal. Chem.* 54, 1065-1071 (1982).
194. Z. Iskandarani and D. J. Pietrzyk, *Anal. Chem.* 54, 2427-2431 (1982).
195. Z. Iskandarani and D. J. Pietrzyk, *Anal. Chem.* 54, 2601-2603 (1982).
196. Hisaaki Itoh and Yoshio Shinbori, *Bunseki Kagaku* 29(4), 239-243 (1980).
197. Hisaaki Itoh and Yoshio Shinbori, *Bunseki Kagaku* 31(4), T39-T43 (1982).
198. Hisaaki Itoh and Yoshio Shinbori, *Chem. Lett.* 12, 2001-2002 (1982).

199. Hisaaki Itoh, Yoshio Shinbori, and Norihiro Tamura, *Bunseki Kagaku* 32(10), 571-575 (1983).
200. J. P. Ivey, *J. Chromatogr.* 267(1), 218-221 (1983).
201. J. P. Ivey, *J. Chromatogr.* 281, 314-318 (1983).
202. J. P. Ivey, *J. Chromatogr.* 287(1), 128-132 (1984).
203. C. J. Jackson, C. Neuberger, and M. Taylor, *Anal. Proc. (London)* 18(5), 201-204 (1981).
204. P. E. Jackson, P. R. Haddad, and S. Dilli, *J. Chromatogr.* 295(2), 471-478 (1984).
205. J. C. Jancar, M. D. Constant, and W. L. Herwig, *J. Am. Soc. Brew. Chem.* 42(2), 90-93 (1984).
206. K. H. Jansen, *GIT Fachz. Lab.* 22(12), 1062,1064-1066, 1069-1071 (1978).
207. Karl-Heinz Jansen, *LaborPraxis* 2(3), 30-34 (1978).
208. A. Jardy and R. Rosset, *Analusis* 7(6), 259-267 (1979).
209. Dennis Jenke, *Anal. Chem.* 53(9), 1535-1536 (1981).
210. Dennis Roger Jenke, *Diss. Abstr. Int. [B]* 44(1) (1983). (Available from Univ. Microfilms Int.)
211. D. R. Jenke, *Anal. Chem.* 56, 2468-2470 (1984).
212. D. R. Jenke, *Anal. Chem.* 56, 2674-2681 (1984).
213. Dennis R. Jenke, Presley K. Mitchell, and Gordon K. Pagenkopf, *J. Chromatogr. Sci.* 21(11), 487-489 (1983).
214. Dennis R. Jenke, Presley K. Mitchell, and Gordon K. Pagenkopf, *Anal. Chim. Acta* 155, 279-285 (1983).
215. Dennis R. Jenke and Gordon K. Pagenkopf, *Anal. Chem.* 54(14), 2603-2604 (1982).
216. D. R. Jenke and G. K. Pagenkopf, *Anal. Chem.* 55, 1168-1169 (1983).
217. Dennis R. Jenke and Gordon K. Pagenkopf, *J. Chromatogr.* 269(3), 202-207 (1983).
218. Dennis R. Jenke and Gordon K. Pagenkopf, *Anal. Chem.* 56(1), 85-88 (1984).
219. Dennis R. Jenke and Gordon K. Pagenkopf, *Anal. Chem.* 56(1), 88-91 (1984).
220. D. R. Jenke and G. K. Pagenkopf, *J. Chromatogr. Sci.* 22(6), 231-233 (1984).
221. Tingan Ji, *Huanjing Baohu (Beijing)* 11, 29-31 (1983).
222. E. L. Johnson, *Am. Lab. (Fairfield, CN)* 14(2), 98,100-102,104 (1982).
223. Ed Johnson, *GIT Fachz. Lab.* 26(3), 241-243 (1982).
224. E. L. Johnson, *Int. Lab.* 12(3), 110,112,114-115 (1982).
225. E. L. Johnson and K. K. Haak, in *Liq. Chromatogr. Environ. Anal.* (J. F. Lawrence, ed.), Humana Clifton, NJ, 1984, pp. 263-300.
226. Thomas Jupille, *LC, Liq. Chromatogr. HPLC Mag.* 1(1), 24-27 (1983).

227. Thomas Jupille, David Burge, and David Togami, *Chromatographia* 16, 312-316 (1982).
228. Thomas Jupille, David Burge, and David Togami, *Res. Dev.* 26(3), 135-139 (1984).
229. T. Jupille, M. Gray, and B. Black, *Am. Lab. (Fairfield, CN)* 13(8), 80,83-86 (1981).
230. Thomas H. Jupille, David W. Togami, and David E. Burge, *Ind. Res. Dev.* 25(2), 151-156 (1983).
231. M. Kan, K. Ohnishi, and M. Shintani, *Yakugaku Zasshi* 104(7), 763-768 (1984).
232. S. M. Kapelner, J. C. Trocciola, and M. S. Freed, in *Ion Chromatogr. Anal. Environ. Pollut.* (James D. Mulik and Eugene Sawicki, eds.), Vol. 2, Ann Arbor Sci., Ann Arbor, MI, 1979, pp. 345-360.
233. Fumio Karasawa, Takashi Ono, and Shinya Mizusawa, *Nippon Shashin Gakkaishi* 43(4), 245-250 (1980).
234. Fumio Karasawa, Takashi Ono, and Shinya Mizusawa, *Nippon Shashin Gakkaishi* 44(2), 96-103 (1981).
235. D. B. Karr, J. K. Waters, and D. W. Emerich, *Appl. Environ. Microbiol.* 46(6), 1339-1344 (1983).
236. Dieter Kasiske and Manfred Sonneborn, *LaborPraxis* 4(4), 76, 78,80,83 (1980).
237. Koichi Katoh, *Bunseki Kagaku* 32(10), 567-570 (1983).
238. Tatsuo Katou and Yoshimichi Hanai, *Yokohama Kokuritsu Daigaku Kankyo Kagaku Kenkyu Senta Kiyo* 10(1), 3-21 (1983).
239. John M. Keller, *Anal. Chem.* 53(2), 344-345 (1981).
240. J. M. Keller and R. R. Rickard, Technical Report, ORNL/TM-7569 (1981). (Available from NTIS.)
241. Wendy Thelaner Kennedy, Walter Bryan Hubbard, and James G. Tarter, *Anal. Lett.* 16(A15), 1133-1148 (1983).
242. Ikuei Kifune and Kikuo Oikawa, *Bunseki Kagaku* 28(10), 587-590 (1979).
243. Ikuhide Kifune and Kikuo Oikawa, *Niigata Rikagaku* 5, 9-14 (1979).
244. W. S. Kim, C. L. Geraci, Jr., and R. E. Kupel, in *Ion Chromatogr. Anal. Environ. Pollut.* (James D. Mulik and Eugene Sawicki, eds.), Vol. 2, Ann Arbor Sci., Ann Arbor, MI, 1979, pp. 171-184.
245. Walter S. Kim, Charles L. Geraci, Jr., and Richard E. Kupel, *Am. Ind. Hyg. Assoc. J.* 41(5), 334-339 (1980).
246. W. S. Kim, J. D. McGlothlin, and R. E. Kupel, *Am. Ind. Hyg. Assoc. J.* 42, 187-190 (1981).
247. P. T. Kissinger, *Anal. Chem.* 49, 447A-456A (1977).
248. E. J. Knudson and K. J. Siebert, *J. Am. Soc. Brew. Chem.* 42(2), 65-70 (1984).
249. William F. Koch, *Anal. Chem.* 51(9), 1571-1573 (1979).

250. William F. Koch, *J. Res. Natl. Bur. Stand. (U.S.)* 84(3), 241-245 (1979).
251. W. F. Koch, *J. Res. Natl. Bur. Stand. (U.S.)* 88(3), 157-161 (1983).
252. William F. Koch and Jeffrey W. Stolz, *Anal. Chem.* 54(2), 340-342 (1982).
253. V. Kordorouba, M. Pelletier, A. Balikungeri, and W. Haerdi, *Chimia* 38(7-8) 253-255 (1984).
254. W. Korth and J. Ellis, *Talanta* 31(6), 467-468 (1984).
255. Gary W. Kramer and Benjamin W. Haynes, Technical Report, RI 8661 (1982).
256. B. Kreilgaard and M. F. Andersen, *Arch. Pharm. Chem. Sci. Educ.* 12(3), 721-726 (1984).
257. S. T. Lai, M. M. Nishina, and L. Sangermano, *HRC CC, J. High Resol. Chromatogr. Chromatogr. Commun.* 7(6), 336-337 (1984).
258. J. D. Lamb, L. D. Hansen, G. G. Patch, and F. R. Nordmeyer, *Anal. Chem.* 53(4), 749-750 (1981).
259. R. P. Lash and C. J. Hill, *J. Liq. Chromatogr.* 2(3), 417-427 (1979).
260. R. P. Lash and C. J. Hill, in *Ion Chromatogr. Anal. Environ. Pollut.* (Eugene Sawicki and James D. Mulik, eds.), Vol. 2, Ann Arbor Sci., Ann Arbor, MI, 1979, pp. 67-74.
261. R. P. Lash and C. J. Hill, *Anal. Chim. Acta* 108, 405-409 (1979).
262. J. Lathouse and R. W. Coutant, in *Ion Chromatogr. Anal. Environ. Pollut.* (Eugene Sawicki, James D. Mulik, and Eva Wittgenstein, eds.), Ann Arbor Sci., Ann Arbor, MI, 1978, pp. 53-64.
263. J. Lathouse and R. E. Heffelfinger, in *Ion Chromatogr. Anal. Environ. Pollut.* (Eugene Sawicki and James D. Mulik, eds.), Vol. 2, Ann Arbor Sci., Ann Arbor, MI, 1979, pp. 59-65.
264. A. Laurent and R. Bourdon, *Ann. Pharm. Fr.* 36, 453-460 (1978).
265. Joseph J. Law, *Power Eng. (Barrington, IL)* 85(6), 94 (1981).
266. Alon Lebel and Teh Fu Yen, *Anal. Chem.* 56(4), 807-808 (1984).
267. Dan Peter Lee, *J. Chromatogr. Sci.* 22, 327-331 (1984).
268. D. P. Lee, *LC, Liq. Chromatogr. HPLC Mag.* 2(11), 828-832 (1984).
269. S. H. Lee and L. R. Field, *Anal. Chem.* 56, 2647-2653 (1984).
270. M. Legrand, M. De Angelis, and R. J. Delmas, *Anal. Chim. Acta* 156, 181-192 (1984).
271. C. W. Lewis and E. S. Macias, *Atmos. Environ.* 14, 185-194 (1980).
272. Mats Lindgren, *Vatten* 36(3), 249-264 (1980).
273. Mats Lindgren, Anders Cedergren, and Jan Lindberg, *Anal. Chim. Acta* 141, 279-286 (1982).

274. A. J. Lipski and C. J. Vairo, *Canadian Res.* 13(1), 45-48 (1980).
275. T. S. Long and A. L. Reinsvold, *Jt. Conf. Sens. Environ. Pollut.* 624-629 (1978).
276. Jeffrey M. Lorrain, Christopher R. Fortune, and Barry Dellinger, *Anal. Chem.* 53(8), 1302-1305 (1981).
277. M. J. McCormick, *Anal. Chim. Acta* 121, 233-238 (1980).
278. Phillip R. McCullough and Jimmy W. Worley, *Anal. Chem.* 51(8), 1120-1122 (1979).
279. K. M. McFadden and T. R. Garland, Technical Report, PNL-SA-7592 (1980). (Available from NTIS.)
280. John C. MacDonald, *Am. Lab. (Fairfield, CN)* 11(1), 45-55 (1979).
281. Hugh Mackie, Susan J. Speciale, Lewis J. Throop, and Taiyin Yang, *J. Chromatogr.* 242(1), 177-180 (1982).
282. Carl J. Mahle and Mani Menon, *J. Urol. (Baltimore)* 127(1), 159-162 (1982).
283. Supachai Maketon and James G. Tarter, *LC, Liq. Chromatogr. HPLC Mag.* 2(2), 124-126 (1984).
284. A. Mangia and M. T. Lugari, *Anal. Chim. Acta* 159, 349-354 (1984).
285. G. H. Mansfield, in *Ion Chromatogr. Anal. Environ. Pollut.* (James D. Mulik and Eugene Sawicki, eds.), Vol. 2, Ann Arbor Sci., Ann Arbor, MI, 1979, pp. 271-288.
286. G. Marko-Varga, I. Csiky, and J. A. Jonsson, *Anal. Chem.* 56(12), 2066-2069 (1984).
287. David W. Mason and Herbert C. Miller, in *Ion Chromatogr. Anal. Environ. Pollut.* (Eugene Sawicki and James D. Mulik, eds.), Vol. 2, Ann Arbor Sci., Ann Arbor, MI, 1979, pp. 193-201.
288. Mitsuhiro Matsumoto, Hiroshi Ichikawa, Kunitoshi Ichimura, Eiji Ueda, and Tatsumitsu Itano, *Zenkoku Kogaiken Kaishi* 6(1), 15-21 (1981).
289. Susumu Matsushita, Yoshimitsu Tada, Nobuyuki Baba, and Keiichi Hosako, *J. Chromatogr.* 259(3), 459-464 (1983).
290. Susumu Matsushita, Yoshimitsu Tada, Katsuo Komiya, and Akia Ono, *Bunseki Kagaku* 32(10), 562-567 (1983).
291. Henryk Matusiewicz and David F. S. Natusch, *Int. J. Environ. Anal. Chem.* 8(3), 227-233 (1980).
292. Sergio Mazzola, Renato Longhi, and Ettore Castiglioni, *Prod. Chim. Aerosol Sel.* 9, 30-31-33,35 (1983).
293. Mani Menon and Carl J. Mahle, *Clin. Chem. (Winston-Salem, N.C.)* 29(2), 369-371 (1983).
294. R. M. Merrill, Technical Report, SAND-82-1929 (1982). (Available from NTIS.)
295. R. Merriweather, Technical Report, K-2031 (1982). (Available from NTIS.)

296. D. P. Miller, *Atmos. Environ.* 18(4), 891-892 (1984).
297. Ted Miller, *Adv. Instrum.* 33(2), 169-176 (1978).
298. T. Miller, *ISA Trans.* 18(2), 59-64 (1979).
299. T. E. Miller, Jr., *InTech* 28(9), 77-79 (1981).
300. Theodore E. Miller, Jr., in *Autom. Stream Anal. Process Control* (D. P. Manka, ed.), Vol. 1, Academic Press, New York, 1982, pp. 1-37.
301. T. E. Miller, Jr., *Adv. Instrum,* 38, 347-364 (1983).
302. A. Mirna, H. Wagner, E. Kloetzer, and E. Fausel, *Lebensmittelchem. Gerichtl. Chem.* 38(1), 18-19 (1984).
303. Constance S. Mizisin, David E. Kuivinen, and Dumas A. Otterson, Technical Report, NASA-TM-78971 E-9743 (1978). (Available from NTIS.)
304. C. S. Mizisin, D. E. Kuivinen, and D. A. Otterson, in *Ion Chromatogr. Anal. Environ. Pollut.* (James D. Mulik and Eugene Sawicki, eds.), Vol. 2, Ann Arbor Sci., Ann Arbor, MI, 1979, pp. 129-139.
305. Munehiko Mizobuchi, Hisako Ohmae, Takeshi Tanaka, Fumiaki Umoto, Kunitoshi Ichimura, Eiji Ueda, and Tatsumitsu Itano, *Zenkoku Kogaiken Kaishi* 7(2), 67-71 (1982).
306. Munehiko Mizobuchi, Eiji Ueda, and Tatsumitsu Itano, *Yosui to Haisui* 23(12), 1437-1440 (1981).
307. Takahiko Mizobuchi, Toshiko Nakaoka, Takeshi Tanaka, Fumiaki Tomoto, Kunitoshi Ichimura, Hideji Ueda, and Taesumitsu Itano, *Nara-Ken Eisei Kenkyusho Nenpo* 15, 52-55 (1981).
308. R. B. Moore, J. E. Wilkerson, and C. R. Martin, *Anal. Chem.* 56, 2572-2575 (1984).
309. Y. Mori and A. Arikawa, *Ebara Infiruko Jiho* 90, 8-13 (1984).
310. Kanemitsu Morimoto and Shinobu Esaka, *Kyoto-Fu Eisei Kosai Kenkyusho Nenpo* 24, 128-131 (1980).
311. C. M. Morrow and R. A. Minear, *Water Res.* 18(9), 1165-1168 (1984).
312. C. O. Moses, D. K. Nordstrom, and A. L. Mills, *Talanta* 31(5), 331-339 (1984).
313. Julia A. Mosko, *Anal. Chem.* 56(4), 629-633 (1984).
314. Shefen Mu and Letian Chen, *Fenxi Huaxue* 11(3), 232-240 (1983).
315. H. Mueller, W. Nielinger, and A. Horbach, *Angew. Makromol. Chem.* 108, 1-8 (1982).
316. K. P. Mueller, *Fresenius' Z. Anal. Chem.* 317(3-4), 345-346 (1984).
317. P. K. Mueller, B. V. Mendoza, J. C. Collins, and E. S. Wilgus, in *Ion Chromatogr. Anal. Environ. Pollut.* (Eugene Sawicki, James D. Mulik, and Eva Wittgenstein, eds.), Ann Arbor Sci., Ann Arbor, MI, 1978, pp. 77-86.
318. Andrew J. Muir, *Sci. Technol. (Surrey Hills, Aust.)* 16(2), 19-23 (1978).

319. J. D. Mulik, E. Estes, and E. Sawicki, in *Ion Chromatogr. Anal. Environ. Pollut.* (Eugene Sawicki, James D. Mulik, and Eva Wittgenstein, eds.), Ann Arbor Sci., Ann Arbor, MI, 1978, pp. 41-52.
320. James D. Mulik, Ralph Puckett, Eugene Sawicki, and Dennis Williams, *NBS Spec. Publ. (U.S.) 464*, 603-607 (1977).
321. J. Mulik, R. Puckett, D. Williams, and E. Sawicki, *Anal. Lett.* 9(7), 653-663 (1976).
322. James D. Mulik and Eugene Sawicki, *Environ. Sci. Technol.* 13(7), 804-809 (1979).
323. J. D. Mulik and Eugene Sawicki, eds., *Ion Chromatographic Analysis of Environmental Pollutants*, Vol. 2, Ann Arbor Science Publishers Inc., Ann Arbor, MI, 1979.
324. J. D. Mulik, G. Todd, E. Estes, R. Puckett, E. Sawicki, and D. Williams, in *Ion Chromatogr. Anal. Environ. Pollut.* (Eugene Sawicki, James D. Mulik, and Eva Wittgenstein, eds.), Ann Arbor Sci., Ann Arbor, MI, 1978, pp. 23-40.
325. G. B. Munier, L. A. Psota-Kelty, and J. D. Sinclair, in *Atmos. Corros. (Pap.-Int. Symp. Atmos. Corros.)*, 275-283 (1982).
326. Kentaro Murano, *Bunseki 12*, 919-924 (1983).
327. Kentaro Murano, Motoyuk Mizuochi, Itsushi Uno, Tsutomu Fukuyama, and Shinji Wakamatsu, *Bunseki Kagaku 32*(10), 620-625 (1983).
328. Yoshikazu Muto, *Ion Chromatography*, Kodansha, Tokyo, Japan, 1983.
329. R. A. Nadkarni, *Proc.-Coal Test. Conf. 3rd*, 135-142 (1983).
330. R. A. Nadkarni and D. M. Pond, *Anal. Chim. Acta 146*, 261-266 (1983).
331. H. Nagashima, K. Kuboyzma, and K. Ono, *Sankyo Kenkyusho Nempo 35*, 59-64 (1983).
332. P. J. Naish, *Analyst (London) 109*, 809-812 (1984).
333. Masaki Nakajima, Keiichi Kimura, and Toshiyuki Shono, *Bull. Chem. Soc. Jpn. 56*(10), 3052-3056 (1983).
334. Kiyoshi Nakamura and Yoshihiro Morikawa, *Bunseki Kagaku 32*(4), 224-228 (1983).
335. Hisako Nakaoka, Fumiaki Umoto, Mitsuo Kasano, Norihiro Ikeda, Kunitoshi Ichimura, Eiji Ueda, and Tatsumitsu Itano, *Bunseki Kagaku 30*(10), T97-T101 (1981).
336. T. Nishino, *Yogyo Kyokaishi 92*(7), 410-412 (1984).
337. Tetsuo Nomura, *Kagaku to Kogyo (Osaka) 52*(12), 448-452 (1978).
338. F. R. Nordmeyer, G. M. Chan, and K. O. Ash, *Clin. Physiol. Biochem. 2*(4), 159-165 (1984).
339. F. R. Nordmeyer, L. D. Hansen, D. J. Eatough, D. K. Rollins, and J. D. Lamb, *Anal. Chem. 52*(6), 852-856 (1980).

340. M. Oehme and H. Stray, *Fresenius' Z. Anal. Chem. 306*(5), 356-361 (1981).
341. Takashi Ohno, Masafumi Kamoi, Fumio Karasawa, and Shinya Mizusawa, *Nippon Sashin Gakkaishi 43*(1), 48-50 (1980).
342. T. Ohno, H. Kobayashi, and S. Mizusawa, *Nippon Shashin Gakkaishi 47*(2), 91-95 (1984).
343. T. Ohno, H. Kobayashi, and S. Mizusawa, *Nippon Shashin Gakkaishi 47*(2), 108-112 (1984).
344. Kikuo Oikawa, *PPM 9*(7), 52-61 (1978).
345. Kikuo Oikawa, *Bunseki 8*, 531-538 (1980).
346. Kikuo Oikawa and Hiroko Saito, *Bunseki 2*, 94-99 (1982).
347. Kikuo Oikawa and Hiroko Saito, *Chemosphere 11*(9), 933-941 (1982).
348. Kikuo Oikawa and Hiroko Saito, *Bunseki Kagaku 32*(10), T105-T109 (1983).
349. Kikuo Oikawa, Hiroko Saito, Shigeaki Sakazume, and Masami Fujii, *Bunseki Kagaku 31*(8), E251-E255 (1982).
350. Kikuo Oikawa, Hiroko Saito, Shigeaki Sakazume, and Masami Fujii, *Chemosphere 11*(9), 953-961 (1982).
351. Tetsuo Okada and Tooru Kuwamoto, *Anal. Chem. 55*(7), 1001-1004 (1983).
352. Tetsuo Okada and Tooru Kuwamoto, *Bunseki Kagaku 32*(10), 595-599 (1983).
353. Tetsuo Okada and Tooru Kuwamoto, *J. Chromatogr. 284*(1), 149-156 (1984).
354. T. Okada and T. Kuwamoto, *Kagaku (Kyoto) 39*, 70-72 (1984).
355. T. Okada and T. Kuwamoto, *Anal. Chem. 56*(12), 2073-2078 (1984).
356. T. Okada and T. Kuwamoto, *Anal. Lett. 17*(A15), 1743-1751 (1984).
357. T. Okada and T. Kuwamoto, *Anal. Chem. 57*, 258-262 (1985).
358. T. Okamoto and Y. Shirane, *Hiroshima-ken Kankyo Senta Kenkyu Hokoku 5*, 39-43 (1983).
359. Franz Ollram, *Wasser Abwasser 27*, 211-227 (1983).
360. R. L. Orwell, D. S. Scurr, A. Smith, and W. G. Robertson, *Fortschr. Urol. Nephrol.. 20*, 263-270 (1982).
361. Dumas A. Otterson, Technical Report, NASA-TM-X-73642 (1977). (Available from NTIS.)
362. D. A. Otterson, in *Ion Chromatogr. Anal. Environ. Pollut.* (Eugene Sawicki, James D. Mulik, and Eva Wittgenstein, eds.), Ann Arbor Sci., Ann Arbor, MI, 1978, pp. 87-98.
363. Dumas A. Otterson, Technical Report, NASA-TM-79020. E-9817 (1978). (Available from NTIS.)
364. J. S. Parigi, *Am. Lab. (Fairfield, CN) 16*(9), 124,126 (1984).
365. D. F. Pensenstadler and M. A. Fulmer, *Anal. Chem. 53*(7), 859A-860A,862A,864A,866A,868A (1981).

366. Jarl M. Pettersen, *Anal. Chim. Acta 160*, 263-266 (1984).
367. D. Pickerell, T. Hook, T. Dolzine, and J. K. Robertson, in *Ion Chromatogr. Anal. Environ. Pollut.* (James D. Mulik and Eugene Sawicki, eds.), Vol. 2, Ann Arbor Sci., Ann Arbor, MI, 1979, pp. 289-294.
368. R. K. Pinschmidt, in *Ion Chromatogr. Anal. Environ. Pollut.* (Eugene Sawicki and James D. Mulik, eds.), Vol. 2, Ann Arbor Sci., Ann Arbor, MI, 1979, pp. 41-50.
369. R. K. Pinschmidt and T. P. Katrinak, in *Ion Chromatogr. Anal. Environ. Pollut.* (James D. Mulik and Eugene Sawicki, eds.), Vol. 2, Ann Arbor Sci., Ann Arbor, MI, 1979, pp. 31-40.
370. M. M. Plechaty, *LC, Liq. Chromatogr. HPLC Mag. 2*(9), 684-688 (1984).
371. C. A. Pohl and E. L. Johnson, *J. Chromatogr. Sci. 18*(9), 442-452 (1980).
372. Christel Pohlandt, *S. Afr. J. Chem. 33*(3), 87-91 (1980).
373. C. Pohlandt, Rep. Natl. Inst. Metall. (S. Afr.) No. 2044, (1980).
374. C. Pohlandt, Rep. Natl. Inst. Metall. (S. Afr.) No. 2132, (1981).
375. C. Pohlandt, Technical Report, NIM-2107 (1981). (Available from NTIS.)
376. C. Pohlandt, Technical Report, NIM-2132 (1982). (Available from NTIS.)
377. C. Pohlandt, *S. Afr. J. Chem. 35*(3), 96-100 (1982).
378. C. Pohlandt, *S. Afr. J. Chem. 37*(3), 133-137 (1984).
379. C. Pohlandt, *S. Afr. J. Sci. 80*(5), 208-209 (1984).
380. R. D. Posner and A. Schoffman, in *Ion Chromatogr. Anal. Environ. Pollut.* (James D. Mulik and Eugene Sawicki, eds.), Vol. 2, Ann Arbor Sci., Ann Arbor, MI, 1979, pp. 51-58.
381. G. S. Pyen and D. E. Erdmann, *Anal. Chim. Acta 149*, 355-358 (1983).
382. G. S. Pyen and M. J. Fishman, in *Ion Chromatogr. Anal. Environ. Pollut.* (Eugene Sawicki and James D. Mulik, eds.), Vol. 2, Ann Arbor Sci., Ann Arbor, MI, 1979, pp. 235-244.
383. Dayong Qi, Changling Qu, and Tianze Zhou, *Huaxue Tongbao 8*, 492-496 (1982).
384. Changling Qu, Dayong Qi, and Tianze Zhou, *Huaxue Shiji 3*, 176-181 (1982).
385. Changling Qu, Dayong Qi, and Tianze Zhou, *Huaxue Shiji 6*(2), 107-110 (1984).
386. E. Rajakyla, *J. Chromatogr. 218*, 695-701 (1981).
387. J. A. Rawa, in *Ion Chromatogr. Anal. Environ. Pollut.* (Eugene Sawicki and James D. Mulik, eds.), Vol. 2, Ann Arbor Sci., Ann Arbor, MI, 1979, pp. 245-269.
388. J. A. Rawa, *ASTM Spec. Tech. Publ. 742*. 92-104 (1981).

389. Judith A. Rawa and Earl L. Henn, *Proc. Int. Water Conf., Eng. Soc. West. Pa. 40th*, 213-219 (1979).
390. J. P. Rawat, M. Iqbal, and H. M. A. Abdul Aziz, *J. Liq. Chromatogr.* 7(8), 1691-1706 (1984).
391. Selwyn J. Rehfeld, Hans F. Loken, Francis R. Nordmeyer, and John D. Lamb, *Clin. Chem. (Winston-Salem, NC)* 26(8), 1232-1233 (1980).
392. P. F. Reigler, Norman J. Smith, and V. T. Turkelson, *Anal. Chem.* 54(1), 84-87 (1982).
393. G. Resch and E. Gruenschlaeger, *Vortr.-VGB-Konf. Chem. Kraftwerk*, 34-39 (1981).
394. G. Resch and E. Gruenschlaeger, *VGB Kraftwerkstech* 62(2), 127-132 (1982).
395. G. Resch and E. Gruenschlaeger, *Gewaesserschutz, Wasser, Abwasser* 67, 265-285 (1983).
396. G. Resch and E. Gruenschlaeger, *Vom Wasser* 62, 207-217 (1984).
397. D. J. Reutter, R. C. Buechele, and T. L. Rudolph, *Anal. Chem.* 55(14), 1468A (1983).
398. G. R. Ricci, L. S. Shepard, G. Colovos, and N. E. Hester, *Anal. Chem.* 53, 610-613 (1981).
399. William E. Rich, *Anal. Instrum.* 15, 113-117 (1977).
400. W. E. Rich, *Instrum. Technol.* 24(8), 47-51 (1977).
401. William Rich, Edward Johnson, Louis Lois, Pokar Kabra, Brian Stafford, and Laurence Marton, *Clin. Chem. (Winston-Salem, NC)* 26(10), 1492-1498 (1980).
402. William E. Rich, Edward Johnson, Louis Lois, Brian E. Stafford, Pokar M. Kara, and Laurence J. Marton, in *Liq. Chromatogr. Clin. Anal.* (P. M. Kabra and L. J. Marton, eds.), Humana, Clifton, NJ, 1981, 393-407.
403. W. Rich, F. Smith, Jr., L. McNeil, and T. Sidebottom, in *Ion Chromatogr. Anal. Environ. Pollut.* (Eugene Sawicki and James D. Mulik, eds.), Vol. 2, Ann Arbor Sci., Ann Arbor, MI, 1979, pp. 17-29.
404. W. E. Rich, J. A. Tillotson, and R. C. Chang, in *Ion Chromatogr. Anal. Environ. Pollut.* (Eugene Sawicki, James D. Mulik, and Eva Wittgenstein, eds.), Ann Arbor Sci., Ann Arbor, MI, 1978, pp. 185-196.
405. W. E. Rich and R. A. Wetzel, *ACS Symp. Ser.* 94, 233-246 (1979).
406. W. E. Rich and R. A. Wetzel, *Actual. Chim.* 6, 51-57 (1980).
407. John Rivielle, Arthur Fitchett, and E. Johnson, *Proc. Int. Water Conf. Eng. Soc. West. Pa.*, 43rd, 458-467 (1982).
408. K. M. Roberts, D. T. Gjerde, and J. S. Fritz, *Anal. Chem.* 53(11), 1691-1695 (1981).
409. W. G. Robertson and D. S. Scurr, *Clin. Chim. Acta* 140(1), 97-99 (1984).

410. W. G. Robertson, D. S. Scurr, A. Smith, and R. L. Orwell, *Clin. Chim. Acta 126*(1), 91-99 (1982).
411. Peter G. Robinson, *Chem. N. Z. 45*(5), 153-154 (1981).
412. Roy D. Rocklin, *LC, Liq. Chromatogr. HPLC Mag. 1*(8), 504-507 (1983).
413. R. D. Rocklin, *LC, Liq. Chromatogr. HPLC Mag. 2*(8), 588-594 (1984).
414. R. D. Rocklin, *Anal. Chem. 56*(11), 1959-1962 (1984).
415. Roy D. Rocklin and Edward L. Johnson, *Anal. Chem. 55*(1), 4-7 (1983).
416. R. D. Rocklin and C. A. Pohl, *J. Liq. Chromatogr. 6*, 1577-1590 (1983).
417. Dennis R. Roden and Dennis E. Tallman, *Anal. Chem. 54*(2), 307-309 (1982).
418. Souji Rokushika, *Kagaku (Kyoto) 37*(7), 557-560 (1982).
419. Souji Rokushika, *Kagaku no Ryoiki, Zokan 138*, 73-85 (1983).
420. Souji Rokushika, Zong Yin Qui, and Hiroyuki Hatano, *J. Chromatogr. 260*(1), 81-87 (1983).
421. S. Rokushika, Z. Y. Qiu, Z. L. Sun, and H. Hatano, *J. Chromatogr. 280*, 69-76 (1983).
422. Souji Rokushika, Zhuo Lian Sun, and Hiroyuki Hatano, *J. Chromatogr. 253*(1), 87-94 (1982).
423. Walter J. Rossiter, Jr., Paul W. Brown, and McClure Godette, *Sol. Energy Mater. 9*(3), 267-279 (1983).
424. R. J. Rucki, *Talanta 27*, 147-156 (1980).
425. J. Rudling, B. O. Hallberg, M. Hultengren, and A. Hultman, *Scand. J. Work, Environ. Health 10*(3), 197-202 (1984).
426. E. Ruseva, *Khim. Ind. (Sofia) 3*, 126-129 (1983).
427. E. Sacher, *IEEE Trans. Electr. Insul. EI-18*(4), 369-373 (1983).
428. Hiroko Saitoh and Kikuo Oikawa, *Bunseki Kagaku 31*(11), E375-E380 (1982).
429. H. Saitoh and K. Oikawa, *Bunseki Kagaku 33*(10), E441-E444 (1984).
430. Hiroko Saitoh, Kikuo Oikawa, Hayao Sakamoto, and Masaakira Kamada, *Chikyu Kagaku (Nippon Chikyu Kagakkai) 16*(2), 43-47 (1982).
431. Hiroko Saitoh, Kikuo Oikawa, Tooru Takano, and Katsura Kamimura, *J. Chromatogr. 281*, 397-402 (1983).
432. Isao Sanemasa, Takemi Mizoguchi, Junko Ohtsuka, Toshio Deguchi, and Hideo Nagai, *Bunseki Kagaku 32*(7), 420-425 (1983).
433. Hisakuni Sato, *Bunseki Kagaku 31*(2), 97-99 (1982).
434. Hisakuni Sato, *Bunseki Kagaku 31*(3), T23-T28 (1982).
435. Hisakuni Sato, *Bunseki Kagaku 32*, 610-614 (1983).
436. Eugene Sawicki, in *Ion Chromatogr. Anal. Environ. Pollut.* (Eugene Sawicki, James D. Mulik, and Eva Wittgenstein, eds.), Ann Arbor Sci., Ann Arbor, MI, 1978, pp. 1-9.

437. E. Sawicki, in *Ion Chromatogr. Anal. Environ. Pollut.* (Eugene Sawicki and James D. Mulik, eds.), Vol. 2, Ann Arbor Sci., Ann Arbor, MI, 1979, pp. 1-15.
438. Eugene Sawicki, J. D. Mulik, and E. Wittgenstein, *Ion Chromatographic Analysis of Environmental Pollutants*, Vol. 1, Ann Arbor Sci., Ann Arbor, MI, 1978.
439. W. G. Sayre and D. C. Constable, *Acqua Aria 2*, 113-116 (1983).
440. Leon J. Schiff, Stephen G. Pleva, and Emory W. Sarver, in *Ion Chromatogr. Anal. Environ. Pollut.* (Eugene Sawicki and James D. Mulik, eds.), Vol. 2, Ann Arbor Sci., Ann Arbor, MI, 1979, pp. 329-344.
441. G. J. Schmidt and R. P. W. Scott, *Analyst (London) 109*, 997-1002 (1984).
442. M. Schnitzler, G. Levay, W. Kuehn, and H. Sontheimer, *Vom. Wasser 61*, 263-276 (1983).
443. F. Schoeller and Franz Ollram, *Oesterr. Wasserwirtsch 35*(3-4), 73-77 (1983).
444. R. Schwabe, T. Darimont, T. Moehlmann, E. Pable, and M. Sommeborn, *Int. J. Environ. Anal. Chem. 14*(3), 169-179 (1983).
445. R. Schwarzenbach, *J. Chromatogr. 251*, 339-358 (1982).
446. Georg Schwedt. *LaborPraxis 8*(1-2), 30,32,34-36,41-42 (1984).
447. Gregory J. Sevenich and James S. Fritz, *Anal. Chem. 55*(1), 12-16 (1983).
448. (a) R. W. Shaw, R. K. Stevens, and W. J. Courtney, pp. 2217-2218, (b) D. A. Hegg and P. U. Hobbs, pp. 2218-2219, (c) L. D. Hansen, N. F. Mangelson and D. J. Eatough, p. 2219, *Anal. Chem. 52*, 2217-2219 (1980).
449. M. Shibukawa and N. Ohta, *Bunseki Kagaku 32*, 557-561 (1983).
450. H. Shintani, K. Tsuji, and T. Oba, *Bunseki Kagaku 33*(7), 347-351 (1984).
451. O. A. Shpigun and Yu. A. Zolotov, *Zavod. Lab. 48*(9), 4-14 (1982).
452. Darryl D. Siemer, *Anal. Chem. 52*(12), 1874-1877 (1980).
453. Darryl D. Siemer and Vergil J. Johnson, *Anal. Chem. 56*(6), 1033-1034 (1984).
454. N. W. Skelly, *Anal. Chem. 54*, 712-715 (1982).
455. J. Slanina, in *Euroanal. 4, Rev. Anal. Chem.*, 4th (L. Niinisto, ed.), Akad. Kiado, Budapest, 1982, pp. 173-182.
456. J. Slanina, F. P. Bakker, P. A. C. Jongejan, L. Van Lamoen, and J. J. Mols, *Anal. Chim. Acta 130*(1), 1-8 (1981).
457. J. Slanina, W. A. Lingerak, J. E. Ordelman, P. Borst, and F. P. Bakker, in *Ion Chromatogr. Anal. Environ. Pollut.* (Eugene Sawicki and James D. Mulik, eds.), Vol. 2, Ann Arbor Sci., Ann Arbor, MI, 1979, pp. 305-317.
458. Rosanne W. Slingsby and John M. Riviello, *LC, Liq. Chromatogr., HPLC Mag. 1*(6), 354-356 (1983).

459. H. Small, in *Ion Chromatogr. Anal. Environ. Pollut.* (Eugene Sawicki, James D. Mulik, and Eva Wittgenstein, eds.), Ann Arbor Sci., Ann Arbor, MI, 1978, pp. 11-21.
460. Hamish Small, *Trace Anal.* 1, 267-322 (1981).
461. H. Small and T. E. Miller, Jr., *Anal. Chem.* 54, 462-469 (1982).
462. H. Small, T. S. Stevens, and W. C. Bauman, *Anal. Chem.* 47, 1801-1809 (1975).
463. Barry W. Smee, G. E. M. Hall, and D. J. Koop, *J. Geochem. Explor.* 10(3), 245-258 (1978).
464. David L. Smith, Walter S. Kim, and Richard E. Kupel, *Am. Ind. Hyg. Assoc. J.* 41(7), 485-488 (1980).
465. Frank C. Smith, Jr., and Richard C. Chang, *CRC Crit. Rev. Anal. Chem.* 9(3), 197-217 (1980).
466. Frank C. Smith, Jr., and Richard C. Chang, *The Practice of Ion Chromatography*, John Wiley and Sons, New York, 1983.
467. J. Smith, Jr., A. McMurtrie, and H. Galbraith, *Microchem. J.* 22(1), 45-49 (1977).
468. Jessie G. Smith, *Anal. Chem. Nucl. Technol., Proc. Conf. Anal. Chem. Energy Technol. 25th*, 63-67 (1982).
469. Robert E. Smith, *Anal. Chem.* 55(8), 1427-1429 (1983).
470. R. E. Smith, Technical Report, BDX-613-2865 (1983). (Available from NTIS.)
471. R. L. Smith and D. J. Pietrzyk, *Anal. Chem.* 56, 1572-1577 (1984).
472. Louise C. Speitel, Joe C. Spurgeon, and Robert A. Filipczak, in *Ion Chromatogr. Anal. Environ. Pollut.* (Eugene Sawicki and James D. Mulik, eds.), Vol. 2, Ann Arbor Sci., Ann Arbor, MI, 1979, pp. 75-87.
473. B. Stauber and H. Wiesmann, *Chem. Rundsch.* 33(17), 3,7 (1980).
474. B. Stauber and H. Wiesmann, *Escher Wyss News* 53(1-2), 119-123 (1980).
475. R. Steiber and R. Merrill, *Anal. Lett.* 12(A3), 273-278 (1979).
476. R. Steiber and R. Merrill, in *Ion Chromatogr. Anal. Environ. Pollut.* (Eugene Sawicki and James D. Mulik, eds.), Vol. 2, Ann Arbor Sci., Ann Arbor, MI, 1979, pp. 99-113.
477. R. Steiber and R. M. Statnick, in *Ion Chromatogr. Anal. Environ. Pollut.* (Eugene Sawicki, James D. Mulik, and Eva Wittgenstein, eds.), Ann Arbor Sci., Ann Arbor, MI, 1978, pp. 141-148.
478. R. K. Stevens, T. G. Dzubay, G. Russwurm, and D. Rickel, *Atmos. Environ.* 12, 55-68 (1978).
479. Timothy S. Stevens, *Ind. Res. Dev.* 25(9), 96-99 (1983).
480. Timothy S. Stevens, James C. Davis, and Hamish Small, *Anal. Chem.* 53(9), 1488-1492 (1981).
481. Timothy S. Stevens, Gary L. Jewett, and Robert A. Bredeweg, *Anal. Chem.* 54(7), 1206-1208 (1982).

482. Timothy S. Stevens and Martin A. Langhorst, *Anal. Chem.* 54(5), 950-953 (1982).
483. T. S. Stevens and T. E. Miller, Jr., *Anal. Chem.* 52, 2023-2026 (1980).
484. Timothy S. Stevens and Hamish Small, *J. Liq. Chromatogr.* 1(2), 123-132 (1978).
485. Timothy S. Stevens, Virgil T. Turkelson, and William R. Albe, *Anal. Chem.* 49(8), 1176-1178 (1977).
486. Robert L. Stevenson and Kervin Harrison, *Am. Lab. (Fairfield, CN)* 13(5), 76,78-81 (1981).
487. Helge Stray, *Kjemi* 4, 30,33,46 (1981).
488. H. H. Streckert and B. D. Epstein, *Anal. Chem.* 56(1), 21-24 (1984).
489. K. Stulik and V. Pacakova, *J. Electroanal. Chem. Interfac. Electrochem.* 129, 1-24 (1981).
490. T. Sunden, A. Cedergren, and D. D. Siemer, *Anal. Chem.* 56, 1085-1089 (1984).
491. Thomas Sunden, Mats Lindgren, Anders Cedergren, and Darryl D. Siemer, *Anal. Chem.* 55(1), 2-4 (1983).
492. Koji Suzuki, Hiroshi Aruga, Hitoshi Ishiwada, Toyoko Oshima, Hidenari Inoue, and Tsuneo Shirai, *Bunseki Kagaku* 32(10), 585-590 (1983).
493. Koji Suzuki, Hiroshi Aruga, and Tsuneo Shirai, *Anal. Chem.* 55(12), 2011-2013 (1983).
494. M. A. Tabatabai and W. A. Dick, in *Ion Chromatogr. Anal. Environ. Pollut.* (Eugene Sawicki and James D. Mulik, eds.), Vol. 2, Ann Arbor Sci., Ann Arbor, MI, 1979, pp. 361-370.
495. M. A. Tabatabai and W. A. Dick, *J. Environ. Qual.* 12(2), 209-213 (1983).
496. Yoji Taguchi, Satoshi Koyanagi, Manabu Ohizumi, Michio Mashima, and Nobuyuki Sakai, *Niigata Daigaku Kogakubu Kenkyu Hokoku* 32, 49-53 (1983).
497. Takejiro Takamatsu, Munetsugu Kawashima, and Mutsuo Koyama, *Bunseki Kagaku* 28(10), 596-601 (1979).
498. Katsushige Takami, Ohkawa Kazunobu, Yoshio Kuge, and Masao Nakamoto, *Bunseki Kagaku* 31(7), 362-367 (1982).
499. Kuniaki Takamine, Shigeru Tanaka, and Yoshikazu Hashimoto, *Bunseki Kagaku* 31(12), 692-696 (1982).
500. B. Takano, M. A. McKibben, and H. L. Barnes, *Anal. Chem.* 56, 1594-1600 (1984).
501. Kazuhiko Tanaka, *Bunseki Kagaku* 31(12), T106-T112 (1982).
502. K. Tanaka, *Bunseki Kagaku* 32, 439-443 (1983).
503. Kazuhiko Tanaka and Yutaka Ishihara, *Mizu Shori Gijutsu* 23(9), 767-771 (1982).
504. K. Tanaka, Y. Ishihara, and K. Kakajima, *Bunseki Kagaku* 32, 626-631 (1983).

505. Kazuhiko Tanaka, Yutaka Ishihara, and Kunio Nakajima, *Mizu Shori Gijutsu* 23(10), 855-859 (1982).
506. K. Tanaka and T. Ishizuka, *J. Chromatogr.* 190, 77-83 (1980).
507. K. Tanaka and T. Ishizuka, *Water Res.* 16(5), 719-723 (1982).
508. K. Tanaka, T. Ishizuka, and H. Sunahara, *J. Chromatogr.* 174, 153-157 (1979).
509. K. Tanaka, T. Ishizuka, and H. Sunahara, *J. Chromatogr.* 177, 21-27 (1979).
510. Shigeru Tanaka, Takayuki Yoshimori, and Yoshikazu Hashimoto, *Bunseki Kagaku* 32(12), 735-739 (1983).
511. Takashi Tanaka and Kazuo Hiiro, *Kankyo Gijutsu* 12(1), 56-59 (1983).
512. Takashi Tanaka, Kazuo Hiiro, Akinori Kawahara, and Shinichi Wakida, *Bunseki Kagaku* 32(12), 771-773 (1983).
513. James Gordon Tarter, *Diss. Abstr. B* 42(2), 620-621 (1981).
514. James G. Tarter, *LC, Liq. Chromatogr. HPLC Mag.* 1(8), 508-509 (1983).
515. James G. Tarter, *Anal. Chem.* 56, 1264-1268 (1984).
516. J. G. Tarter, *J. Liq. Chromatogr.* 7(8), 1559-1566 (1984).
517. Fernando Tateo, Maria Luisa Faleschini, and Marcello Fossati, *Ind. Conserve* 57(1), 30-33 (1982).
518. S. B. Tejada, R. B. Zweidinger, J. E. Sigsby, Jr., and R. L. Bradow, in *Ion Chromatogr. Anal. Environ. Pollut.* (Eugene Sawicki, James D. Mulik, and Eva Wittgenstein, eds.), Ann Arbor Sci., Ann Arbor, MI, 1978, pp. 111-124.
519. Shigeru Terabe, Kiyoshi Yamamoto, and Teiichi Ando, *Kenkyu Hokoku - Asahi Garasu Kogyo Gijutsu Shoreikai* 39, 131-134 (1981).
520. K. The and R. Roussel, *Light Met. (Warrendale, PA)*, 115-125 (1984).
521. F. J. Trujillo, M. M. Miller, R. K. Skogerboe, H. E. Taylor, and C. L. Grant, *Anal. Chem.* 53(12), 1944-1946 (1981).
522. Yuichi Tsuchitani, *Bunseki* 9, 603-605 (1979).
523. V. T. Turkelson and M. Richards, *Anal. Chem.* 50(11), 1420-1423 (1978).
524. S. Y. Tyree, Jr., Jan M. Stouffer, and Mark Bollinger, in *Ion Chromatogr. Anal. Environ. Pollut.* (Eugene Sawicki and James D. Mulik, eds.), Vol. 2, Ann Arbor Sci., Ann Arbor, MI, 1979, pp. 295-304.
525. United Kingdom, Dept. of the Environment, *Methods Exam. Water Assoc. Mater.* (1984).
526. Yoshio Urishiyama, Toshio Fukuzaki, Takahide Haga, and Yoshio Ichikawa, *Niigata Rikagaku* 9, 45-48 (1983).
527. M. J. Van Os, J. Slanina, C. L. DeLigny, and J. Agterdenbos, *Anal. Chim. Acta* 156, 169-180 (1984).

528. M. J. Van Os, J. Slanina, C. L. DeLigny, W. E. Hammers, and J. Agterdenbos, *Anal. Chim. Acta 144*, 73-82 (1982).
529. J. Vialle, *TrAC, Trends Anal. Chem. (Pers. Ed.) 3*(2), 61-65 (1984).
530. Helena Viikamo, *Kem.-Kemi 6*(4), 190-191 (1979).
531. Dutt V. Vinjamoori and Chaur-Sun Ling, *Anal. Chem. 53*(11), 1689-1691 (1981).
532. Puligandla Viswanadham, Donald R. Smick, Jerome J. Pisney, and Walter F. Dilworth, *Anal. Chem. 54*(14), 2431-2433 (1982).
533. H. Waki and Y. Tokunaga, *J. Chromatogr. 201*, 259-264 (1980).
534. H. Waki and Y. Tokunaga, *J. Liq. Chromatogr. 5*, 105-119 (1982).
535. Chung Yu Wang, Scott D. Bunday, and James G. Tarter, *Anal. Chem. 55*(9), 1617-1619 (1983).
536. Chung Yu Wang and James G. Tarter, *Anal. Chem. 55*(11), 1775-1778 (1983).
537. Wang Nang Wang, Yeong Jgi Chen, and Mou Tai Wu, *Analyst (London) 109*(3), 281-286 (1984).
538. W. B. Wargotz, *Proc. Int. Symp. Contam. Control. 4th*, 291-297 (1978).
539. W. B. Wargotz, *Surf. Contam.: Genesis, Detect., Control, (Proc. Symp.) 2*, 877-895 (1979).
540. R. G. Warren, *Food Technol. Aust. 33*(6), 300,302 (1981).
541. Isao Watanabe, Ryoichi Tanaka, and Takashi Kashimoto, *Shokuhin Eiseigaku Zasshi 22*(3), 246-247 (1981).
542. Isao Watanabe, Ryoichi Tanaka, and Takashi Kashimoto, *Shokuhin Eiseigaku Zasshi 23*(2), 135-141 (1982).
543. Yoshiko Watanabe and Shinobu Esaka, *Kyoto-fu Eisei Kogai Kenkyusho Nenpo 24*, 124-127 (1980).
544. V. T. Wee and J. M. Kennedy, *Anal. Chem. 54*, 1631-1633 (1982).
545. Joachim Weiss, *CLB, Chem. Labor Betr. 34*(7), 293-297 (1983).
546. Joachim Weiss, *CLB, Chem. Labor Betr. 34*(8), 342-345 (1983).
547. Joachim Weiss, *CLB, Chem. Labor Betr. 34*(11), 494-495,498-500 (1983).
548. Joachim Weiss, *CLB, Chem. Labor Betr. 35*(2), 59-60,62-66 (1984).
549. L. C. Westwood and E. L. Stokes, in *Ion Chromatogr. Anal. Environ. Pollut.* (Eugene Sawicki and James D. Mulik, eds.), Vol. 2, Ann Arbor Sci., Ann Arbor, MI, 1979, pp. 141-156.
550. Roy Wetzel, *Environ. Sci. Technol. 13*(10), 1214-1217 (1979).
551. Roy Wetzel, *Ind. Res. Dev. 24*(4), 92-95 (1982).
552. R. A. Wetzel, C. L. Anderson, Helmut Schleicher, and G. D. Crook, *Anal. Chem. 51*(9), 1532-1535 (1979).
553. B. B. Wheals, *J. Chromatogr. 262*, 61-76 (1983).

554. J. W. Whittaker and P. R. Lemke, *J. Pharm. Sci. 71*(3), 334-338 (1982).
555. Richard J. Williams, *J. Chromatogr. Sci. 20*(12), 560-565 (1982).
556. Richard J. Williams, *Anal. Chem. 55*(6), 851-854 (1983).
557. M. J. Willison and A. G. Clarke, *Anal. Chem. 56*(6), 1037-1039 (1984).
558. John P. Wilshire, *LC, Liq. Chromatogr. HPLC Mag. 1*(5), 290-293 (1983).
559. John P. Wilshire and Willia A. Brown, *Anal. Chem. 54*(9), 1647-1650 (1982).
560. Stephen A. Wilson and Carol A. Gent, *Anal. Lett. 15*(A10), 851-864 (1982).
561. Stephen A. Wilson and Carol A. Gent, *Anal. Chim. Acta 148*, 299-303 (1983).
562. S. A. Wilson and E. S. Yeung, *Anal. Chim. Acta 157*, 53-63 (1984).
563. S. A. Wilson, E. S. Yeung, and D. R. Bobbitt, *Anal. Chem. 56*, 1457-1460 (1984).
564. Jerry W. Wimberley, *Anal. Chem. 53*(11), 1709-1710 (1981).
565. Jerry W. Wimberley, *Anal. Chem. 53*(13), 2137-2138 (1981).
566. P. Witier and P. Advielle, *Bull. Liaison Lab. Ponts Chaussées 130*, 121-126 (1984).
567. D. J. Woo and J. R. Benson, *Am. Lab. (Fairfield, CN) 16*, 50, 52-54 (1984).
568. R. Wood, L. Cummings, and T. Jupille, *J. Chromatogr. Sci. 18*, 551-558 (1980).
569. Weh S. Wu, Douglas K. Arai, Mark A. Nazar, and David K. Leong, *Am. Ind. Hyg. Assoc. J. 43*(12), 942-945 (1982).
570. Lijuan Xiang, Xiulin Yang, and Zhenyuan Hu, *Youji Huaxue 5*, 353-357 (1983).
571. M. Yamamoto, H. Yamamoto, Y. Yamamoto, S. Matsushita, N. Baba, and T. Ikushige, *Anal. Chem. 56*(4), 832-834 (1984).
572. Takashi Yasuoka, Jiro Takano, Shunmei Mitsuzawa, Hiroko Saitoh, and Kikuo Oikawa, *Bunseki Kagaku 32*(10), 580-584 (1983).
573. Shigeaki Yonemori and Makoto Noshiro, *Asahi Garasu Kenkyu Hokoku 31*(1), 17-22 (1981).
574. Xinsen Zhang and Qing Song, *Huanjing Kexue 4*(6), 47-50 (1983).
575. Yu. A. Zolotov, A. A. Ivanov, and A. A. Shpigun, *Zh. Anal. Khim. 38*(8), 1479-1483 (1983).
576. Yu. A. Zolotov, O. A. Shpigun, and L. A. Bubchikova, *Dokl. Akad. Nauk SSSR 266*(5), 1144-1147 (1982).
577. Yu. A. Zolotov, O. A. Shpigun, and L. A. Bubchikova, *Fresenius' Z. Anal. Chem. 316*(1), 8-12 (1983).

578. Yu. A. Zolotov, O. A. Shpigun, L. A. Bubchikova, and E. A. Sedel'nikova, *Dokl. Akad. Nauk SSSR* 263(4), 889-892 (1982).
579. Fritz Zuercher, Technical Report, EUR 7623, *Anal. Org. Micropollut. Water*, 272-276 (1982).
580. R. B. Zweidinger, S. B. Tejada, J. E. Sigsby, Jr., and R. L. Bradow, in *Ion Chromatogr. Anal. Environ. Pollut.* (Eugene Sawicki, James D. Mulik, and Eva Wittgenstein, eds.), Ann Arbor Sci., Ann Arbor, MI, 1978, pp. 125-139.

Index

A

Acetate ion
   determination of, 383-384
   variable response, 5
Acrylate polymer resin material, 45
Alcohols by ion chromatography exclusion, 180-182
Alkaline earth determination using zinc eluant, 275
Amines
   determination of, 386
   determination of simple, 39
Amino acids by ion chromatography exclusion, 174, 181, 183
Ammonia
   determination of, 386
   by ion chromatography exclusion, 173, 186
Amperometric detection
   of amino acids, 114-116
   applications, 122
   auxiliary electrodes, 109
   basic principles, 106
   carbon electrodes, 118, 277

[Amperometric detection]
   cell design, 106-107
   copper electrodes, 116
   current-potential plot, 103
   general, 54-56
   glassy carbon electrodes, 113
   hydrodynamic voltammograms, 104
   use in ion chromatography exclusion, 174
   large area coulometric cell, 108
   limiting current, 105
   mercury based electrodes, 244
   reference electrodes, 104
   sensitivity, 102
   silver electrodes, 111-112
   thin layer cell, 107
   tubular cadmium electrode, 116
   voltammogram, 103
   wall jet cell, 108
Anion columns
   pellicular (see pellicular columns)
   sample loading capacity, 7
   surface area, 7
Arsenic, determination of, 382

Atomic spectroscopic detection
  atomic absorption, 144
  atomic emission, 144
  ion-interaction reagents, 269
  organometallics, 279, 281-282

B

Basic organic compounds by ion chromatography exclusion, 183
Beverage analysis (see food analysis)
Biological applications, 375
  by ion chromatography exclusion, 176-177
Biomedical analysis (see pharmaceutical analysis)
Borate ion, determination of, 383
Bromide ion, determination of, 377

C

Calcium, determination of, 386
Carbonate ion
  determination of, 382
  by ion chromatography exclusion, 177, 179, 183-185
Cation columns
  exchange in single column ion chromatography, 38
  types, 13
Cell constant, 89
Charge exclusion (see Donnan exclusion)
Chloride ion
  in biological samples, 377
  in environmental samples, 376
  in industrial samples, 376
Citric acid, use in eluants, 43
Columns (see specific type such as anion, cation, etc.)

Column clean-up procedures
  use of EDTA, 41
  use of nitric acid, 41
Column parameters
  affinity, 26
  capacity, 44
  efficiency, 51
  flow, 25
  pellicular column crosslinkage, 9
  pressure, 25
  pore size, 45
Complexing agents, monitoring of, 64
Conductive ion pairs (see eluant suppression)
Conductivity detection
  active thermostating systems, 240
  background conductivity, 95-96
  baseline drift, 99-100
  bipolar pulse techniques, 99, 242
  bulk property, 2
  capacitance effects, 99
  cell current, 99
  cell design requirements, 97-100
  cell volume, 97-98
  conductance changes, 91
  detector noise, 93
  detector response equations, 89-91
  electrode materials, 98
  eluant conductivity, 2
  enhancement of sensitivity, 93-94
  indirect methods, eluants, 241
  use in ion chromatography exclusion, 173-174
  metal separations, 274
  non-linear response, 101-102
  paired flow cell, 100
  permittivity detection, coupling with, 242

Index

[Conductivity detection]
  solute specific, 2
  temperature effect, 37, 100
Contamination by leaching metals, 287
Coulometric detection
  based on pH changes, 244-245
  indirect post-column detection, 277-278
  ion chromatography exclusion, 174
Crosslinkage in ion chromatography exclusion columns, 169, 172
Cyclodextrin bonded phases for metal separations, 296

D

Debye-Falkenhagen effect, 337
Detectable species
  anions by eluant suppressed ion chromatography, 12-13
  cations by eluant suppressed ion chromatography, 18
Detection limits
  anions by eluant suppressed ion chromatography, 12-13
  by anion exclusion, 49
  cations by eluant suppressed ion chromatography, 19
  definition, 237-238
  by ion chromatography exclusion, 174-175
  metal ions by conductivity detection, 274
  by single column ion chromatography, 35-37
  by UV detection, 130
  by UV detection, indirect method, 249-250
Differential detection, 88
  of metal ions, 288
Diode array detectors, 337
Direct current plasma atomic emission, 145
Direct detection methods, definition of, 88
Donnan Exclusion, 5-6, 46, 158, 224
Donnan Exclusion Chromatography (see ion chromatography exclusion)
Driving strength, 26, 41
Dynamic ion exchange chromatography (see ion interaction reagents)

E

Electrochemical detection (see amperometric detection)
Electronic suppression ion chromatography (see single column ion chromatography)
Electron spin resonance detection of metal ions, 288
Eluants, nonaqueous, 25
Eluant suppressed ion chromatography
  comparison with single column ion chromatography, 84-85
  eluants for anion determination, 10
  eluants for cation determination, 14-16
Eluant suppression
  anion exchange membranes, 204
  band dispersion, 195
  carbonate dip elimination, 235
  carbonate regenerants, 17
  carbon dioxide post-suppression, 232-233
  cation exchange fibers, 195
  conductive ion pairs, 5
  critical micelle concentration, 228-229
  diffusive transport through membrane, 226
  effect of pK, 5
  electrolytic post-column conversion, 230-231

[Eluant suppression]
  exchange efficiency, 206
  fiber based, 6
  filament filled membranes, 207
  filament insertion methods, 209-210
  forbidden ions, 225
  helical membrane suppressors, 207
  hollow fiber band dispersion, 207-208
  in ion chromatography exclusion, 166, 174
  by ion exchange hollow fibers, 194-195
  leak proof connections, 208, 220
  liquid ion exchangers, 229
  mass transfer of permeated ion, 214
  mass transfer of regenerant ion, 214
  mass transport, 198
  mass transport from membrane wall
    carbon dioxide, 234-235
    exchange efficiency, 205
    regenerant penetration, 205-206
  mass transport through membrane wall
    band dispersion, 218
    diffusion coefficients, 199-200
    flux, 200
    selectivity coefficients, 198-199, 202
    wall transport limited, 201
  mass transport to membrane wall
    Dean number, 204
    helix design, 204
  membrane based, 6, 195
  membrane lifetime, 224

[Eluant suppression]
  membrane poisoning, 224
  membrane resistance to transport, 233-234
  membrane tube characteristics, 221-223
  multiple strand design, 195-197
  packed bead suppressor, 212
  packed columns, 194
  penetration of counter ion, 230
  penetration of forbidden ion, 226-229
  performance characteristics, 217
  permeability characteristics, 198, 229
  polymeric regenerants, 228
  porous polymer tubes, 220
  principles, 1-6
  radial compression, 218
  radiation grafting, 220
  reactions, 3-4
  regenerants for anion determination, 10-11
  regenerants for cation determination, 16-17
  regenerant selection, 225
  restrictions on, 4
  single bead-string reactor, 210-212
  solubility, 5
  water dip elimination, 235
Environmental analysis
  air, 373
  arsenates in, 77
  nitrogen oxides, 76-77
  soils, 373
  sulfur, 74
  water, 372-373
  water by ion chromatography exclusion, 183
Equilibration times, 8
Equivalent conductance, 26
Ethanolamine, 39
Ethylenediamine, use as eluant, 41

F

Fluorescence detector
  indirect methods, 143, 250
  in ion chromatography exclusion, 174
  of metal ions, 287
Fluoride ion
  in environmental samples, 375
  in industrial samples, 376
Food analysis
  additives, 73
  anions, 69-73
  by ion chromatography exclusion, 178-179
Formate ion, determination of, 383-384

G

Glycolate-borate as eluant, 33
Gormley-Kennedy equation, 195, 204
Gradient elution, 6, 310
  using carbonate based eluants, 334-335
  using electrochemical detection, 336
  using ion interaction reagents, 336-337

H

Heavy metal separations, 273
Hydroxide ion, use as eluant, 35
p-Hydroxybenzoic acid, use as eluant, 32-33

I

Indirect detection methods
  definition, 88, 236
  see also specific detectors, indirect
Indirect photometric detection (see UV detection, indirect)
Inductively coupled plasma systems, 144-145
Industrial applications, 374
  water analysis by ion chromatography exclusion, 179, 183, 186
Injection volumes, 25
Inorganic acids by ion chromatography exclusion, 183-185
Inorganic bases by ion chromatography exclusion, 183, 185
Inorganic compound analysis by ion chromatography exclusion, 165, 183
Iodide ion, determination of, 378
Ion chromatography exclusion, 1, 24
  adsorption, 159
  anion fiber suppressor in, 166, 168
  application, 175-186
  columns, 169-172
  comparison with alternate techniques, 165
  detectors (see specific detector type)
  distribution coefficients, 162
  elevated temperature separation, 174
  eluants, 172-173
  mechanism of separation, 159
Ion chromatography exclusion-ion chromatography, 166-167, 178, 184-185
Ion exchange chromatography, 24
Ion exclusion chromatography (see ion chromatography exclusion)
Ion exclusion partition chromatography (see also ion chromatography exclusion)
  elution behavior, 160-162

Ionic chromatography, definition, 192
Ion interaction reagents, 24
　ammonium acetate, 54
　dynamic ion exchange model, 253-255
　electrical double layer, 256
　eluants, 51, 254-255, 257, 259-261, 263-264, 267-272, 285, 292-293
　ion-pair model, 255-257
　separations, 259-261, 270-272, 285, 292-293
　surfactants in, 53
　transition metal detection system, 269-270
Ion moderated partition chromatography (see ion chromatography exclusion)
Ion pair chromatography (see ion interaction reagents)

J

Juice analysis by ion chromatography exclusion, 178-179, 181, 184

L

Limiting equivalent conductance, 89, 91-92
Limit of detection (see detection limit)
Limit of quantitation, definition of, 238
Lithium, determination of, 385

M

Magnesium ion, determination of, 386
Manufacturers of equipment, 86
Metal separations
　applications, 276-277
　[Metal separations]
　on bonded silica based resins, 307-310
　on chelating resins, 296-300
　on crown ether based resins, 300-307
　on exchange resins, 290
　normal phase chromatography, 294
Methanol as eluant component, 39
Minimum detectable concentration, 60
Mobile-phase ion chromatography, 1

N

Negative peak, 29, 38
Nicotinic acid as eluant, 34
Nitrate ion
　in environmental samples, 380
　in industrial samples, 380
Nitrite ion
　determination of, 381
　variable response, 5
Non-suppressed ion chromatography (see single column ion chromatography)
Normal phase chromatography, 25

O

On-line operation, 388
Organic acid analysis
　determination of, 384
　by ion chromatography exclusion, 161, 163-164, 174-175, 177-179, 181
Organic base analysis by ion chromatography exclusion, 165, 169, 183
Organic modifiers
　acetonitrile, 47
　in ion chromatography exclusion, 173

Index

P

Peak width, relation to injection volume, 25
Pellicular column
　functional groups, 9
　hydrophobicity, 9
　latex, 8
　silica based, 44
Permeability of pellicular columns, 8
Permittivity detector, 242
Pharmaceutical analysis
　cosmetics, 69
　dosage forms, 67
　excipients/preservatives, 68-69
　by ion chromatography exclusion, 175, 177, 183
　microchemical, 68
　quantitative analysis, 67
Phenate ion, use as eluant, 33
Phenol analysis by ion chromatography exclusion, 174, 181
Phosphate ion
　in environmental samples, 381
　in industrial samples, 381
　by ion chromatography exclusion, 173, 186
Phthalic acid, use as eluant, 29-32
Plating bath analysis, 62
　by ion chromatography exclusion, 180, 185
　for metals, 64
Polarographic detection, 108-109
　complexation reactions, 110
Post-column reactors
　active reactors
　　gas pressurized, 329
　　operational characteristics, 330
　　air segmented, 321
　　angle of impingement, 147

[Post-column reactors]
　applications, 148-149, 331-334
　color forming reactions, 147
　decomposition reactions, 320
　design requirements, 321
　detection enhancement, 318-320
　fluorescence detection, 151
　gaseous reagents, 330
　hybrid membrane reactors, 330
　liquid segmented, 321-322
　membrane reactors, 326
　mixing noise, 322
　mixing Tee, 323-325
　packed-bed reactors, 147-148, 150
　passive reactors
　　mass transport, 328
　　membrane types, 327
　　sample loss, 327
　reagent introduction, 320-321
　reagents, 320-321, 331-333
　serpentine tubes, 322
　single bead-string reactors, 322
Potassium ion, determination of, 385
Potentiometric detection
　applications, 123
　chloride ion selective electrodes, 243
　copper electrodes, 121-125, 243-244
　definition, 118
　detection modes, 125
　indirect detection of halides, 242
　using ion interaction reagents, 270
　using ion selective electrodes, 119-120
　Nernst equation, 116, 124-125
　silver/silver chloride electrodes, 120, 243
　silver/silver salicylate electrodes, 120-121

Process bath analysis
 major components, 63
 trace components, 64
Pulping liquors, thiosulfate in, 75

R

Radioactivity detector for metal ions, 287-288
Rapid analysis of nitrate and sulfate, 32
Refractive index detection
 differential detection, 130-131
 direct methods, 130
 indirect methods, 140-143
 eluants, 251
 use in ion chromatography exclusion, 174
 sensitivity, 55
Regenerants (see eluant suppression)
Replacement ion chromatography, 144, 251, 253
Resistivity detection (see conductivity detection, indirect)
Reverse phase chromatography, 24-25
 metal separations, 294-296

S

Salicylate, use as eluant, 33
Sample concentration, 25, 59-60
 applications, 317-318
 breakthrough, 61
 dialytic preconcentration, 318
 extractive methods, 311-312
 of metal ions, 297-300
 on-column methods, 313-317
 use of precolumn, 313-316
 on stationary phase, 312-314
Sample preparation, 390
Schoniger flask for microchemical analysis, 68

Selectivity
 guidelines, 27-28
 of pellicular columns, 9
Selenium, determination of, 380
Sensitivity, definition of, 237
Separation of fluoride-borate, 32
Shrinking of pellicular columns, 8
Silica gel, use as substrate, 45
Silicate ion, determination of, 382
Signal enhancement, 235
Single column ion chromatography
 comparison with eluant suppressed ion chromatography, 84-85
 detectors in, 23
 effect of eluant pH, 26-27, 29-32
 eluants for anion measurements, 29-35
 eluants, rules for selection, 26-28
 equilibrium reactions, 28-29
 inorganic acid determination, 30
Sodium, determination of, 385
Spectroscopic detection (see also UV detection)
 types, 126
Strontium, determination of, 386
Sugar analysis by ion chromatography exclusion, 174, 180-181
Sulfate ion
 in environmental samples, 378-379
 in industrial samples, 379
Sulfite ion, determination of, 379
Suppressor column (see eluant suppression)
Suppressorless ion chromatography (see single column ion chromatography)
Swelling of pellicular columns, 8
System peak, 28-29, 54, 139-140

Index

T

Tartaric acid, use as eluant, 43
Tellerium, determination of, 380
Temperature compensation, 62
Transition metals
  as contaminants, 41
  determination of, 387
  determination of as oxo-anions, 383

U

UV detection
  direct, 56-57
  indirect, 58-59
  alkylquaternaryammonium compounds, 247
  cations, 134-135, 246
  detector response equations, 132-134
  eluants, 245-246
  for ion interaction reagents, 262-263, 269
  matrix effects, 138-139
  method for anions and cations, 136-138
  without standards, 134, 247-249

[UV detection]
  ion chromatography exclusion, 51, 173
  ion interaction reagents, 261
  metal-chloro complexes, 128
  metal ions, 286
  metal ions by ion interaction reagents, 269
  simultaneous direct and indirect method, 247
  species detectable by direct modes, 127
Urine analysis, 375
  by ion chromatography exclusion, 175-176, 178

V

Vacancy chromatography (see UV detection, indirect)
Van der Waals forces, 8

W

Water analysis by ion chromatography exclusion, 179, 186
Water dip, 5

Z

Zwitterionic buffers, 94

/543.0893I64>C1/

DATE DUE

8-8-90 SLS		
JUN 2 5 1991		
DEC 1 0 1996		
9/9/98 Cook MPL		

Demco, Inc. 38-293